D0853278

AN INTRODUCTION TO THE CHEMISTRY AND BIOCHEMISTRY OF PYRIMIDINES, PURINES AND PTERIDINES

AN INTRODUCTION TO THE CHEMISTRY AND BIOCHEMISTRY OF PYRIMIDINES, PURINES AND PTERIDINES

DEREK T. HURST

School of Chemical and Physical Sciences,
Kingston Polytechnic

JOHN WILEY & SONS

Chichester · New York · Brisbane · Toronto

British Library Cataloguing in Publication Data:

Hurst, Derek T
An introduction to the chemistry and biochemistry
of pyrimidines, purines and pteridines.
1. Pyrimidines 2. Pteridines 3. Purines
I. Title
547'.593 QD401 79–40736

ISBN 0 471 27647 2

Typeset by Preface Ltd., Salisbury, Wilts. and
Printed in Great Britain by Page Bros. (Norwich) Ltd., Norwich

Preface

There are a few text-books of heterocyclic chemistry but in general they pay comparatively little attention to the pyrimidines and the fused ring analogues the purines and pteridines. Yet these are very interesting compounds and play very important roles in biochemistry and in medicinal chemistry. Although texts on biochemistry deal with the biochemical aspects of these compounds they also have scant coverage of the chemistry of the pyrimidines, purines, and pteridines. The John Wiley Monographs on the Chemistry of Heterocyclic compounds are comprehensive and are valuable reference books which include volumes on the pyrimidines and the purines but not, as yet, on the pteridines.

This book is an attempt to cover the basic chemistry of the pyrimidines, purines and pteridines at a level suitable for undergraduate and research students of chemistry, biochemistry and pharmaceutical chemistry and to provide an introduction to the biochemistry and biochemical uses of these compounds at a level suitable for undergraduate and postgraduate chemists. There are several good books on the biochemistry of the nucleic acids, and this book does not intend to cover this ground in such detail but only to indicate the place of the nitrogen heterocycles in this field.

The literature on the various aspects of such compounds—from molecular biology to medicinal chemistry, to basic organic chemistry—is immense and the coverage in this book cannot but be incomplete. However, it is hoped that obvious points have not been overlooked and that what errors there are are of omission not of commission.

As the weeks pass, so more papers are published on aspects of the pyrimidines, purines and pteridines, and already some of the information given here will be out of date, but it is hoped that this book provides a reasonable survey of the topics at present (April 1979).

This book is dedicated to all those, whether included in references or not, whose work has made the writing of such a book possible.

Kingston-upon-Thames
April 1979

DEREK T. HURST

Contents

Chapter 1

Introduction to Nitrogen Heteroaromatic Chemistry

(A) HISTORICAL INTRODUCTION

Nitrogen-containing heterocyclic compounds featured prominently in early studies of chemistry and they were closely associated with the development of 'organic' chemistry, which was concerned with the study of materials isolated from living sources, whilst 'inorganic' chemistry was concerned with the study of inanimate materials.

As early as 1776 a purine derivative (uric acid) was isolated, in a pure form, from urinary calculi (stones) by Scheele.[1] The first pyrimidine derivative (alloxan) to be isolated was obtained in 1818 by the oxidation of uric acid with nitric acid (Brugnatelli,[2,3] whilst pyrrole was first isolated by Runge[4] from bone oil in 1834, and in 1846 Anderson[5] obtained a pyridine derivative (picoline) from coal tar. Pteridines were not isolated until 1889 when Gowland Hopkins[6,7] first recognised that the pigments obtained from the wings of certain butterflies were compounds which showed similarities to polyhydroxypurines such as uric acid.

Such discoveries were being made during the 'classical' period of chemistry during which the idea that 'organic' and 'inorganic' materials, obtained from living and inanimate sources, were inherently different was refuted, and when ideas concerning the structure and composition of compounds were put on a logical basis, and the structure and special nature of the 'aromatic' class of compounds related to benzene was recognized. These ideas were of great importance in helping to solve the structures of the nitrogen heterocyclic compounds, and the discovery of such compounds in organic matter initiated studies to discover, and explain, the role of such compounds in living organisms.

At about the same time that Kekué[8] proposed his structure for benzene, Miescher[9] first separated 'nuclein' from pus cells from discarded bandages, whilst in 1889 Altmann[10] introduced the term *nucleic acid* and developed general methods for the isolation of nucleic acids from many sources. The discovery of the presence of pyrimidines and purines in nucleic acids soon followed, for example cytosine was isolated from calf thymus nucleic acid in 1894 by Kossel and Neumann.[11] These discoveries have led to the development of a new area of scientific study—molecular biology—and has led to important discoveries such as the structure of nucleic acids, the nature of the genetic material, the explanation of mutation, and a variety of other important consequences.

During this period developments were being made in other fields which, although apparently unconnected with discoveries such as the presence of pyrimidines and purines in nucleic acids, and the occurrence of uric acid, guanine, etc., in the excrement of some animal species, were later shown to have the common feature of involvement of the nitrogen heterocycles. Experiments on animals and empirical cures of certain diseases in man showed the need for certain dietary constituents, termed *accessory growth factors* by Gowland Hopkins. The condition known as beriberi had been long known, and in 1897 Eijkmann, a prison officer in Java, showed that prisoners who were normally fed on polished rice, and amongst whom beriberi was prevalent, had this condition greatly alleviated if rice polishings were added to their diet. Later Funk[12] isolated the active component of rice polishings as a crystalline compound which he termed a *vitamine*. This observation was one of several which led to the discovery of a number of other essential growth factors, now called *vitamins*, several of which are derivatives of the purines, pyrimidines and pteridines. These discoveries have also stimulated interest in the study of these heterocycles.

Yet another field in which these heterocycles found application was also discovered in this early period of chemistry, Baeyer[13] had prepared barbituric acid in 1863, but it was not until 1903 that the hypnotic properties of 5,5-diethylbarbituric acid were discovered by Fischer and von Mering.[14] The discovery of 'Veronal' initiated considerable activity in the synthesis of other derivatives of barbituric acid, a number of which have found valuable use as hypnotics, tranquillizers and treatments for epilepsy.

During the twentieth century considerable research effort has been directed towards investigating the structure, properties, and synthesis of the nitrogen heterocycles. The structures, roles, and metabolic pathways of those found in living organisms have, for the most part, been elucidated, and many synthetic derivatives of these heterocycles have been shown to have chemotherapeutic applications in a variety of cases. A number of such compounds have also found use as agricultural chemicals and pesticides. The detailed structures of large and complex natural products containing these heterocycles, for example Vitamin B_{12} and some nucleic acids, have also been solved and such studies continue to attract great interest. The nucleotide sequences of some tRNAs and the three-dimensional structure of such compounds have recently been reported, and interest, not only in molecular biology but in other aspects of nitrogen heterocycles, seems to continue unabated.

This book attempts to provide the background chemistry of the pyrimidines, purines, and pteridines, and to provide an account of the derivatives of these compounds which play important roles in living processes. It indicates the major metabolic pathways of such compounds and also indicates the role of nucleic acid in directing protein synthesis. Points of interest concerning the use of certain pyrimidines, purines, and related compounds, as anti-bacterial agents, anti-cancer agents, and various other physiological activities will also be indicated. It is hoped that this book will provide the reader with the foundations for the understanding of the chemistry, biochemistry, and utility, of the nitrogen heterocycles pyrimidine, purine, and pteridine, which play such an important part in the living process.

(B) AROMATICITY: THE STRUCTURE AND NUMBERING OF THE NITROGEN HETEROCYCLES

The concept of *aromaticity* and the attributes of *aromatic character* have attracted much attention since the discovery and early investigations of benzene and its derivatives in the middle of the last century. A number of books have been written on aromaticity and some of these are listed in the reference section at the end of the chapter. All organic chemistry texts include a section on aromaticity, also reference can be made to the report of a symposium[15] on aromaticity (Sheffield, 1966).

Badger[16] has defined aromaticity in the following terms:

> 'An unsaturated cyclic or polycyclic molecule or ion (or part of a molecule or ion) may be classified as aromatic if all the annular atoms participate in a conjugated system such that, in the ground state, all the π electrons (which are derived from atomic orbitals having axial orientation to the ring) are associated in bonding molecular orbitals in a closed (annular) shell'.

Other writers have used different criteria for obtaining a definition of aromaticity, but a simple description which applies to most cases is as follows:

> An aromatic compound contains a cyclic system of atoms with a cyclic, conjugated, π bond system of $(4n + 2)$ π electrons (the Hückel number).

Such a system of π bonds is caused by the overlap of unhybridized p (or d) atomic orbitals of the adjacent atoms in the ring. Such a situation results in a *delocalization energy*, i.e. a compound in which there is a continuous cyclic overlap of p orbitals has less energy (is more stable) than that calculated for such a compound if there were no such cyclic conjugation. For a compound to show a high degree of aromaticity it is necessary that there is a maximum overlap of the p orbitals, i.e. the ring should approach planarity, and also the compound should have a high delocalization energy.

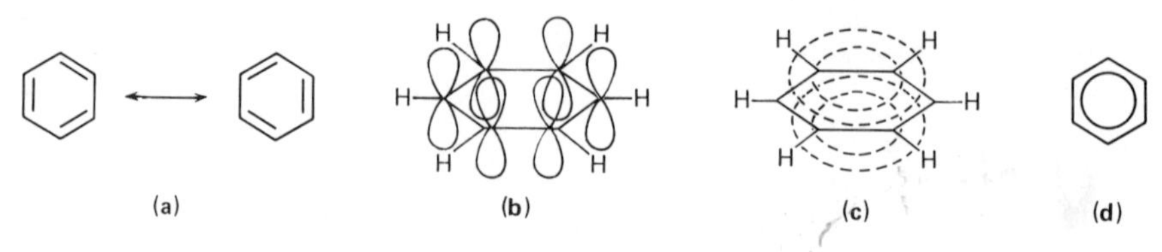

Fig. 1.1 Benzene molecule: (a) Kekulé formulae. (b) Showing sp^2 hybrids and unhybridized p orbitals. (c) Delocalized π bond system. (d) Usual notation

Benzene, the model aromatic compound, has a planar ring system of six carbon atoms each having three sp^2 hybrid orbitals, two of which are used for σ bonding to the adjacent carbon atoms and the third bonds to hydrogen, and each having an unhybridized p orbital, containing one electron, perpendicular to the plane of the ring and resulting in a delocalised π bond system having $n = 1$ in the Hückel expression.

The usual properties associated with aromaticity are:

(a) a tendency for aromatic compounds to undergo ring substitution reactions rather than addition reactions which are characteristic of alkenes;
(b) a tendency for the cyclic system to be more stable to both oxidation and reduction than a typical alkene;
(c) the existence of a 'delocalization energy';
(d) bond lengths in aromatic rings are intermediate, between normal σ bonded and normal π bonded atoms;
(e) characteristic absorbance in the u.v. or visible region of the spectrum;
(f) the exhibition of a *ring current* when an aromatic compound is placed in a magnetic field, this accounting for the low-field absorbance of hydrogen atoms attached to such rings in the ^1H n.m.r. spectrum. (This has been used as a definition of an aromatic compound as 'a cyclic system which exhibits a diamagnetic ring current and in which all of the ring atoms are involved in a single conjugated system'.)

Aromaticity is not an absolute but a relative property, and compounds can show aromatic character to a greater or lesser extent. Thus benzene is taken as the aromatic standard and, for example, some aromatic compounds are oxidized much more readily than benzene whilst others are more resistant to oxidation than benzene. Some compounds have a greater tendency to undergo additions to the ring whilst others have little or no tendency for such additions, and some compounds have a high delocalization energy relative to benzene, whilst others have a lower delocalization energy. In some compounds there is also a greater localization of the π electrons in certain bonds within a ring; thus some bonds within such a ring tend more towards a true alkenic carbon–carbon bond and others tend more towards a single-bonded character for the carbon–carbon bond.

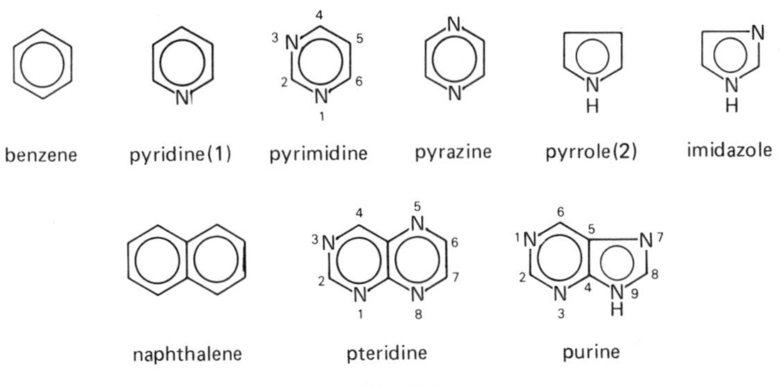

Fig. 1.2.

In the heterocyclic compounds of the pyridine (1) type, having −N= in place of −CH= in the benzenoid ring, the annular nitrogen atom adopts the sp^2 hybrid atomic orbital configuration. Two of the sp^2 hybrid orbitals overlap with adjacent carbon atoms to form σ bonds, whilst the third sp^2 hybrid orbital contains the 'lone pair' of electrons, and the unhybridized p orbital contains one electron available for the cyclic, conjugated, six π electron system.

In compounds of the pyrrole (2) type −NH− takes the place of the two-carbon unit −CH=CH− in benzene. The nitrogen atom is also in the sp^2 configuration, but in this case the third sp^2 hybrid orbital overlaps with a hydrogen atom and it is the lone pair of electrons which occupies the unhybridized p orbital. In this case the nitrogen atom contributes two electrons to the cyclic, conjugated, six π electron system. The compounds pyridine, pyrimidine, pyrazine, pyrrole, and imidazole in Fig. 1.2 are thus six π electron aromatic heterocycles. The heterocycles purine and pteridine shown in Fig. 1.2 are ten π electron systems comparable with the aromatic ten π electron compound naphthalene.

The numbering system which will be used in this book for these heterocycles is indicated in Fig. 1.2, and although it is not entirely in accordance with the recommendations of the Chemical Society,[17] the system and orientation of the rings used is one which is widely used.

(C) 'π EXCESSIVE' AND 'π DEFICIENT' HETEROCYCLES

Whereas in benzene the dispositions of the delocalized π molecular orbitals are symmetrical and the π *electron density* at each atom in the ring is the same and equal to one, the presence of an atom other than carbon in the ring causes distortion of the π orbitals so that the π electron density is not the same at each site in the ring. Nitrogen has a higher electronegativity than carbon, and when present as −N= in an aromatic ring nitrogen distorts the π electron distribution by its inductive ability, so

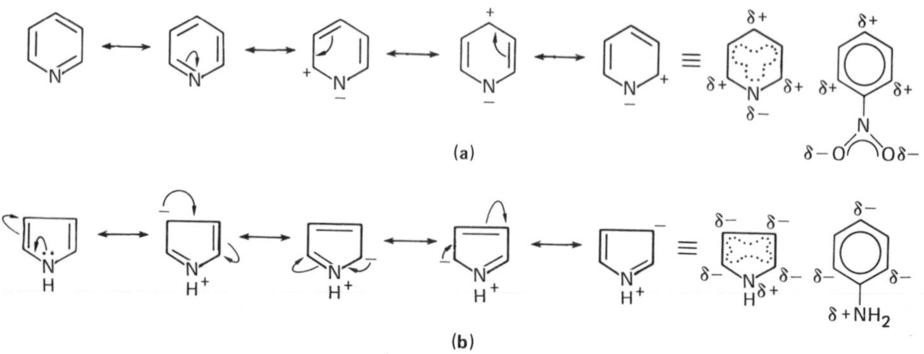

(a)

(b)

Fig. 1.3. (a) Mesomeric forms of pyridine and comparison with nitrobenzene. (b) Mesomeric forms of pyrrole and comparison with aniline

that the π electron density at nitrogen is increased relative to one, and the electron densities at the carbon sites in the ring are decreased relative to the carbon sites in benzene.

In the case of pyrrole and similar compounds which have the $-NH-$ group in place of one of the $-CH{=}CH-$ groups of benzene, the inductive effect of nitrogen is counteracted by the fact that nitrogen provides two electrons for the cyclic, conjugated, π bond system. In effect the nitrogen atom is losing a share of its electron pair to the rest of the ring. Thus in such a compound the π electron density at nitrogen is less than two (the number of π electrons nitrogen is in this case contributing), whilst the π electron densities at the carbon sites are greater than that at the carbon atoms in benzene.

Compounds of the pyridine type in which case the π electron density at carbon sites is less than that of benzene are termed π *deficient* heterocycles, whilst those of the pyrrole type in which the carbon sites have a greater π electron density than that of benzene are termed π *excessive* heterocycles.

The presence of an annular $-N{=}$ in a heterocyclic compound can be compared with the effect of the presence of a nitrogroup in benzene at such a position, and the presence of the $-NH-$ of pyrrole can be compared with the effect of an amino acid attached to the benzene ring.

The effect of the presence of nitrogen in an aromatic ring in these two situations can be shown, using the valence bond theory, by considering the possible mesomeric forms for such rings (Fig. 1.3).

In the the pyrimidine ring the annular nitrogen atoms are situated in the 1,3 positions, which results in a combination of their effects causing marked π electron deficiency at the 2,4, and 6 positions whilst the 5 position is also probably slightly electron deficient due to induction, although it is similar to a normal benzenoid position (Fig. 1.4).

In purine and pteridine each of the C positions is π electron deficient,

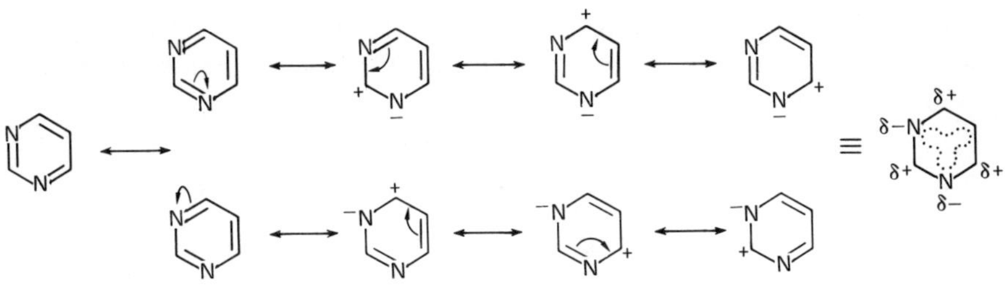

Fig. 1.4. Mesomeric forms of pyrimidine

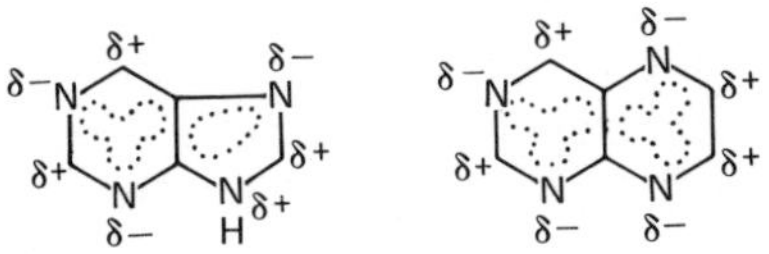

Fig. 1.5.

the effect being particularly marked in the case of pteridine, whilst in purine the effect is less due to the presence of a pyrrole-type –NH– group in the five-membered ring (Fig. 1.5).

Values for the π electron densities in these heterocycles have been calculated by several methods, including the use of valence bond theory and the use of self-consistent field molecular orbital theory. The results differ in detail and they must be regarded as approximations only, since different figures are obtained by varying some of the parameters involved and different approaches also give different results. However, the results obtained do show the same trends and indicate that in compounds related to pyridine sites adjacent to the annular nitrogen atom are relatively π electron deficient, whilst sites meta to nitrogen are normal and the sites in a pyrrole-type ring are relatively π excessive. Some values which have been obtained for some of the nitrogen heterocycles are given in Fig. 1.6.

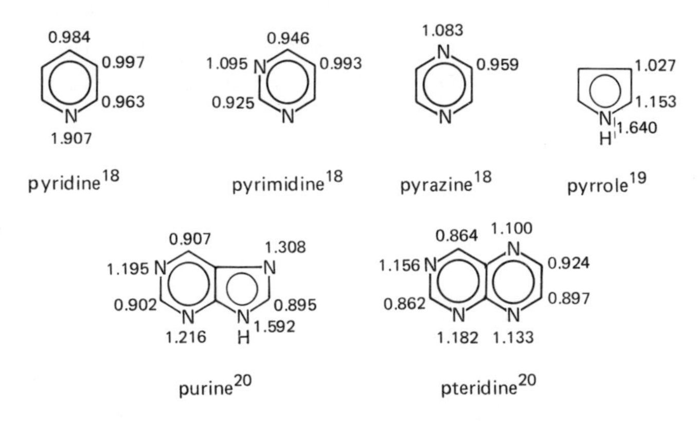

Fig. 1.6 π electron densities of some nitrogen heterocycles

The unequal π electron distribution in the heterocyclic rings greatly affects the physical and chemical properties of these compounds and these are discussed below.

(D) GENERAL PROPERTIES OF NITROGEN HETEROCYCLES

All of the nitrogen heterocycles are basic, the base strengths of some of the parent compounds being listed in Table 1.1.

Pyridine, and related compounds, are basic due to the presence of the lone pair of electrons in an sp^2 hybrid orbital of the nitrogen atom, this electron pair not being involved in bonding and thus being available for sharing with an appropriate electron acceptor. Protonation thus does not affect the aromaticity of the ring. Pyridine ($pK_a \sim 5.25$) is a stronger base than aniline ($pK_a \sim 4.63$), which has a lowered basicity due to the conjugation of the lone pair electrons of the exocyclic nitrogen with the aromatic ring, but it is a much weaker base than aliphatic amines ($pK_a \sim 10$). This has been attributed to the greater 's' character of the nitrogen lone pair electrons in pyridine[21]. The presence of the extra nitrogen atoms in the polyaza heterocycles decreases their basicity relative to pyridine due to the electron-attracting power of the other nitrogen atoms and to unfavourable charge localization in the protonated forms relative to the unprotonated forms of these heterocycles.

The pK_a of pteridine is anomalous. It would be expected to be a much weaker base. However, due to covalent hydration (see section 3C(i) and (ii)f) the basicity of N3 is greatly increased.

In the case of pyrrole the nitrogen lone pair electrons are not available for sharing with an electron acceptor as they are involved in the cyclic conjugated π bond system. The aromaticity of the ring is lost on protonation, and pyrrole is a very weak base, being protonated on C2 and not on the nitrogen atom (ref. 19, p. 255).

The presence of the –NH– group in pyrrole and purine gives these compounds weakly acidic properties (comparable with methanol, $pK_a \sim 15.5$) since the proton can be removed without disruption of the aromatic character. However, these compounds are weaker acids than water and their salts are immediately hydrolysed by water.

Table 1.1

Compound	Basic pK_a	Acidic pK_a	Reference
Pyrrole	−3.80	16.5	22 (pp. 60, 61)
Pyridine	5.23 (5.27)	—	22 (p. 146)
Pyrimidine	1.3	—	18 (p. 28)
Purine	2.4	8.9	23 (p. 26)
Pteridine	4.1	—	20 (p. 367)
(Aniline	4.63)	—	—

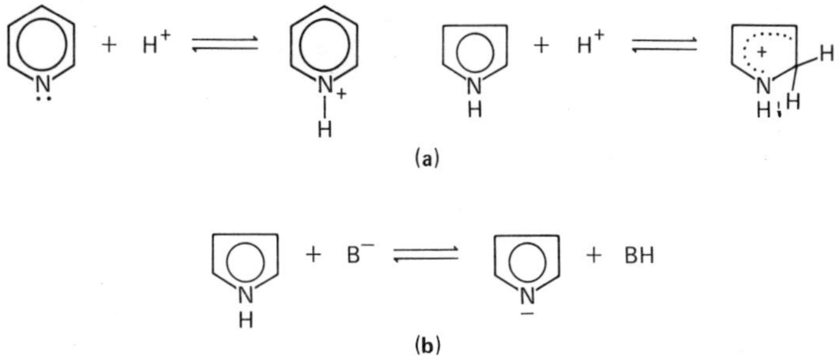

(a)

(b)

Fig. 1.7. Protonation of pyridine and pyrrole. (b) Proton loss from pyrrole

The π deficient heterocycles are deactivated to electrophilic substitution relative to benzene and such reactions usually proceed slowly and require the use of high temperatures. The presence of the annular nitrogen atom lowers the π electron availability of the ring and also destabilizes the charged intermediate formed in electrophilic attack due to unfavourable localization energies in the formation of the cation. In addition to these effects co-ordination of the lone pair of the annular nitrogen with the electrophile occurs. This has the effect of decreasing the concentration of free base, hence lowering the rate of formation of the intermediate, whilst electrophilic attack on the complex is not favoured due to electrical repulsion between the similarly charged species and due to unfavourable localization energies in the charged intermediate.

Relative to benzene all the positions of the pyridine ring are deactivated to electrophilic attack, but those *meta* to the nitrogen atom are least affected, so that when this position is vacant, the 3-position is the site of electrophilic substitution. A comparison of the stabilities of the

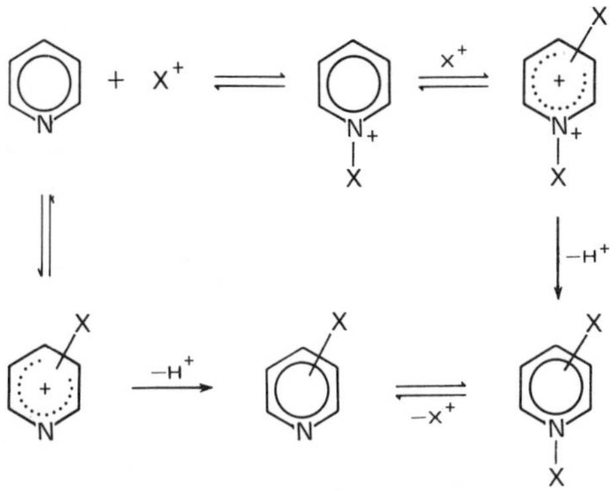

Fig. 1.8. Electrophilic attack on pyridine and the pyridinum ion

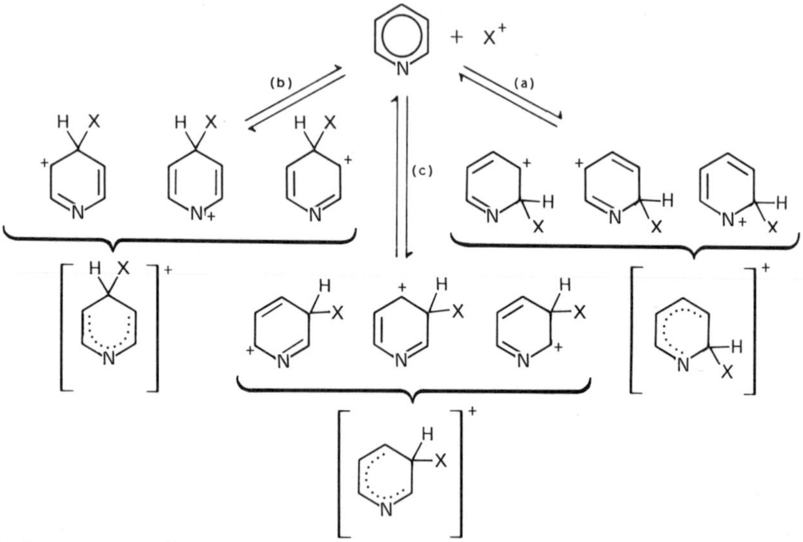

Fig. 1.9. Electrophilic attack on pyridine: (a) 2-position; (b) 4-position; (c) 3-position. Scheme (c) gives most favourable change localizations $-\overset{+}{N}-$ is an unfavourable form

intermediates for attack at the 2–, 3–, and 4–positions of the neutral pyridine molecule can be made by considering the mesomeric forms of the charged intermediate. The intermediate having the largest number of favourable forms is preferred. (Fig. 1.9).

The presence of electron-releasing substituents on the pyridine ring partly off-sets the deactivation to electrophilic substitution by increasing the π electron availability in the ring and by stabilizing the charged intermediate. Pyridines having such substituents thus undergo electrophilic substitutions more readily than pyridine, whilst the presence of electron-attracting substituents accentuates the electron deficiency of the pyridine ring and destabilizes the intermediate. As a result pyridines having electron-attracting substituents rarely undergo electrophilic substitutions.

The presence of extra nitrogen atoms in the polyaza-heterocycles acts in a similar way to the presence of electron-attracting groups, and such compounds are even more deactivated to electrophilic substitution than pyridine, and these reactions occur only if the polyaza-heterocycles have appropriate counter-acting electron-releasing groups if they occur at all.

In contrast to electrophilic attack nucleophiles attack pyridine derivatives with comparative ease. Localization energies for the possible intermediates indicate[24] that attack at positions 2– and 4– are preferred to attack at position 3. (Fig. 1.10). Thus in the π deficient heterocycles the positions adjacent to the anullar nitrogen atoms are readily attacked by nucleophiles whilst those *meta* to the nitrogen atoms are relatively inert and resemble a benzene carbon site.

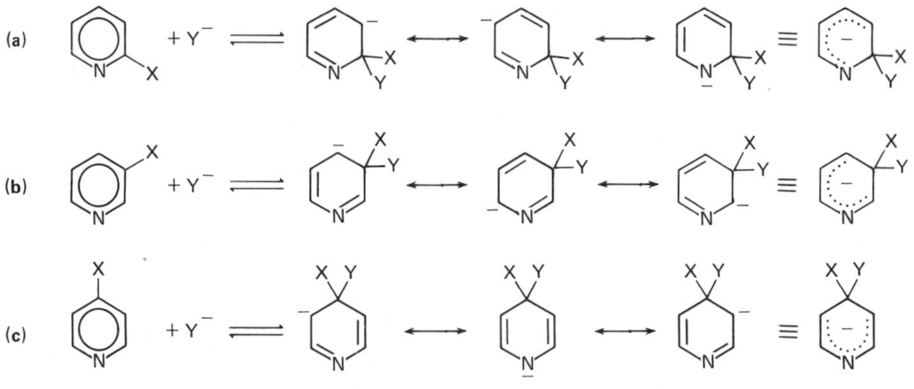

(a) (b) (c)

Fig. 1.10. Nucleophilic attack on pyridine: (a) 2-position; (b) 3-position; (c) 4-position, –N̄– is a favoured form

The presence of the nitrogen atom in π deficient heterocycles aids the decarboxylation of the *ortho* and *para* carboxylic acids and also activates alkyl groups at such positions to oxidation, nitrosation and Knoevenagel-type condensation reactions with aromatic aldehydes. But oxidation of the ring system of the π deficient heterocycles occurs less readily than the benzene ring whilst hydrogenation of such rings occurs more readily than the hydrogenation of benzene. The π excessive and the π deficient heterocyclic rings are more susceptible to addition reactions than are benzene rings.

The properties of the π excessive heterocycles of the pyrrole type are quite different from those of the π deficient type. The acidity of pyrrole has been mentioned above. Electrophilic attack occurs readily in such compounds, whilst comparatively little is known about free-radical attack or nucleophilic attack of pyrroles. In many cases electrophilic attack on the pyrrole ring occurs so readily that polysubstituted products are the only ones to be isolated, and in acid solutions polymerization of pyrroles readily occurs. Each position of the pyrrole ring is activated towards electrophilic attack, but it seems that attack at the 2–position is favoured.

A comprehensive account of the chemistry of the pyrrole and pyridine ring systems has been given by Schofield.[22]

(E) TAUTOMERISM IN THE NITROGEN HETEROCYCLES

Tautomerism is the ability of a compound which has a π bond and a mobile atom, such as hydrogen, in a conjugated position to exist in two or more different forms in which the positions of the π bond and the hydrogen atom are changed. The *tautomers* are in thermodynamic equilibrium and this is a special case of isomerism. Such tautomerism is well known in the case of amides as well as other carbonyl compounds. (Fig. 1.11).

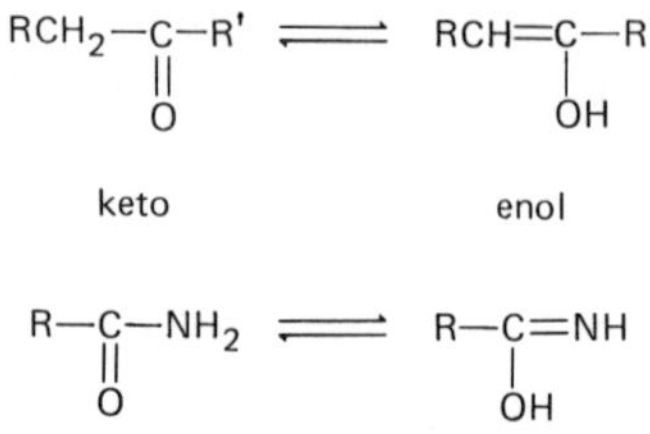

$$RCH_2{-}\underset{\underset{O}{\|}}{C}{-}R' \rightleftharpoons RCH{=}\underset{\underset{OH}{|}}{C}{-}R'$$

keto enol

$$R{-}\underset{\underset{O}{\|}}{C}{-}NH_2 \rightleftharpoons R{-}\underset{\underset{OH}{|}}{C}{=}NH$$

Fig. 1.11. Tautomeric forms of a ketone and an amide

The possibility of the existence of tautomerism in the nitrogen heterocycles had been postulated for some time, but the chemical investigations were not satisfactory and clear evidence which showed the predominant tautomer in each case was not obtained until the i.r., u.v., and n.m.r. spectroscopic methods were developed. The tautomers of some pyridine derivatives are shown in Fig. 1.12.

Much i.r., u.v., and n.m.r. spectroscopic evidence has now been accumulated which indicates that the 2- and 4-hydroxypyridines exist almost entirely in the lactam (amide) forms rather than the lactim (imino) forms, but the 3-hydroxypyridines show no such tendency for tautomerisation and are true hydroxy compounds. The tautomers of 3-hydroxypyridine in which the aromaticity of the ring is lost, and the zwitterionic structure, are unfavourable. However, the amido forms for the 2- and 4-hydroxypyridines are favoured and such compounds are pyridones and not 'hydroxypyridines' and this is reflected by the chemical characteristics of these compounds.

The case of the hydroxypyrroles is complex and there is no clear picture of which form predominates. The position of the equilibrium seems to depend on the solvent, the conditions, and also on the other substituents in

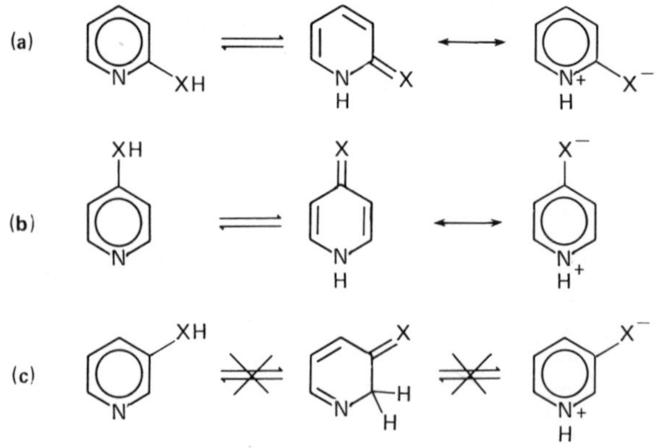

(a) N XH ⇌ N H X ⟷ N+ H X⁻

(b) XH N ⇌ X N H ⟷ X⁻ N+ H

(c) XH N ⇌̸ X N H H ⇌̸ X⁻ N+ H

Fig. 1.12. Some possible tautomers of pyridine derivatives (X = O, S, NH, CH_2)

(a) [pyrrole-2-XH ⇌ tautomer with exocyclic =X]

(b) [pyrrole-3-XH ⇌ tautomer with exocyclic =X]

Fig. 1.13. Possible tautomers of pyrrole derivatives

the ring (see ref. 22, p. 104). Each tautomer having an exocyclic π bond has lost the aromaticity of the ring, but nevertheless such tautomers do occur (Fig. 1.13).

The mercaptopyridines (X = S) also exist predominantly in the thione rather than the thiol form, and are thus similar to the hydroxy compounds, but the amino (X = NH) and alkyl (X = CH_2) pyridines show no appreciable degree of tautomerism and seem to exist entirely in the amino and alkyl forms.

Similar observations have also been made in the case of the pyrimidines, purines, and pteridines and the predominant tautomers in many compounds have been established. The existence of particular tautomeric forms in certain cases is of extreme importance in the biochemical properties of such compounds and this topic will be discussed further in the relevant sections of subsequent chapters. A detailed account of the tautomerism of a wide variety of heterocycles has recently been published[25] and a comprehensive review of the tautomerism of pyrimidines of biological importance is also available.[26]

A very comprehensive account of the physical properties and physical studies of heterocyclic compounds including accounts of u.v., i.r., n.m.r., and mass spectra amongst others, is given in the series *Physical Methods in Heterocyclic Chemistry*.[27]

REFERENCES

1. C. W. Scheele, 'Examen Chemicum Calculi Urinari', *Opuscula*, **2**, 73 (1776).
2. L. Brugnatelli, *Giornale di fisica, Chimica, et storia naturale (Pavia)* deacada seconda, **I**, 117 (1818).
3. L. Brugnatelli, *Ann. Chim.*, (2), **8**, 201 (1818).
4. E. Runge, *Ann. Phys.*, **31**, 65 (1834).
5. T. Anderson, *Annalen*, **60**, 86 (1846).
6. F. G. Hopkins, *Nature*, **40**, 335 (1889); **45**, 197, 581 (1892).
7. F. G. Hopkins, *Trans. Roy. Soc.*, (B), **186**, 661 (1895).
8. F. A. Kekulé, *Ber.*, **2**, 362 (1869); *Annalen*, **162**, 77 (1872).
9. F. Miescher, *Hoppe-Seyler's Medicinische—Chemische Untersuchungen*, A. Hirschwald, Berlin, 441 (1871).

10. R. Altmann, *Archiv für Anatomie und Physiologie, Physiol. Supple.*, 524 (1889).
11. A. Kossel and A. Neumann, *Ber.*, **27**, 2215 (1894).
12. C. Funk, *J. Physiol.*, **45**, 489 (1912–13); **46**, 173 (1913).
13. A. Baeyer, *Annalen*, **127**, 199 (1863); **130**, 129 (1864).
14. E. Fischer and J. von Mering, *Therap. d. Gegenwart,* **44**, 97 (1903); **45**, 145 (1904).
15. *Aromaticity, An International Symposium, Sheffield, 1966*, The Chemical Society, Sp. pub. No. 21, 1967.
16. G. M. Badger, *Aromatic Character and Aromaticity*, Cambridge University Press (1969).
17. *Handbook for Chemical Society Authors*, The Chemical Society, Sp. pub. No. 14 (1960).
18. From: D. J. Brown, *The Pyrimidines*, Wiley–Interscience, New York and London p. 27 (1962).
19. From: M. H. Palmer, *The Structure and Reactions of Heterocyclic Compounds*, Arnold, London, p. 255 (1967).
20. From: G. M. Badger, *The Chemistry of Heterocyclic Compounds*, Academic Press, New York and London p. 366 (1961).
21. C. K. Ingold, *Structure and Mechanism in Organic Chemistry*, Bell, London, p. 174 (1953).
22. From: K. Schofield, *Hetero-aromatic Nitrogen Compounds——Pyrroles and Pyridines*, Butterworths, London (1967).
23. From: J. H. Lister, *Fused Pyrimidines, Part II, Purines*, Wiley—Interscience, New York, London, Sydney, Toronto (1971).
24. R. D. Brown and M. L. Heffernan, *Austral. J. Chem.*, **9**, 83 (1956).
25. J. Elguero, C. Marzin, A. R. Katritzky, and P. Linda. The tautomerism of heterocycles in *Advances in Heterocyclic Chemistry*, Supplement I, Academic Press, New York, San Francisco, London (1976).
26. J. S. Kwiatkowski and B. Pullman, *Advances in Heterocyclic Chemistry*, **18**, 199 (1975).
27. A. R. Katritzky (ed.), *Physical Methods in Heterocyclic Chemistry*, Vols. I–VI, Academic Press, London and New York (1963–1974).

Some suggested text-books for further reading:

R. M. Acheson, *An Introduction to the Chemistry of Heterocyclic Compounds*, 3rd edn., Wiley–Interscience, New York, London, Sydney, Toronto (1976).

R. C. Elderfield (ed.), *Heterocyclic Compounds*, Vols. 6 and 7, Wiley, New York and London (1957, 1961).

Progress in Nucleic Acid Research and Molecular Biology, Academic Press, New York, San Francisco, London, Vols. 1–19. (1963–1976).

Advances in Heterocyclic Chemistry, Academic Press, New York, San Francisco, London. Vols. 1–20 (1963–1976).

H. R. Mahler and E. H. Cordes, *Biological Chemistry*, Harper and Row New York, Evanston, San Francisco and London, 2nd edn., (1971).

R. T. Morrison and R. N. Boyd, *Organic Chemistry*, 3rd edn., Allyn and Bacon, Boston, (1973).

Chapter 2

The Chemistry of the Pyrimidines

(A) INTRODUCTION

Pyrimidine is the compound 1,3- (or *meta*) diazine, and may be regarded as being derived from benzene by the replacement of two *meta* −CH= groups by −N=. It is the parent heterocycle of a very important group of compounds which have been studied for many years, much of the work having been instigated because of the occurrence of pyrimidine derivatives in living systems and because some derivatives were found to have biological activity.

In this account the pyrimidine ring will be represented as illustrated in Fig. 2.1, with the numbering system shown.

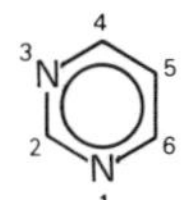

Fig. 2.1. The pyrimidine ring

The 'hydroxy' and 'mercapto' pyrimidines will be written as hydroxy and mercapto derivatives and not as oxo or thione forms, irrespective of evidence for the existence of a particular, predominant, tautomeric form, unless there is a need to represent an alternative tautomer to illustrate a particular point or to facilitate the understanding of a particular point.

Such a system makes the nomenclature of these pyrimidines simpler and is more convenient for the representations of the pyrimidine ring, and should not lead to any confusion as long as the points concerning the tautomerism of the hydroxy and mercapto groups in nitrogen heterocycles are appreciated (Section 1A).

This is the system which has been used in the excellent, definitive, accounts of pyrimidine chemistry by D. J. Brown.[1,2] The confusion which may arise by using a system other than this can be shown by considering the compound 4-hydroxypyrimidine (Fig. 2.2).

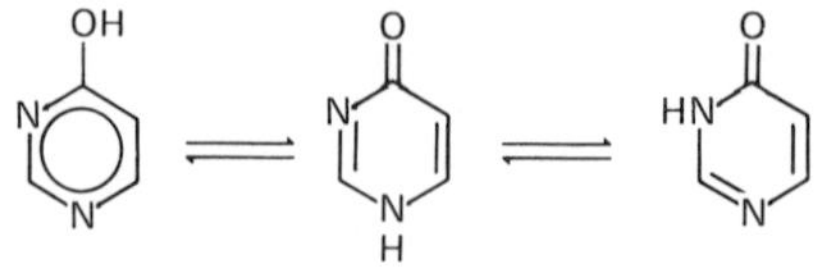

Fig. 2.2. 4-hydroxypyrinidine

This compound could be called 4-hydroxypyrimidine, 4-pyrimidol, 4-pyrimidinol, 4-pyrimidone, 4-pyrimidinone, 4(1)-pyrimidone, 4(3)-pyrimidone, 1,4-dihydro-4-oxo-pyrimidine, 3,4-dihydro-4-oxopyrimidine, and other names still, and this is only using the English language. Similar problems arise with the nomenclature of mercaptopyrimidines and when the pyrimidine ring is partially reduced as well as having such a tautomeric group. Use of the full systematic name can give a false impression of the real oxidation state of the ring.

(B) THE FIVE PRINCIPLES OF PYRIMIDINE CHEMISTRY

Brown[1] has listed five simple principles of pyrimidine chemistry which are useful in understanding the properties and reactions of pyrimidines and which often enable predictions to be made concerning the properties of any pyrimidine derivative. These principles are a summary of some of the general properties of π deficient nitrogen heterocycles, but together account for the unique basis of pyrimidine chemistry.

(i) The 'active' 2-, 4-, and 6-positions

The 2-, 4-, and 6-positions of the pyrimidine ring have a marked π electron deficiency (see Section 1C) due to the presence of the 1,3-annular nitrogen atoms and thus these positions resemble the 2-, and 4-positions of pyridine and the 2-, 4-, and 6-, positions of nitrobenzene. In the case of pyrimidine the 1,3- positioning of the nitrogen atoms results in reinforcement of their effects so that the resultant is greater than in the isomers pyrazine and pyridazine in which the effects of the nitrogen atoms partly antagonize one another.

Such 'active' positions are readily attacked by nucleophiles, for example resulting in ready removal of halogens in such positions by water, amines, etc. Carboxy groups situated at these active positions are labile, and methyl groups are readily oxidized, undergo ready reaction with aromatic aldehydes, and in some cases undergo C-nitrosation. It is very rare for electrophilic substitution to occur at such an active position.

(ii) The 'aromatic' 5-position

The 5-position of the pyrimidine ring most closely resembles a true 'aromatic' position, being made only slightly π electron deficient by induction. As a result, substituents in such a position resemble the benzene analogues in their reactivity, for example halogens are more resistant to nucleophilic displacement, methyl groups are less readily attacked, and carboxy groups are less labile than such groups when present in the 2-, 4-, or 6-position of the pyrimidine ring. However, if electrophilic substitution occurs at all, then the 5-position is the site of attack. A few cases of electrophilic attack at the 6-position are known in pyrimidines having several electron-releasing substituents and the 5-position occupied.

(iii) The effect of electron-releasing substituents

The introduction of electron-releasing substituents—hydroxy, amino, mercapto, etc.—counteracts the π electron deficiency of the pyrimidine ring such that it now more closely approximates to the 'true aromatic' ring. Electrophilic substitution is facilitated, and the 2-, 4-, and 6-positions are deactivated towards nucleophilic attack.

(iv) The effect of electron-attracting substituents

The π electron deficiency of the pyrimidine ring is further enhanced by the introduction of electron-attracting substituents and pyrimidines having such substituents are more susceptible to nucleophilic attack and readily form covalent hydrates. For example, 2,4-dichloro-5-nitropyrimidine(**1**) reacts very readily with ammonia at 0°C to give 4-amino-2-chloro-5-nitropyrimidine (**2**). The displacement of the second chlorine occurs much more slowly as the newly introduced electron-releasing amino group antagonizes the electron-attracting effect of the nitro group and is best carried out above 100°C when the diamino product, (3), is obtained (see ref. 1, p. 8).

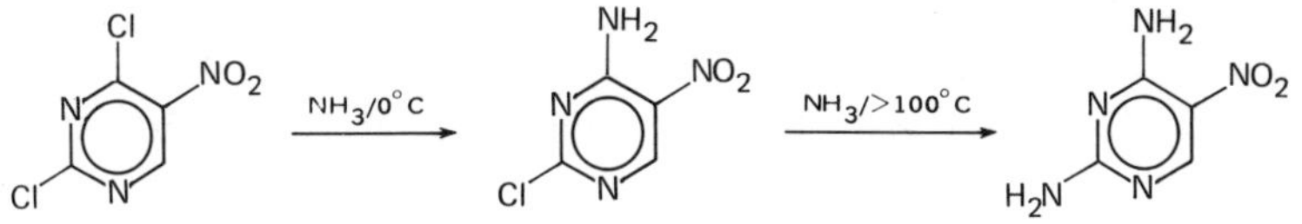

An interesting apparent nucleophilic displacement of hydride ion occurs when 4,6-dichloro-5-nitropyrimidine (**4**) reacts with the diethylmalonyl anion to give the product (**5**).[3] However, the actual mechanism of the reaction is unknown.

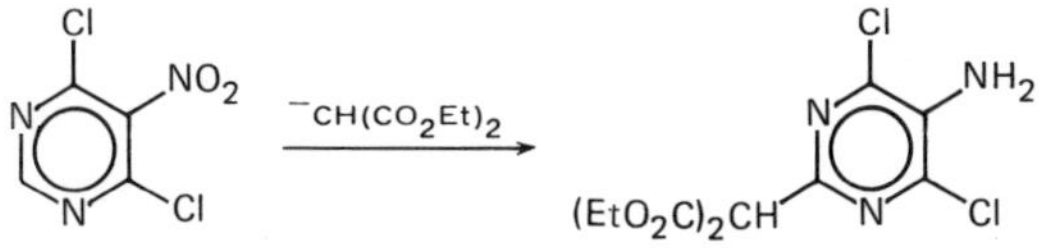

(v) Tautomerism of substituents

The fact that hydroxy and mercapto groups, when situated *ortho* to the nitrogen of nitrogen heterocycles, have the oxo and thione forms as the predominant tautomer has been mentioned in Section 1E. Such tautomerism is also possible in derivatives of pyrimidine and it has been established that 2-, 4-, and 6-hydroxy- and mercaptopyrimidines exist in the oxo and thione forms respectively, whilst the aminopyrimidines do not exist to any appreciable extent in the imino form (Fig. 2.3). The 5-hydroxy- and mercaptopyrimidines show no tendency to form the oxo or thione tautomers.

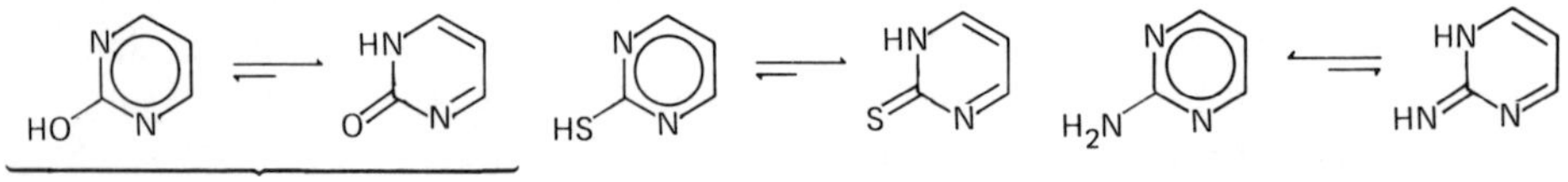

(6)

Fig. 2.3. Tautomerism of 2-substituted pyrimidines

Thus N- as well as O-alkylation is possible in compounds such as 2-hydroxypyrimidine (**6**).

The tautomerization of such groups can also involve the 5-position, although this seems to occur only when there is no annular nitrogen atom available. Thus barbituric acid (2,4,6-trihydroxypyrimidine; Fig. 2.4), which was at one time thought to exist in the dioxomonohydroxy form (**7b**), has now been confirmed to have the trioxo form (**7c**) as the predominant tautomer by spectroscopic and X-ray crystallographic studies.[4]

Further points about barbituric acid and the barbiturates will be made later (see Section 10B).

Some discussion has arisen concerning the tautomerism of 4,6-dihydroxypyrimidine (Fig. 2.5) (see ref. 2, p. 396), but it seems that the major tautomer is an hydroxy-oxo form (**8b**) with a small equilibrium concentration of the dioxo form (**8c**). Proposals for the existence of significant contributions from zwitterionic (**8d**) or di-zwitterionic (**8e**) forms have not found favour.[5]

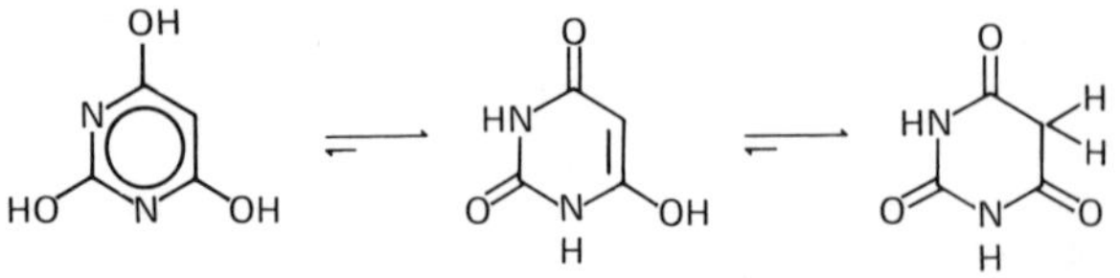

(7a) (7b) (7c)

Fig. 2.4. Some tautomeric forms of barbituric acid

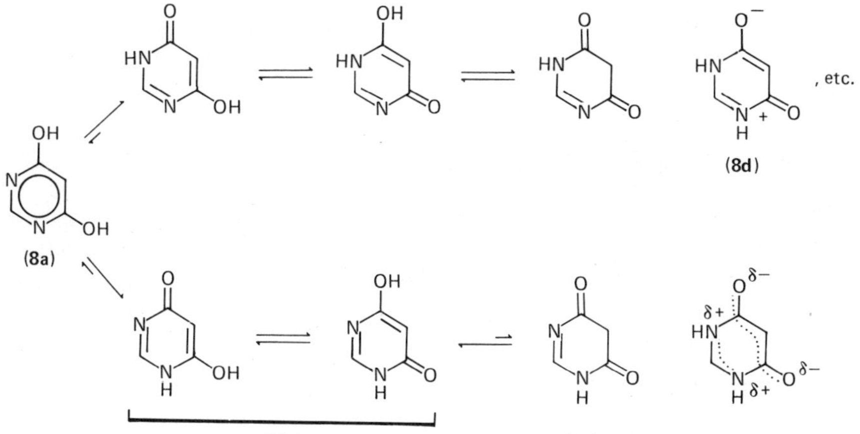

Fig. 2.5. Tautomeric forms of 4,6,-dihydroxypyrimidine

The use of u.v. and of n.m.r. spectroscopy to study the tautomerization of substituted pyrimidines has been very rewarding and has given valuable results, although there is still much work to be carried out in this field.

The importance of the existence of particular tautomeric forms in certain cases for understanding the structures and roles of nucleic acids is referred to elsewhere (see Chapter 7).

(C) SYNTHESIS OF THE PYRIMIDINE RING

The pyrimidine ring can be synthesized from acyclic precursors or can be obtained by ring expansion, isomerization, or degradation of other ring systems. Syntheses starting from acyclic precursors are the most common and, as synthetic methods, are probably the most useful, although reactions producing pyrimidines from other ring systems are of mechanistic and theoretical interest for the most part.

The term 'principal' or 'common' synthesis is used to describe the most widely used of pyrimidine syntheses, that is, the condensation of an N—C—N with a C—C—C fragment. Of the other possible reactions, the condensation of a C—C—C—N fragment with a C—N fragment is also useful, whilst reactions involving an N–C–C–C–N condensation with a C fragment, and other possible condensations, are limited in their applicability.

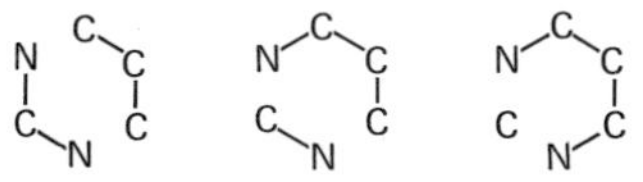

Fig. 2.6. Some possible condensations to form the pyrimidine ring

In the following discussion most attention will be devoted to the principal synthetic method, although examples of other types of pyrimidine synthesis will be given.

(i) The Principal Synthetic Method

In general this approach to pyrimidine synthesis involves the condensation of an amidine or an urea with a 1,3-bifunctional three-carbon fragment. These functional groups may be either carbonyl groups or nitriles (Fig. 2.7).

Table 2.1 lists the usual types of N–C–N fragment, and Table 2.2 lists some common three carbon fragments.

Fig. 2.7. The formation of N-C bonds

Table 2.1

N–C–N fragment		2-substituents in pyrimidine ring
NH_2CONH_2	urea	Hydroxy
NH_2CSNH_2	thiourea	Mercapto
$NH_2C(=NH)NH_2$	guanidine	Amino
$NH_2C(OR)=NH$	e.g. *O*-methylurea	Substituted oxy
$NH_2C(SR)=NH$	e.g. *S*-methylurea	Substituted thio
$NH_2C(R)=NH$	e.g. acetamidine	Alkyl, aryl

Table 2.2

C–C–C fragment	Typical product With XC:NH.NH_2
1,3-Dialdehydes, e.g. 1,1,3,3-tetraethoxypropane	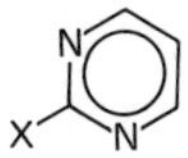
1,3-Diketones, e.g. acetylacetone	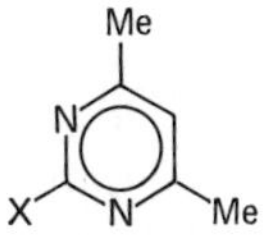
β-Ketoesters, e.g. ethylacetoacetate	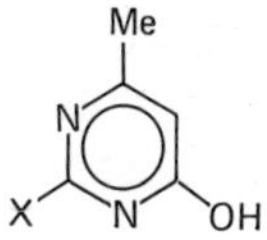
β-Cyanoesters, e.g. ethyl cyanoacetate	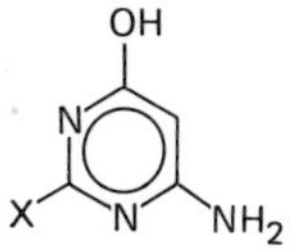
1,3-Diesters, e.g. diethyl malonate	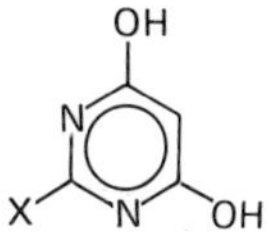
1,3-Dinitriles, e.g. malonitrile	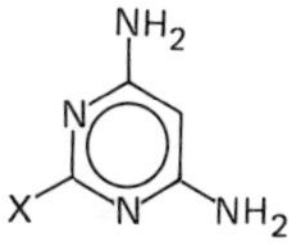

Thus condensation between an NH_2 group and an aldehyde group results in a pyrimidine which is unsubstituted at position 4(6), condensation with a ketone give an alkyl or aryl substituent at position 4(6), an acid, ester, or amide gives an hydroxy substituent, and a nitrile an amino substituent at these positions. The use of formamidine as the N–C–N fragment would produce a pyrimidine unsubstituted at position 2, but owing to the ease of hydrolysis of formamidine under normal condensation conditions the yield of the pyrimidine, if any is formed at all, is often very poor. In practice many combinations of the above types of compound do not condense together to form the desired pyrimidine but, nevertheless, by a judicious choice of intermediates and reaction conditions, a very wide range of pyrimidines can be obtained by such reactions. Further reactions of the products obtained should lead to the formation of any required pyrimidine but, because of the limitations of many reactions, and the restrictions imposed by the chemistry of the pyrimidine ring, several simple substituted pyrimidines have still to be synthesized.

The conditions under which the N−C−N and C−C−C fragments can be condensed to form pyrimidines vary with the compounds concerned and a variety of conditions has been used. Some compounds react together without solvent and without catalyst, for example, guanidine carbonate and acetylacetone give 2-amino-4,6-dimethylpyrimidine (**9**) on heating together at 100°C for about 30 min.[6] However, such condensation reactions usually require the presence of a solvent, although the catalyst may be either an acid or a base, the choice depending on the intermediates being used. For example, urea and acetylacetone react to form 2-hydroxy-4,6-dimethylpyrimidine (**10**) on refluxing with concentrated hydrochloric acid in ethanol.[7]

Malic acid (**11**) in fuming sulphuric acid reacts to form formylacetic acid (**12**) which condenses in the presence of urea to form uracil (2,4-dihydroxypyrimidine) (**13**).[8]

$NH_2-C^+(NH_2)_2 \ \tfrac{1}{2}(CO_3)^{2-}$ + $CH_3.CO.CH_2.CO.CH_3$ —heat, 100°C→ (9) [2-amino-4,6-dimethylpyrimidine: H_2N, Me, Me]

$NH_2.CO.NH_2$ + $CH_3.CO.CH_2.CO.CH_3$ —HCl/EtOH, reflux→ (10) .HCl [2-hydroxy-4,6-dimethylpyrimidine: HO, Me, Me]

$CO_2H.CH_2.CHOH.CO_2H$ (11) —fuming H_2SO_4→ $CO_2H.CH_2.CHO$ (12) ⇌ $CO_2H.CH{=}CHOH$ —urea→ (13) [HO, OH pyrimidine]

But the majority of these condensation reactions seem to proceed best in the presence of bases which may be either organic or inorganic, for example acetamidine condenses with acetylacetone in aqueous potassium carbonate to give 2,4,6-trimethylpyrimidine (**14**);[9] sodio nitromalondialdehyde (**15**) condenses with benzamidine in the presence of Triton B (40% aqueous trimethylammonium hydroxide) to give 5-nitro-2-phenylpyrimidine (**16**).[10] However, the most common method of carrying out a cyclization to form a pyrimidine ring is to use sodium ethoxide in refluxing ethanol. For example, acetamidine reacts with diethyl malonate to give 2-methyl-4,6-dihydroxypyrimidine (**17**).[11] Some typical examples of the pyrimidine principal synthetic method are given in Table 2.3, but extensive lists of these reactions are given in two books by D. J. Brown.[1,2]

Table 2.3 Some examples of the principal synthetic method for pyrimidines

Three-carbon fragment	N–C–N Fragment	Conditions	Product
Ethyl benzoylacetate	*S*-Methylthiourea	Alkali at RT for 48 h	Ph, N, MeS, N, OH
Benzoylacetone	Urea	Ethanolic HCl/reflux for 34 h	Ph, N, HO, N, Me
Ethyl benzoylacetate	Acetamidine	Alkali at RT for 14 days	Ph, N, Me, N, OH
Diethylmalonate	Guanidine	Ethanolic sodium ethoxide reflux for 0.5 h	OH, N, H_2N, N, OH
1,1,3,3-Tetraethoxypropane	Urea	Ethanolic HCl, reflux for 6 h	N, HO, N
Bromomucic acid	Benzamidine	Aqueous Triton B, RT overnight	CO_2H, Br, N, Ph, N

Table 2.3 continued

Three-carbon fragment	N–C–N Fragment	Conditions	Product
Ethyl cyanoacetate	Urea	Ethanolic sodium ethoxide, reflux, 3 h	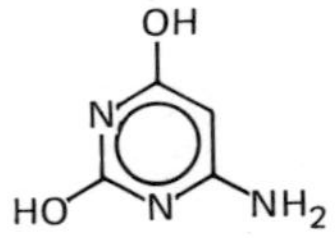
Malonitrile	Guanidine	Ethanol, reflux	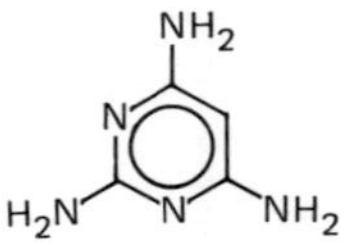
Malonitrile*	Formamidine	Sodium ethoxide in ethanol, RT for 24 h	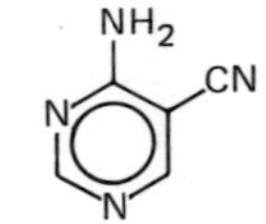
Ethyl ethoxymethylene cyanoacetate	*O*-Methylurea	Sodium methoxide in methanol, RT, 12 h	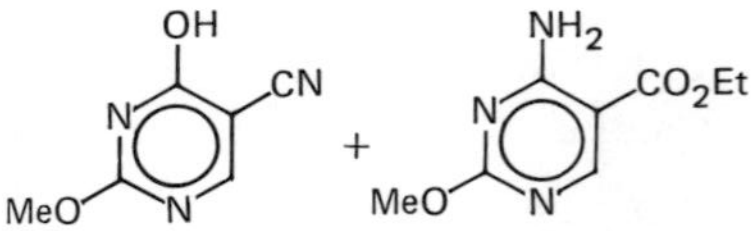
Acetylacetone	*N*-Methylurea	Ethanolic HCl, reflux 3 h	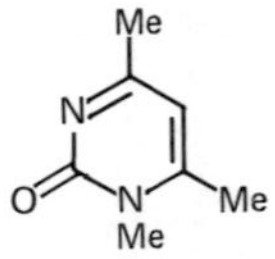

*Malonitrile usually undergoes abnormal reactions (see ref. 1, p. 72), and this is not a true principal synthesis

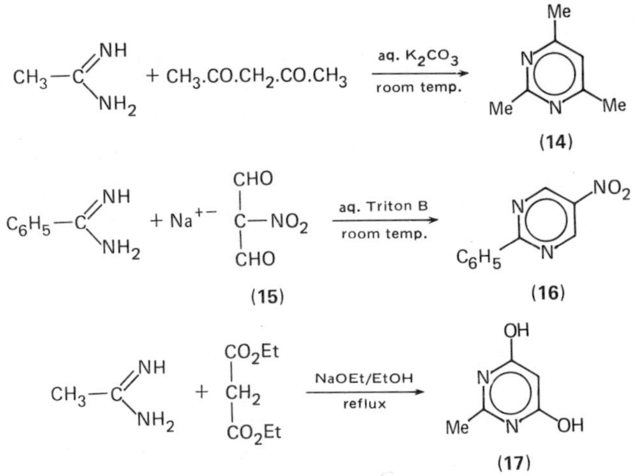

(14)

(15)

(16)

(17)

(ii) Other Synthetic Methods

The synthesis of the pyrimidine ring from C–C–C–N and C–N fragments is exemplified by the reaction between ethyl aminomethylene malonitrile **(18)** and ethyl acetimidate to give 4-amino-5-cyano-2-methylpyrimidine **(19)**.[12] Several similar reactions have been reported, but in the case of condensations involving ethyl aminomethylene cyanocetate **(20)** or similar compounds, pyrimidines with either 5-cyano-4-hydroxy or 4-amino-5-ethoxycarbonyl groups may be obtained, depending on the reactive grouping involved in ring closure. It has been reported[12] that the reaction of **(20)** with ethyl acetimidate gives 4-amino-5-ethoxycarbonyl-2-methylpyrimidine **(21)** whilst ethyl benzoimidate gives 5-cyano-4-hydroxy-2-phenylpyrimidine **(22)**.

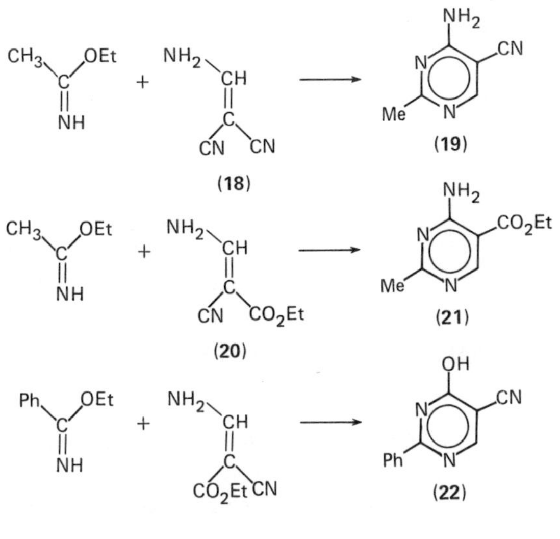

(18)

(19)

(20)

(21)

(22)

The condensation of malondiamide (**23**), an N–C–C–C–N fragment, with esters or amides as one carbon fragments, results in the formation of 4,6-dihydroxypyrimidines and is useful in some cases, for example 4,6-dihydroxypyrimidine itself may be conveniently made by such a synthesis. This type of reaction, first described by Remfry,[13] was further investigated by Hull[14] and has been termed the Remfry–Hull synthesis. Malondiamidine (**24**) can also be used in this type of condensation to give 4,6-diaminopyrimidines.

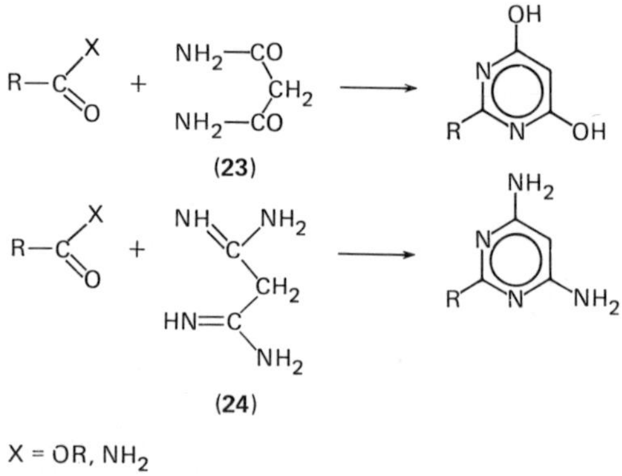

(23)

(24)

X = OR, NH_2

A variety of other reactions can lead to the pyrimidine ring system but most of them are not of general applicability and are of limited use. A comprehensive account of such reactions is also given in the books of D. J. Brown[1,2] which have been referred to earlier.

Of several methods of obtaining the parent compound pyrimidine, probably the most useful involves reaction between 1,1,3,3-tetraethoxypropane (the ethyl acetal of malondialdehyde) and formamide.[15] The mechanism of the reaction is not fully established but may involve an intermediate such as (**25**).

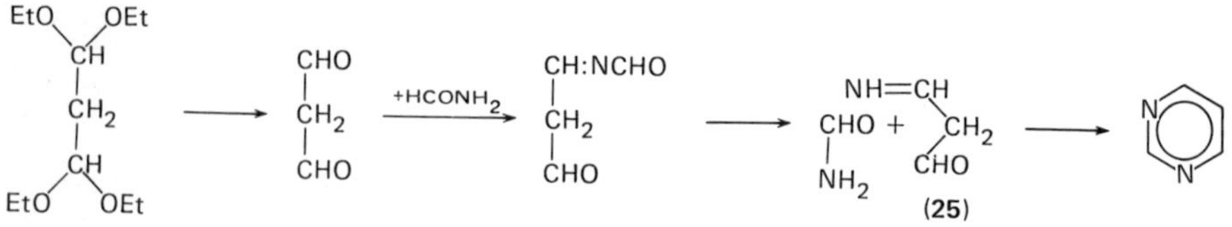

+HCONH$_2$

(25)

(D) GENERAL PROPERTIES, IONIZATION, AND SPECTRA OF PYRIMIDINES

Pyrimidine itself is a water-soluble colourless hygroscopic solid with m.p. 22.5°C, dipole moment 2.13 D, and being a weak base [1,2] (pK_a 1.31 and *ca*. -6.3). The general properties of the ring are considerably modified by the presence and orientation of substituents. In general pyrimidines having alkyl, aryl, alkoxy, alkylthio, or halogen substituents are liquids or low melting solids and are more soluble in organic than aqueous solutions. The introduction of polar groups causes a marked increase in melting point and a decrease in the solubility in organic solvents. The presence of groups such as amino, hydroxy, mercapto, initially increases the water solubility of pyrimidines, but polysubstitution by such groups causes a sharp decrease in the solubility in all solvents. The reason for this is that such compounds are capable of forming strong intermolecular hydrogen bonds.

(i) Ionization of Pyrimidines

Pyridine is a strong base with pK_a 5.2, but the electron-attracting power of the second nitrogen atom in pyrimidine means that in comparison pyrimidine is a much weaker base, having pK_a 1.31. This value is comparable with that of 3-nitropyridine (pK_a 0.8) which again demonstrates the analogy between an annular nitrogen atom and a nitro group at a similar site in an aromatic compound. In the same way that the basicity of anilines is increased by the introduction of electron-releasing substituents, the introduction of an electron-releasing group causes an increase in the basicity of a pyrimidine: for example, 4-methylpyrimidine[16] has pK_a 2.0, 4-methoxypyrimidine[17] pK_a 2.5, and 4-acetamidopyrimidine[18] pK_a 2.76.

In the case of the aminopyrimidines, the effect of the amino group is very much greater than might be expected when present in the 2 or 4 position, and causes a much greater increase in the basicity of pyrimidine than the introduction of a 5-amino group. The values quoted for 2-. 4-. and 5-aminopyrimidine respectively are pK_a 3.54, 5.71, and 2.60. This is due to the extra delocalization possible in the cation in the cases of 2- and

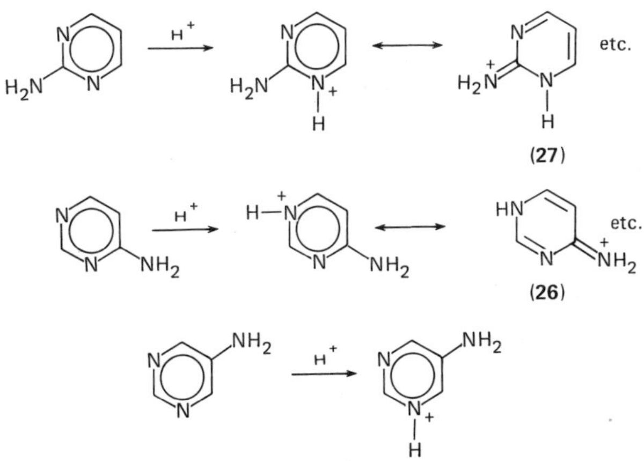

Fig. 2.8. Protonation of the aminopyrimidines

4-amino pyrimidine, and hence an increase in stability of the ion relative to the unprotonated species, Fig. 2.8. This effect is not possible in the case of 5-aminopyrimidine, which shows only a slight increase in basicity relative to pyrimidine.

The 4-amino group has a more pronounced effect than the 2-amino group since the *p*-quinonoid form **(26)** is more favoured than the *o*-quinonoid form **(27)** and the cation thus formed is more stable than in the case of 2-aminopyrimidine. The introduction of more amino groups further enhances the basic character of a pyrimidine to an extent governed by the ability of the group to aid further delocalization in the cation.

The simple hydroxy derivatives of pyrimidine also show an increased basicity relative to pyrimidine, the pK_a values for 2-, 4-, and 5-hydroxypyrimidine respectively being 2.24, 1.85, 1.87. However, in the case of the 2- and 4-isomers the effect is not that of an electron-releasing effect by the hydroxy group but, since they exist as the pyrimidone tautomers (Section 2B(v)), their enhanced basicities are due to delocalization aided by the second annular nitrogen atom (Fig. 2.9).

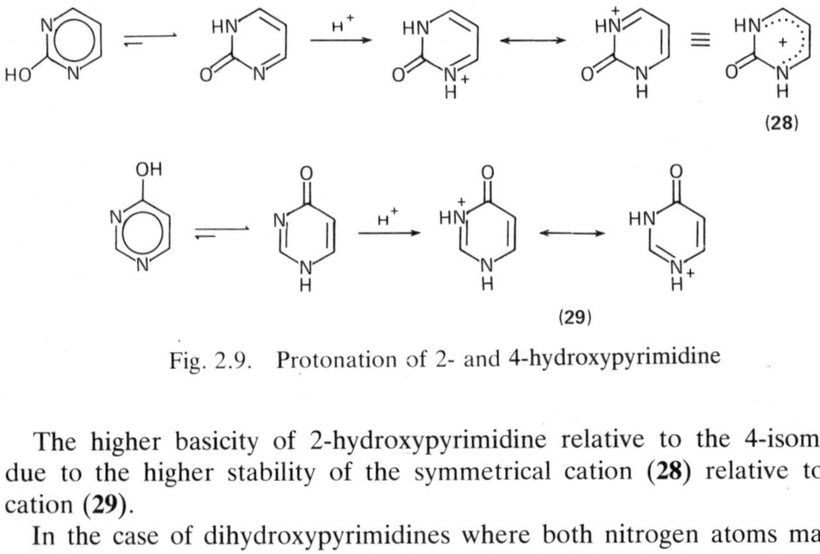

(28)

(29)

Fig. 2.9. Protonation of 2- and 4-hydroxypyrimidine

The higher basicity of 2-hydroxypyrimidine relative to the 4-isomer is due to the higher stability of the symmetrical cation **(28)** relative to the cation **(29)**.

In the case of dihydroxypyrimidines where both nitrogen atoms may be involved in lactam tautomerization, the basicity is lower than for pyrimidine. Uracil (2,4-dihydroxypyrimidine **(30)**), for example, has pK_a -3.38.

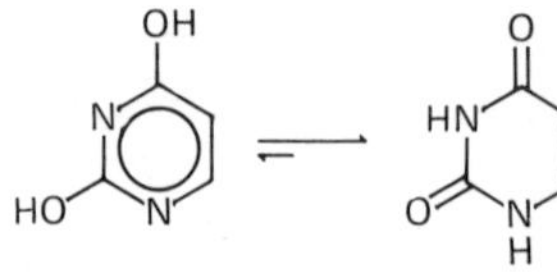

(30)

Table 2.4

Compound	Acidic pK_a
Phenol	9.95
p-Nitrophenol	7.14
2,4-Dinitrophenol	4.4
1,3,5-Trinitrophenol	0.38
2-Hydroxypyridine	11.62
3-Hydroxypyridine	8.72
4-Hydroxypyridine	11.09
2-Hydroxypyrimidine	9.17
4-Hydroxypyrimidine	8.59
5-Hydroxypyrimidine	6.58
2-Mercaptopyrimidine	7.14
4-Mercaptopyrimidine	6.87
2,4-Dimercaptopyrimidine	6.35, 12.57
4,6-Dimercaptopyrimidine	3.60, 9.70
2-Carboxypyrimidine	2.85
Benzoic acid	4.20
p-Nitrobenzoic acid	3.44
o-Nitrobenzoic acid	2.17

N-Methylation of hydroxypyrimidines also lowers their basicity due to the limitation of delocalization and lack of symmetry of the cation, causing a reduction in its stability.

The introduction of electron-attracting substituents into the pyrimidine ring causes the expected result of a further diminution in basicity, for example 2-amino-5-cyanopyrimidine, pK_a 0.66; 2-amino-5-nitro-pyrimidine, pK_a 0.35. However, it has been found that in a few cases the introduction of an *electron-releasing* group can also reduce basicity, for example 4,5-diaminopyrimidine has pK_a 6.04; 4,5-diamino-6-methylpyrimidine has pK_a 5.93; 4-amino-6-methoxypyrimidine, pK_a 4.02; 4-amino-2-methoxypyrimidine, pK_a 5.3; cf. 4-aminopyrimidine, pK_a 5.71. This unexpected result has not yet been adequately explained.

In addition to the expected acid character of the pyrimidine carboxylic acids, hydroxy and mercaptopyrimidines can also ionize as acids. The acidic pK_a of some pyrimidines and comparable pyridines and phenols are given in Table 2.4.

Using the analogy of a cyclic =N– being comparable in its effect to a nitro group present on the ring at a similar position, it might be expected that the hydroxypyrimidines would have similar pKa values to nitrophenols and dinitrophenols respectively. However, this is not the case for the 2- and 4-hydroxypyridines or pyrimidines because these exist as the lactam tautomers and the acidity of such compounds should be more comparable to that of lactams, although a somewhat higher acidity should be evident because of the further delocalization possible for the anion, Fig. 2.10.

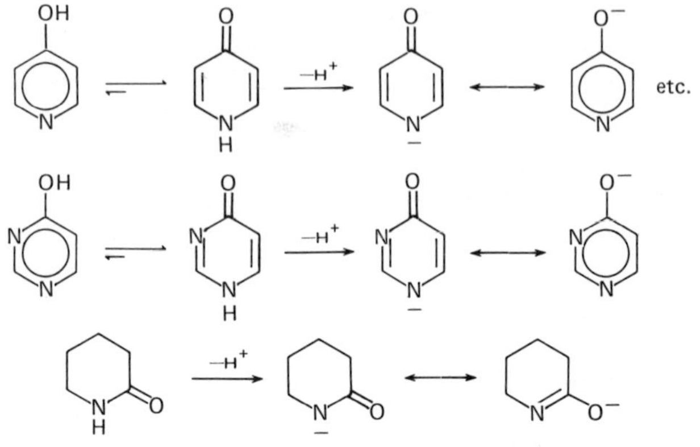

Fig. 2.10. Ionization of 4-hydroxypyridine, 4-hydroxypyrimidine and δ-lactam

In fact 2- and 4-hydroxypyridine are weaker acids than phenol, 2- and 4-hydroxypyrimidine are slightly more acid than phenol and the pyridine analogues due to the presence of the second nitrogen atom, but they do not approach the acidity of the nitrophenols. 3-Hydroxypyridine and 5-hydroxypyrimidine, which are true phenols, have pK_a values which are much closer to their nitrophenol counterparts.

The introduction of electron-attracting or electron-releasing substituents into hydroxypyrimidines has the expected effects, but the introduction of further hydroxy groups needs some consideration. The pK_a values for some polyhydroxypyrimidines are given in Table 2.5.

The introduction of an hydroxy group to 2- or 4-hydroxypyrimidine to give uracil (2,4-dihydroxypyrimidine) reduces the acidity slightly since both hydroxy groups are present as the lactam tautomer and the lactam group has a slight and weakening effect on the other group. A similar effect is observed in the case of 4,5-dihydroxypyrimidine when the lactam group reduces the acidity of the 5-hydroxy group slightly, but in the case of 4,6-dihydroxypyrimidine the di-lactam structure is not possible and the favoured form is (**31**), in which the second hydroxy group is phenolic. The

Table 2.5 (Data from refs. 1 and 2)

Pyrimidine	Acidic pK_a	Basic pK_a
2,4-Dihydroxy-	9.46, 13.2	−3.38
4,5-Dihydroxy-	7.48, 11.61	1.99
4,6-Dihydroxy-	5.4	0.26
2,4,5-Trihydroxy-	8.11, 11.48	—
2,4,6-Trihydroxy-	3.9, 12.5	—
2,4,5,6-Tetrahydroxy	2.83, 11	—

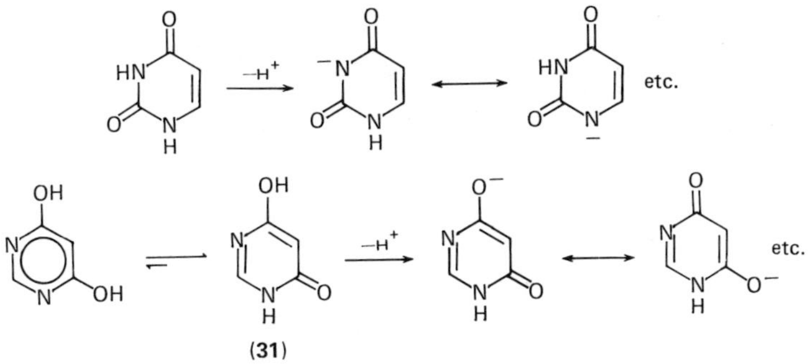

(31)

Fig. 2.11. Ionization of 2,4- and 4,6-dihydroxypyrinidine

anion is stabilized by delocalization and a reasonably high acidity is observed.

2,4,5-Trihydroxypyrimidine has a pK_a similar to that of 4,5-dihydroxypyrimidine, but the pK_a of barbituric acid (2,4,6-trihydroxypyrimidine) is considerably lowered, as is that of 2,4,5,6-tetrahydroxypyrimidine. The reason for the comparatively high acidity of barbituric acid has been attributed to the loss of a proton from the 5-position and stabilization of the anion (Fig. 2.12), but this has been questioned and a satisfactory explanation has not yet been given.

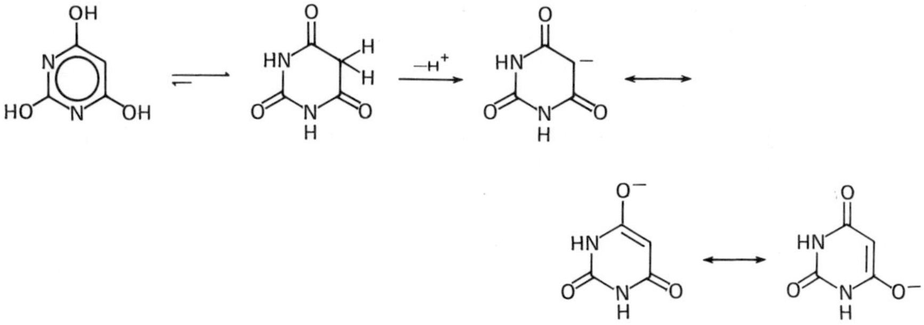

Fig. 2.12. Ionization of barbituric acid

(ii) Spectra of Pyrimidines

(a) Ultraviolet spectra

Like benzene and other aromatic compounds, pyrimidines have characteristic u.v. absorption spectra. However, pyrimidines are basic and, in aqueous solution, can exist as neutral or cationic species, whilst pyrimidines having substituents such as carboxy, hydroxy, or mercapto

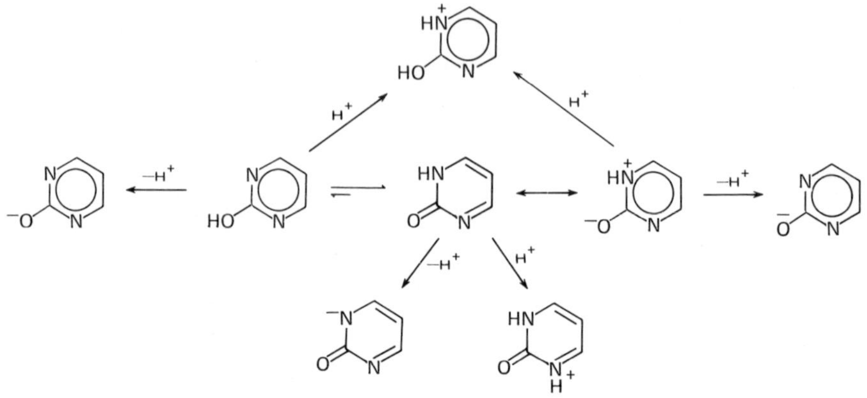

Fig. 2.13.

groups can exist in anionic forms, in other tautomeric neutral forms, or in cationic forms. The possible forms of 2-hydroxypyrimidine are shown in Fig. 2.13.

Thus the spectrum of a pyrimidine in an aqueous solution of arbitrary pH may well represent a combination of the spectra of more than one form of the compound, and the spectrum obtained for any pyrimidine will be pH dependent. In measuring the u.v. spectrum of a pyrimidine in aqueous solution it is therefore always desirable to state the pH of the solution, and for a more complete characterization, to measure the pK_a value of the pyrimidine and to record the spectrum of each ionic form separately, having adjusted the pH of the solution to at least two units above or below the pK_a values for the pyrimidine.

The u.v. spectrum of pyrimidine consists of two bands, a π–π^* band at 243 nm, similar to the 250 nm aromatic absorption of benzene, and an n–π^* band at higher wavelength—271 nm in water at pH 7, 280 nm in ethanol, and 298 nm in cyclohexane. The n–π^* absorption band is due to an electronic transition from a nitrogen lone-pair non-bonding orbital to an empty π orbital of the ring. The band moves to lower wavelength on changing from a non-polar to a more polar solvent since the lone-pair electrons on the annular nitrogen atoms are involved in hydrogen bonding in a polar solvent and extra energy is therefore required to enable the electronic transition to occur.

The presence of electron-releasing substituents in a pyrimidine causes an hypsochromic shift of the n–π^* band (shift to shorter wavelength), whilst an electron-attracting substituent causes a bathochromic shift (to longer wavelength), but neither type of substituent affects the intensity of the band. Both types of substituent cause a bathochromic shift of the π–π^* band, but cause changes of intensity of the band depending on the nature and position of the substituent.

When several substituents are present in the ring the spectrum of the pyrimidine shows approximately additive effects due to the individual substituents, but when more than one substituent capable of tautomerism is present the additive rules are not followed. This is because such compounds are not simple analogues of benzene, and the spectra of pyrimidines having strongly electron-attracting, potentially tautomeric substituents have been more satisfactorily interpreted by comparing them with the spectrum of the benzyl anion.[19]

Studies of the u.v. spectra of the amino-, hydroxy-, and mercaptopyrimidines have been used to establish the tautomeric form of these compounds which predominate in solution, and such studies have been in agreement with the results obtained using n.m.r. spectroscopy. For example, Fig 2.14 indicates the u.v. spectra of 4-hydroxypyrimidine **(32)**, 4-methoxypyrimidine **(33)**, 1,4-dihydro-1-methyl-4-oxopyrimidine **(34)**, and 1,6-dihydro-1-methyl-6-oxopyrimidine **(35)**. The similarity between the spectrum of **(32)** and those of **(34)** and **(35)**, and the difference from

(32) (33) (34) (35)

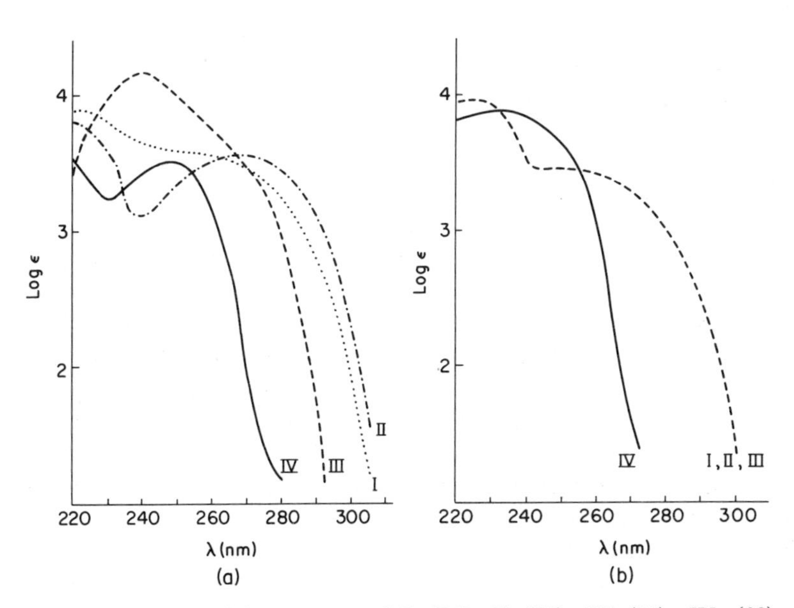

Fig. 2.14. Ultraviolet spectra of I, **(32)**; II **(35)**; III **(34)**; IV, **(33)**. (a), neutral species in water; (b), cations in water (data from ref. 17). Reproduced by permission of The Chemical Society

the spectrum of (**33**) is immediately apparent. Similar studies have shown that the amino form of aminopyrimidines, and the thione form of 2- and 4-mercaptopyrimidines, are the favoured forms.

(b) Infrared spectra

Because of the importance of pyrimidine derivatives in chemistry and biochemistry and because of their wide occurrence, studies of the i.r. spectra of pyrimidines have been extensive. Infrared spectroscopy has been used for identification and structural confirmation purposes as well as for the study of the tautomeric nature of substituted pyrimidines. A number of detailed spectroscopic studies have also been carried out and total assignments have been reported for several pyrimidines including pyrimidine itself and the 2,4,6-trifluoro and 2,4,6-trichloro derivatives.

A brief account of the i.r. spectra of pyrimidines is given by Brown[1] and an account of the i.r. spectra of nucleic acid components is given in Zorbach and Tipson.[20] More comprehensive reviews of the applications of i.r. spectroscopy to heterocyclic compounds in general are given in Katritzky.[21]

Of particular interest has been the study of the tautomerism of substituted pyrimidines and the results of these studies have been in agreement with those obtained from u.v. and n.m.r investigations. The i.r. spectra of 4- and 2-hydroxypyrimidine both in chloroform solution and in the solid state show C=O and N−H stretching vibration absorption bands[22,23] confirming the presence of the lactam tautomer as the predominant form. Such studies have also concluded that the aminopyrimidines exist in the true amino form. the spectra of aminopyrimidines in chloroform or carbon tetrachloride solution showing N−H stretching bands at 3400 and 3500 cm^{-1} respectively.[24,25] These absorption bands correspond to the symmetric and antisymmetric stretching vibrations of the primary amino group in aromatic amines.

The 2- and 4-hydroxypyrimidines show no strong absorptions in the free or H−bonded O−H stretching region. However, they do show C=O stretching vibrations at ~1640–1670 cm^{-1}. Barbituric acids (2,4,6-trihydroxypyrimidines) show C=O stretching vibrations at 1700–1750 cm^{-1}.

The C=N and C=C stretching vibrations of the pyrimidine ring occur at about 1600 cm^{-1} and the C=O, C=N, and C=C vibrations often couple with one another, resulting in complicated vibrations in this part of the spectrum.

The 2- and 4-hydroxypyrimidines show N−H stretching bands for the lactam NH at about 3100 cm^{-1}, this being a rather broad absorption band whilst the N−H stretching vibrations of aminopyrimidines occur at about 3400 cm^{-1}. The C–NH_2 stretching vibration of cytosine and its analogues occurs at ~ 1275 cm^{-1}.

The C–H stretching bands of pyrimidines usually occur in the 3000–3100 cm^{-1} region and are often hidden under the broad, strong absorption bands of N–H. The out-of-plane bending mode bands of C–H and N–H also occur at about the same part of the spectrum (NH ~ 825, CH ~ 800).

Extensive studies of the vibrations of the pyrimidine nucleus have been carried out and the spectra of many types of substituted pyrimidine have been recorded (see refs. 1, 21 and many others) but a detailed account of such spectra is not intended here and the reader is recommended to consult one of the many excellent books on the interpretation of i.r. spectra where a detailed account of the bands shown by particular atomic groupings will be given.

One point of interest which should be noted is that in most cases the characteristic C≡N stretch band of nitriles (~ 2200 cm^{-1}) when attached to the pyrimidine ring does not appear or is, at best, of very weak intensity.[26, 27]

(c) *Nuclear magnetic resonance spectra*

In recent years the application of n.m.r. spectroscopy, in particular ^{1}H magnetic resonance, has been very valuable in pyrimidine chemistry. As pyrimidine has a ring current, the hydrogen atoms attached to the ring absorb at low field, in the aromatic region of the spectrum. As the 4- and 6-positions of pyrimidine are equivalent, the hydrogen atoms at these positions absorb at the same frequency, and pyrimidine shows the ^{1}H n.m.r. spectrum of a broad singlet due to C–H at position 2 (9.26 δ) a doublet due to the 4 and 6 hydrogen atoms (8.78 δ), and a triplet of doublets due to the 5 hydrogen (7.36 δ), Fig. 2.15. The para coupling

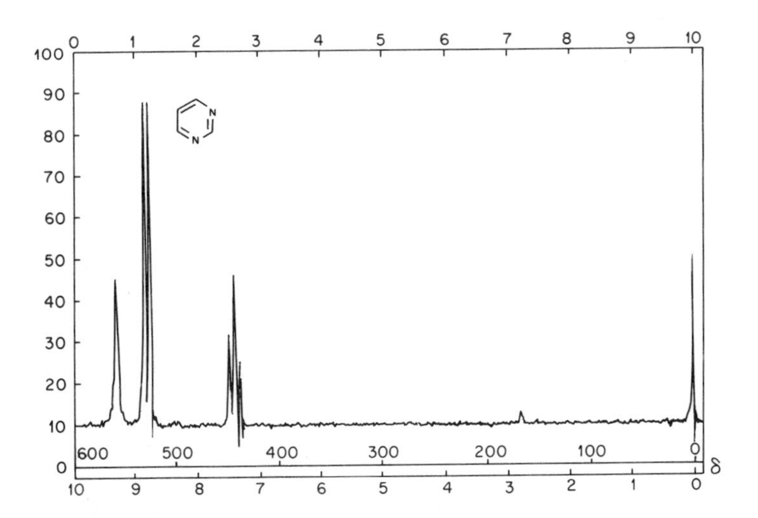

Fig. 2.15. ^{1}H n.m.r. spectrum of pyrimidine. Reproduced by permission of Aldrich Chemical Co., Inc.

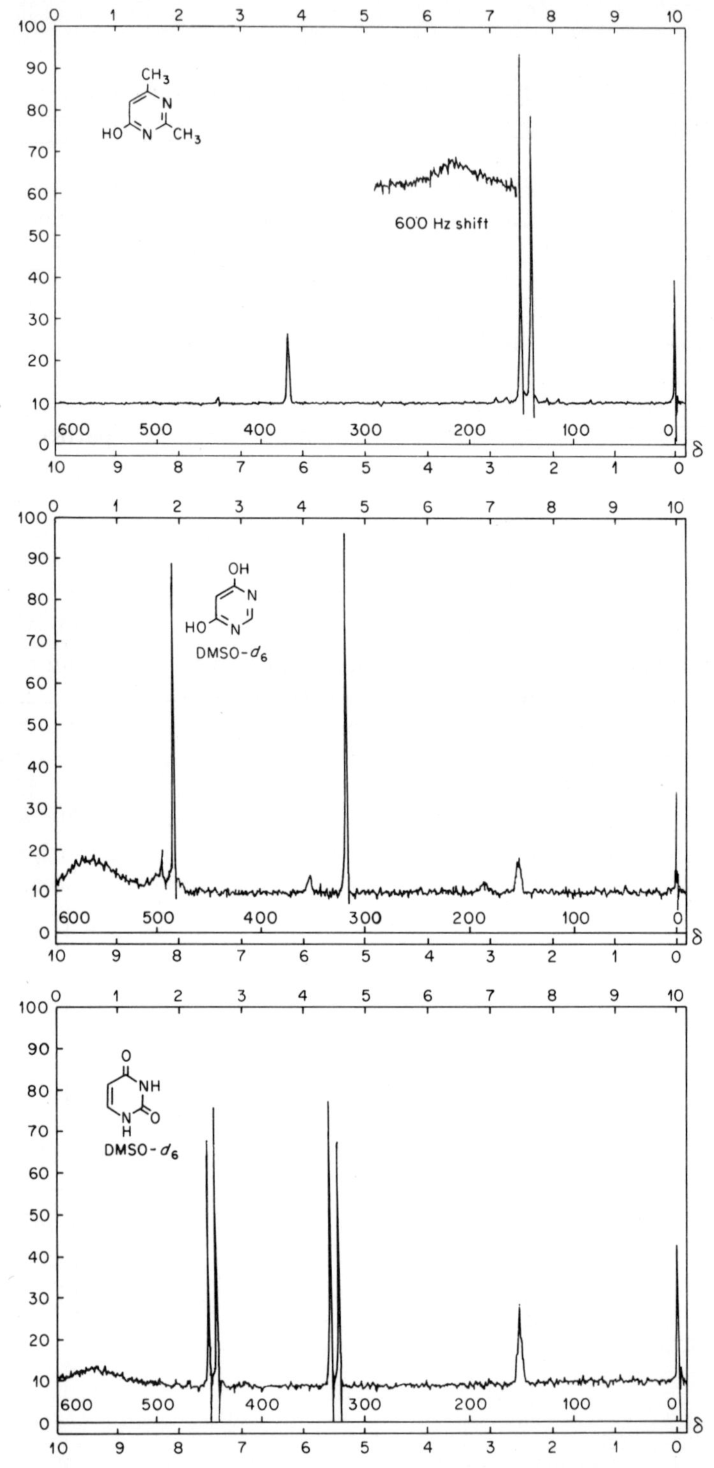

Fig. 2.16. [1]H n.m.r. spectra of some simple pyrimidines. Reproduced by permission of Aldrich Chemical Co., Inc.

constant $J_{2,5}$ is about 1.5 Hz and the splitting of H-2 not observed due to the broadening brought about by the presence of the two adjacent N atoms.

There is an overall deshielding effect on the pyrimidine hydrogen atoms due to the electron-attracting character of the nitrogen atoms, this effect being greatest at the 2-position (lowest chemical shift), then the 4-and 6-positions, whilst the 5-position is least affected. The introduction of electron-attracting groups causes the remaining ring hydrogen atoms to appear at a lower chemical shift. Similar effects are produced for the resonances of hydrogens attached to substituents at sites in the ring, for example CH_3 or NH_2 hydrogen atoms. The magnitude of the chemical shift effect upon the other ring positions, but not its direction, depends on the site of a substituent in the ring. The spectra of some typical pyrimidines are given in Fig. 2.16, and some values of typical chemical shifts and coupling constants are given in Table 2.6.

^1H n.m.r. studies of pyrimidines have been particularly valuable in elucidating problems of the tautomeric nature of pyrimidines. A typical example is that discussed by the late T. J. Batterham (see ref. 2, Chapter XIII, and ref. (5), the problem of the structure of the predominant form of 4,6-dihydroxypyrimidine (Fig. 2.17). This problem has also been discussed by Elguero *et al.*[28]

Table 2.6 ^1H n.m.r. spectral data from some pyrimidines (bracketed numbers refer to methyl groups; solvents are deuterated; data from Ref. 5)

Substituents	Solvent	δ_2	δ_5	$\delta_{4,6}$	$J_{2,4}$	$J_{2,5}$	$J_{4(6),5}$
$2-NH_2$	$(CD_3)_2CO$	—	6.56	8.27	—	—	—
	DMSO	—	6.54	8.22	—	—	—
2−CN	$(CD_3)_2CO$	—	7.86	9.04	—	—	5.10
2−Cl	$(CD_3)_2CO$	—	7.55	8.78	—	—	4.90
2−OH	$(CD_3)_2CO$	—	6.40	8.33	—	—	5.30
	DMSO	—	6.34	8.24	—	—	—
$2-OCH_3$	$(CD_3)_2CO$	(3.93)	7.07	8.60	—	—	4.8
	$CDCl_3$	(4.09)	7.09	8.72	—	—	6.0
$4-NH_2$	DMSO	8.39	6.44	8.04	—	1.25	6.15
4−OH	DMSO	8.17	6.29	7.86	0.7	1.05	6.75
$4-OCH_3$	$CDCl_3$	8.39	6.44	8.65(4.99)	—	—	7.0
5−CN	$(CD_3)_2CO$	9.28	—	9.00	—	—	—
5−OH	$(CD_3)_2CO$	8.69	—	8.39	—	—	—
	DMSO	8.70	—	8.37	—	—	—
$5-OCH_3$	$(CD_3)_2CO$	8.78	—	8.50	—	—	—
$2-OH,4-NH_2$	DMSO	—	5.62	7.36	—	—	7.2
4,6−diOH	DMSO	8.09	5.32	—	—	—	—
$4,6-diOCH_3$	D_2O	8.32	6.22	(3.96)	—	—	—
$4-OH,6-OCH_3$	DMSO	8.12	5.57	(3.80)	—	—	—
	D_2O	8.19	5.83	(3.90)	—	—	—
1−Me,4=O,6−OMe	D_2O	9.15	6.04	(3.53,N 2.86)			
1−Me,4−OMe,6=O	D_2O	8.31	5.88	(3.91,N 3.53)			

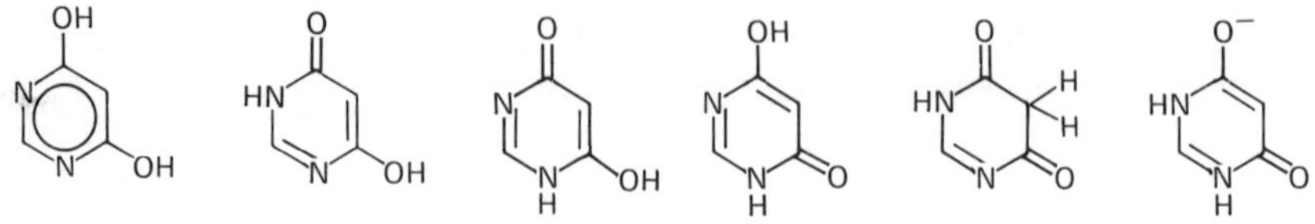

Fig. 2.17.

The first ^{1}H n.m.r. study[29] which was carried out in dimethyl sulphoxide solution showed the presence of only two singlets for the ring hydrogen atoms, indicating that the predominant form should be of type **(36)**–**(39)** or **(41)**–**(42)** but not the di-oxo form **(40)**. But u.v. spectroscopic evidence[30] had indicated that the di-oxo form was the predominant tautomer in aqueous solution. The addition of 50% deuterium oxide to a solution of 4,6-dihydroxypyrimidine in dimethylsulphoxide caused no change in the spectrum, indicating that the same species were present in both solutions. By comparison of the spectrum of 4,6-dihydroxypyrimidine with the spectra of the methylated derivatives **(43)**–**(46)**, it was concluded that **(37)** and **(39)** were the main species present with a very small amount of the di-oxo form **(40)**. These suggestions have been criticized,[31,32] but Batterham supported the ^{1}H n.m.r. interpretations. The chemical shifts for 4,6-dihydroxypyrimidine and the methylated derivatives are given in Table 2.6.

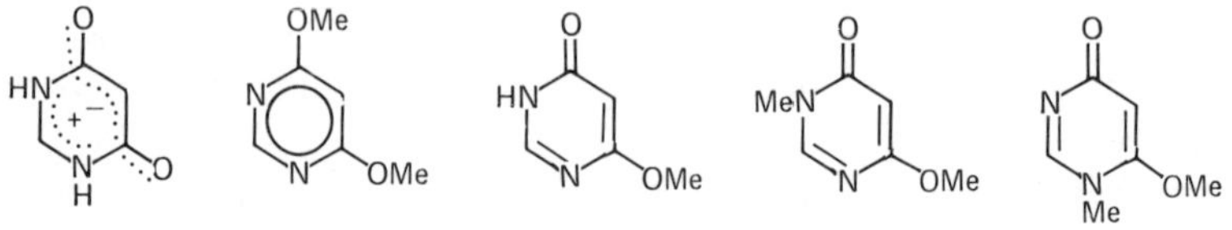

For discussion of ^{13}C and ^{15}N spectra of pyrimidines, the reader is directed to specialist texts and the original literature.

(d) Mass Spectra

The mass spectra of many pyrimidines have now been measured, although comparatively little correlation between spectra has been attempted. The mass spectrum of pyrimidine (Fig. 2.18) shows a peak due to the molecular ion but the base peak is at *m/e* 26 due to ionized acetylene. The loss of HCN from the molecular ion followed by the loss of a second HCN seems to be the principal mode of fragmentation. Substituted pyrimidines frequently have the molecular ion as the base peak and the loss of HCN or HCNO as a major fragmentation pathway. The proposed pathways[33] for the fragmentation of 2-amino-4,6-dimethyl- and 6-amino-2,4-dimethylpyrimidine are shown in Figs. 2.19 and 2.20.

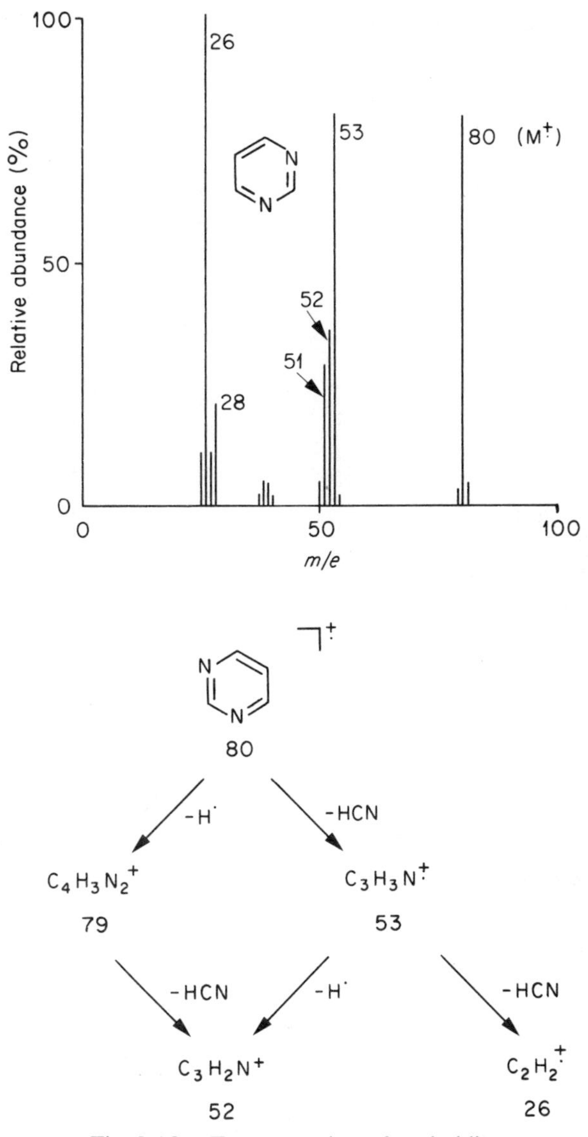

Fig. 2.18. Fragmentation of pyrimidine

The main fragmentation of pyrimidine nucleosides occurs at the C–N bond between the sugar and base with hydrogen transfer to the base. The further fragmentation patterns of such compounds include peaks due to fragmentation of the base, sugar, and some peaks due to losses from the nucleoside in which the glycosidic bond has not been cleaved. Examples of such spectra are given in refs. 34 and 35, and in original literature to which the reader is referred for the detailed spectra of individual pyrimidines.

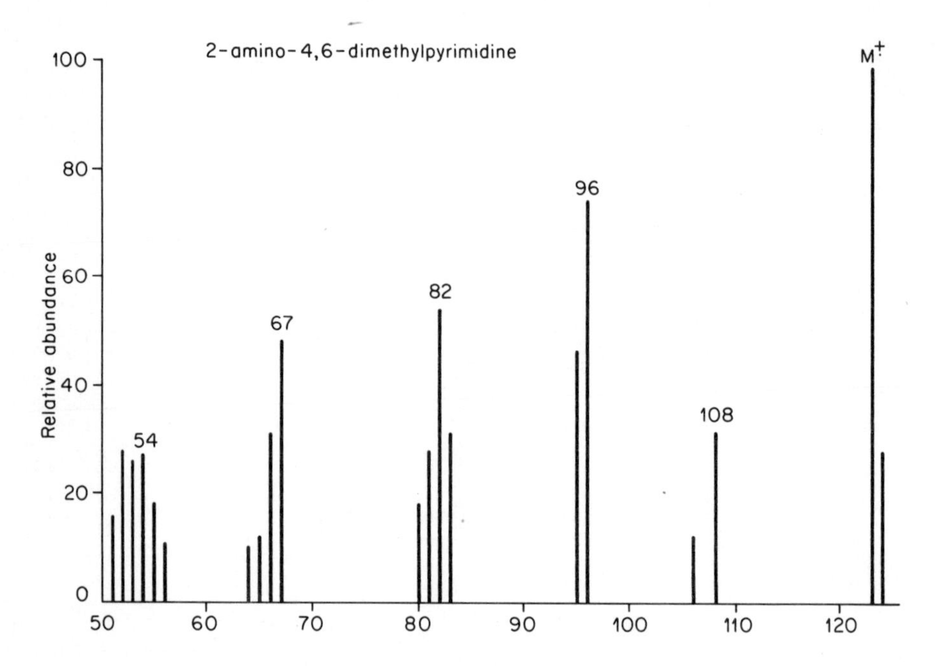

Fig. 2.19. Fragmentation of 2-amino-4,6-dimethylpyrimidine

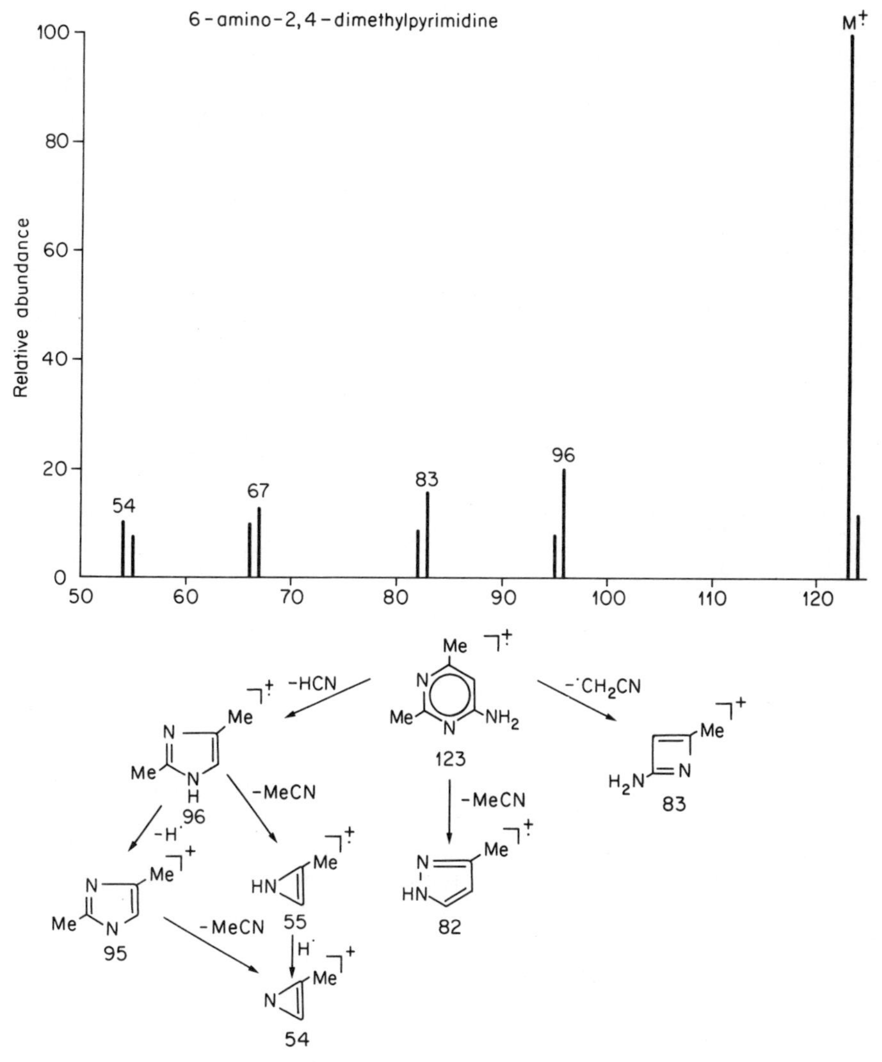

Fig. 2.20. Fragmentation of 4-amino-2,6-dimethylpyrimidine

(E) REACTIONS OF PYRIMIDINES

(i) Electrophilic substitution

The general reaction scheme for a normal aromatic electrophilic substitution is shown below

$$+ \overset{\delta+}{X}-\overset{\delta-}{Y} \rightleftharpoons \quad (H\ X,\ +) \quad + Y^- \longrightarrow \quad (X) \quad + H-Y$$

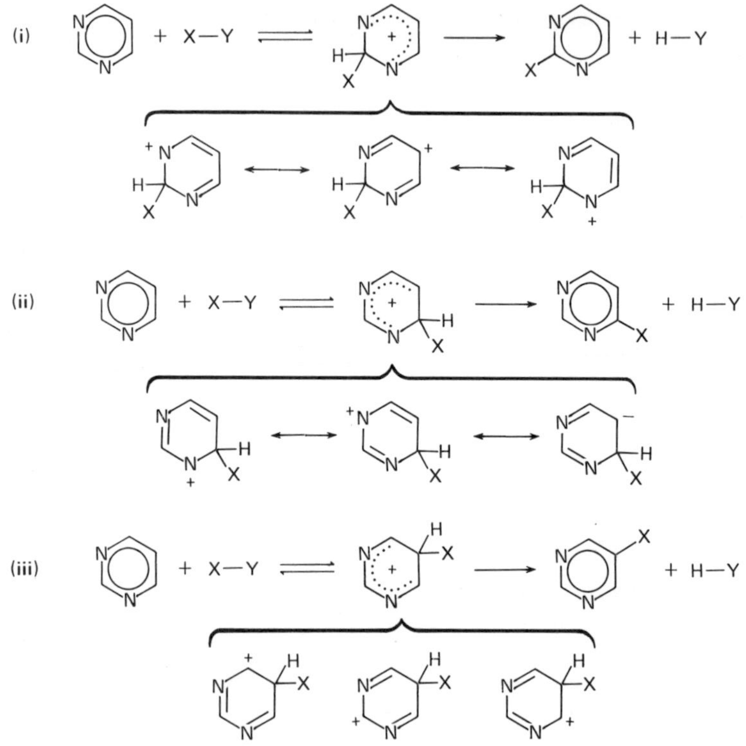

Fig. 2.21. Possibilities for electrophilic substitution of pyrimidine

In the case of pyrimidine there are three possible sites of attack—position 2, position 4(6) and position 5. The schemes for such reactions are illustrated in Fig. 2.21.

Inspection of the delocalization in the charged intermediates in Fig. 2.21 clearly shows that electrophilic substitution at position 5 is very much more favoured than such attack at positions 2 or 4, since the mesomeric forms in which sp_2 hybridized nitrogen carries a positive charge are very unfavourable energetically, and it is only for 5-attack that such forms do not contribute. There is also a greater ease of approach of the electrophile to the 5-position, as this is the only carbon site which is not significantly electron deficient (see Section 2B(*ii*)).

In fact only very few cases have been reported, so far, of electrophilic substitution in the pyrimidine ring at a position other than the 5-position. Isobarbituric acid (2,4,5-trihydroxypyrimidine), **(47)** and 2-amino-4,5-dihydroxypyrimidine **(48)** have been nitrosated and diazo coupled at the 6 position,[36,37] and 4,5-dihydroxypyrimidine **(49)** has been diazo coupled at position 6.[37]

However, even electrophilic attack at position 5 in pyrimidines is

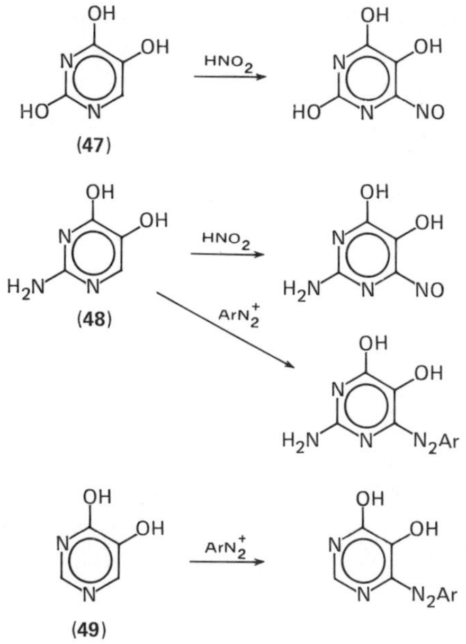

(47)

(48)

(49)

unfavoured (cf. 1,3-dinitrobenzene) and, in general, at least one strongly electron-releasing group must be present before the reaction is successful. The only recorded electrophilic substitution in pyrimidine itself is the bromination of pyrimidinium chloride, but this is probably not a normal electrophilic substitution and may involve an intermediate perbromide similar to that which has been established in the bromination of pyridine. The expected effects are observed (Section 2B (*iii*) (*iv*)), viz. the introduction of an electron-attracting group further deactivates the ring towards electrophilic attack whilst the introduction of electron-releasing groups enhances reaction.

(*a*) *Halogenation*

Typically pyrimidines having at least one electron-releasing substituent may be halogenated in position 5 by chlorine or bromine in warm aqueous solution, or by bromine in acetic acid. *N*-chloro and *N*-bromosuccinimide have also been used for the 5-halogenation of some pyrimidines.

The direct introduction of the iodogroup is frequently unsuccessful, but in some cases the reaction may be carried out using iodine in aqueous alkali, or in methanol, and iodine monochloride has also been used to iodinate some pyrimidines, although in the case of 4,6-dihydroxypyrimidine this reagent unexpectedly gave the 5-chloro derivative.[38] However, the use of *N*-iodosuccinimide has greatly facilitated the direct 5-iodination of pyrimidines.

Pyrimidines can also be directly fluorinated at the 5-position, the first record of this reaction being the conversion of 2,4,6-trifluoropyrimidine to 2,4,5,6-tetrafluoropyrimidine using silver difluoride in hot triperfluorobutylamine.[39] However, more recent work[40,41] has shown that ready electrophilic fluorination of a number of pyrimidine derivatives can be effected using the reagent trifluoro-oxy methane (CF_3OF).

(b) Nitration

Pyrimidine and its simple alkylated derivatives do not undergo nitration, but the nitration of pyrimidines having two or more electron-releasing groups may be readily carried out. The nitration of mono hydroxy and aminopyrimidines may also be effected, but by using different reaction conditions. So far no nitropyrimidines other than 5-nitropyrimidines have been reported in the literature (except for those having nitrogroups present in substituents). However, several have now been made by substituent modification (E. C. Taylor, personal communication, March 1979).

2,4-Dihydroxypyrimidine (uracil) may be nitrated using fuming nitric acid at 100°C,[42,43] but when alkyl or amino groups are present less vigorous conditions are needed for successful nitration. For example, 2,4-dihydroxy-6-methylpyrimidine is best nitrated[44,45] using one molar equivalent of nitric acid in sulphuric acid at 40–50°C.

Substituents at the 4- and 6-positions seem to have a more pronounced activating effect than 2-substituents and 4,6-dihydroxypyrimidine may be nitrated[46,47] in acetic acid below 20°C.

2 and 4-hydroxypyrimidine have been nitrated[48] using potassium nitrate

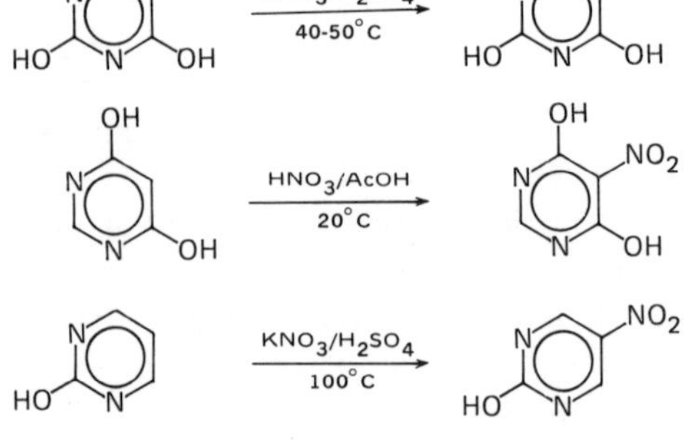

Fuming HNO_3, 100°C

HNO_3/H_2SO_4, 40-50°C

HNO_3/AcOH, 20°C

KNO_3/H_2SO_4, 100°C

Fig. 2.22. The nitration of some hydroxypyrimidines

Table 2.7 Some reactions between pyrimidines and nitrous acid.

1. *Ring nitrosation*

(a) 5-position

NH_2 N N NH_2 → NH_2 NO N NH_2

NOTE

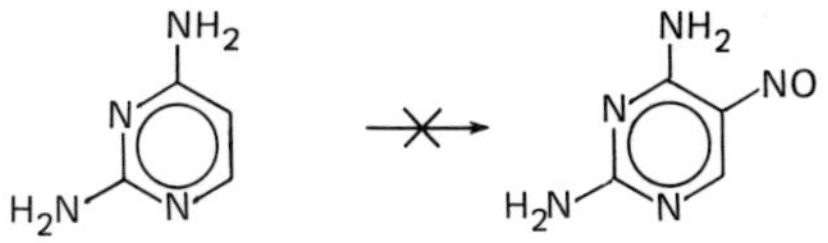

NH_2 N H_2N N ↛ NH_2 NO N H_2N N

(b) 6-position

OH N OH H_2N N → OH N OH H_2N N NO

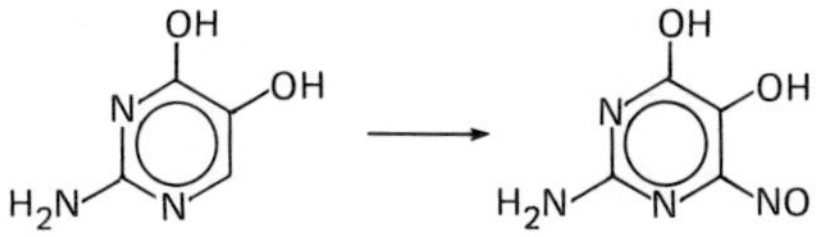

2. *Substituent nitrosation*

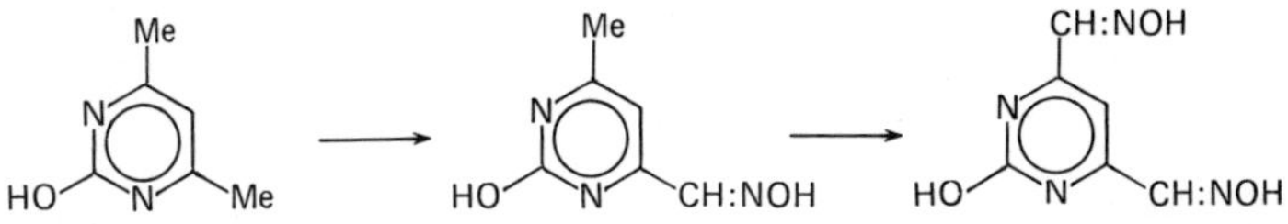

Me N HO N Me → Me N HO N CH:NOH → CH:NOH N HO N CH:NOH

3. *Substituent oxidation*

(a)

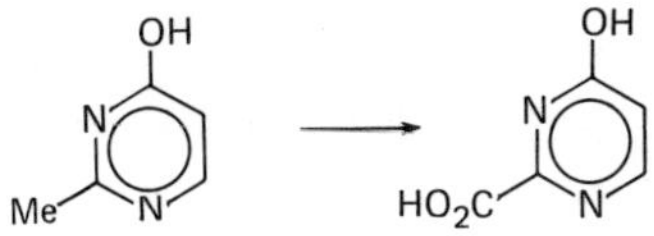

OH N Me N → OH N HO_2C N

(b)

OH N NO Ph N NH_2 → OH N NO_2 Ph N NH_2

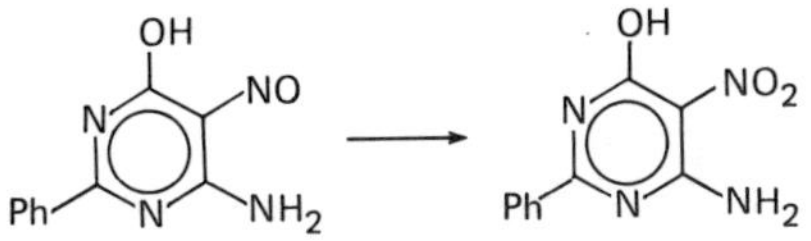

4. *Deamination of a 2-,4-, or 6-amino group*

(a) Replacement by OH

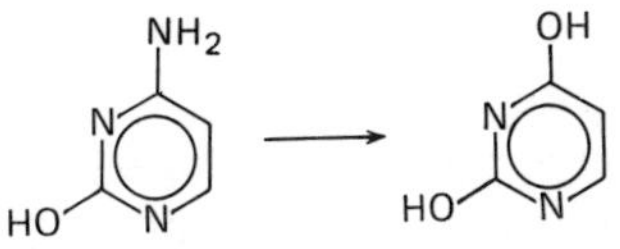

NH_2 N HO N → OH N HO N

(b) Replacement by halogen

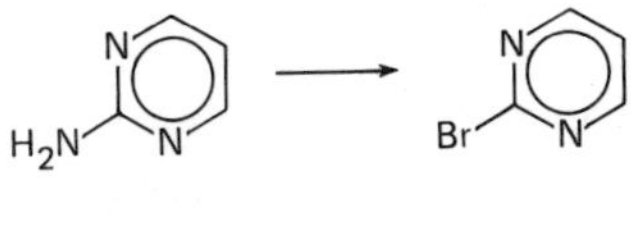

N H_2N N → N Br N

Table 2.7 Continued

5. *Diazotization of a 5-amino group*

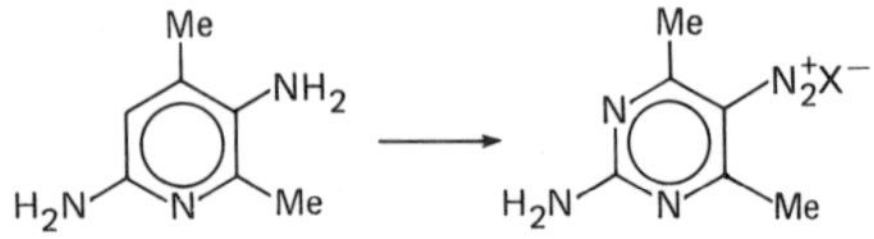

6. *Formation of cyclic compounds*

(a)

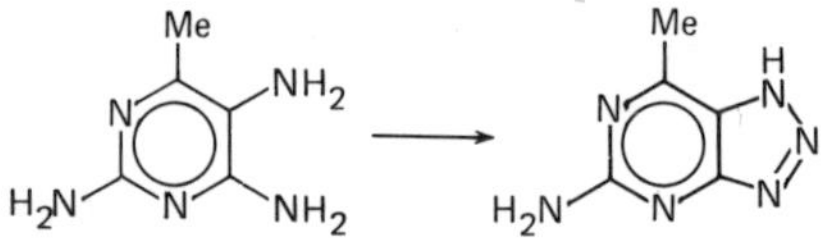

(b)

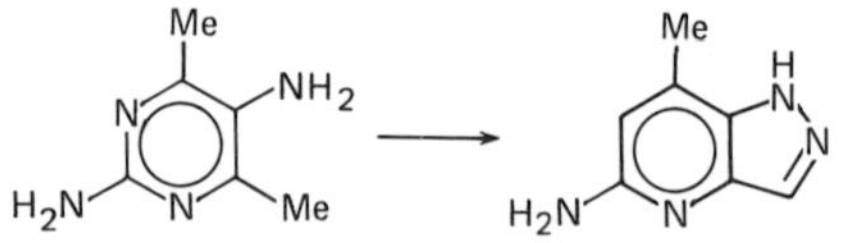

in concentrated sulphuric acid at 100°C, conditions which have also been used for the nitration of 2,4-diaminopyrimidine.[49]

Care must be exercised in the nitration of pyrimidines since in some cases reactions can also occur with substituents. For example, thio (and alkylthio) groups may be converted to hydroxy groups, methyl groups may be converted to carboxy groups, and amino groups may be either hydrolysed or converted to nitroamino groups.

(c) Nitrosation

Except for the examples given earlier, the nitrosation of pyrimidines gives 5-nitroso derivatives, but only pyrimidines having at least two electron-releasing groups will nitrosate, and again 4,6-disubstituted pyrimidines are more susceptible to attack than are 2,4-disubstituted pyrimidines. For example, whereas 4,6-dihydroxy- (and 4,6-diamino) pyrimidine give 5-nitrosoderivatives, the 2,4-disubstituted compounds apparently do not nitrosate.

The greater activation in the case of the 4,6-dihydroxypyrimidines can be explained with reference to the preferred tautomeric form of 4,6-dihydroxypyrimidine (Section 2C), which shows that one hydroxy group must be present as a phenolic hydroxyl giving a greater activation than the two keto groups of 2,4-dihydroxypyrimidine. But this explanation does not apply to the case of the aminopyrimidines.

The nitrosation of pyrimidines is far from being the 'general' reaction it was thought to be, since a variety of other reactions may also occur and 5-nitrosation only seems to occur in favoured situations. Examples of other reactions which can occur include deamination or diazotization of an amino group, nitrosation of a secondary amino group, attack at a side-chain alkyl group, oxidation of a mercaptopyrimidine to a bipyrimidinyl disulphide, cyclization reactions, or a combination of such reactions. Examples of the types of reaction which have been reported to occur between pyrimidines and nitrous acid are given in Table 2.7.

(d) *Diazo-coupling*

Pyrimidines may be coupled with diazotized amines to give 5-arylazo derivatives but the reactions seems to be general only when at least two electron-releasing groups are present. Many examples are known,[1,2] and the best conditions under which the coupling occurs have been surveyed.[50,51] However, when only one electron-releasing group is present the reaction is frequently unsuccessful and the products have not been fully investigated. Some mercapto pyrimidines do not couple at the 5-position but react at the sulphur atom. For example, 2-thiouracil **(50)** reacts with diazotized *p*-chloroaniline to give[52] compound **(51)**. This observation indicates that the products reported for other diazo-coupling reactions of mercaptopyrimidines must be accepted with reserve.

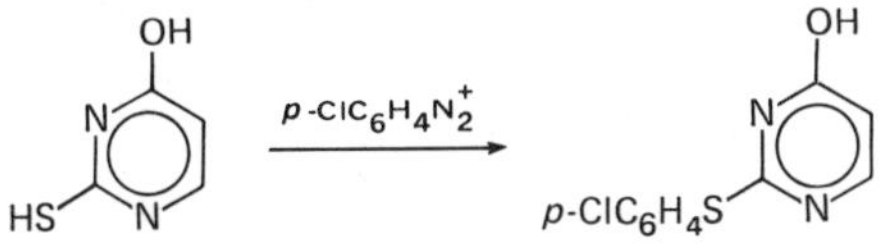

$$\xrightarrow{p\text{-}ClC_6H_4N_2^+}$$

(50) (51)

It is also possible that the diazo-coupling of some methylpyrimidines may occur via the methyl groups and not at the 5-position, such reactions being similar to some nitrosation reactions or the coupling of diazo compounds with active methylene compounds. (D.T.H. current work, cf. *J. Chem. Soc.* (Perkin I), 1688, 1985(1977)).

It is of interest that thymine, 2,4-dihydroxy-5-methylpyrimidine is reported to couple with diazotized sulphanilic acid[53] and that this is the basis of a colorimetric method for the determination of thymine,[54] but the exact nature of the coupled product has not yet been ascertained.

The nitration, nitrosation, or diazo-coupling reactions of 4-aminopyrimidines to give the 5-substituted products are of great importance since reduction of these products leads to 4,5-diaminopyrimidines which are important precursors to purines and pteridines. Diazo-coupling reactions are of particular importance since they are carried out under very mild conditions and sometimes pyrimidines having labile or reactive substituents may be successfully diazo-coupled. For example, even some pyrimidine nucleosides undergo diazo-coupling.

(e) *Sulphonation*

Sulphonation and chlorosulphonation of pyrimidines seems to be a comparatively little explored reaction but does not seem to be generally applicable nor can the products be predicted with certainty. For example, uracil reacts with chlorosulphonic acid at 110°C to give 5-chlorosulphonyl-2,4-dihydroxypyrimidine (**52**).[55–57] However, a similar treatment of cytosine or 6-methylcytosine gives the sulphonic acids (**53a**) and (**53b**) respectively.[57,58]

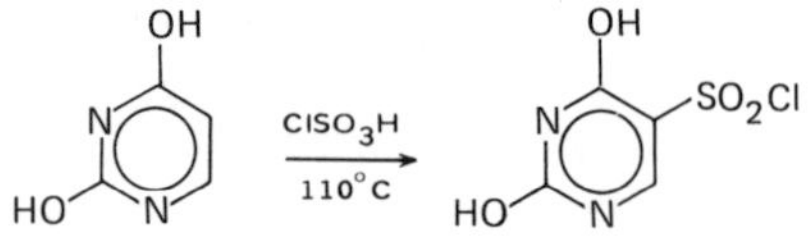

$ClSO_3H$, 110°C

(**52**)

$ClSO_3H$

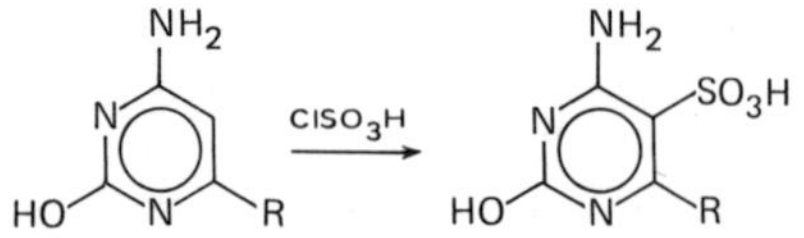

(**53**) **a**: R = H
b: R = Me

Several other aminopyrimidines have also been reported to give pyrimidine-5-sulphonic acids on reaction with chlorosulphonic acid.

Direct sulphonation of 2-aminopyrimidine with fuming sulphuric acid at 180°C gives only a low yield (223°C) of 2-hydroxypyrimidine-5-sulphonic acid and none of the expected 2-amino product.[59]

(f) *Other reactions*

Some other electrophilic substitution reactions of pyrimidines are also of interest. The oxidation of pyrimidines, having at least two electron-releasing groups, with potassium persulphate in alkaline solution (the Elb's persulphate oxidation) leads to pyrimidine-5-hydrogen sulphates which may be hydrolysed to give the corresponding 5-hydroxypyrimidine.

The Reimer–Tiemann reaction ($CHCl_3$ + KOH) for the synthesis of aromatic phenolic aldehydes can also be applied to the synthesis of pyrimidine-5-aldehydes, and the Mannich reaction (HCHO + R_2NH) has also been used in some cases. These reactions also seem to require the presence of at least two electron-releasing groups, including a 4-hydroxy group. Examples of the above types of reaction are shown in Fig. 2.23.

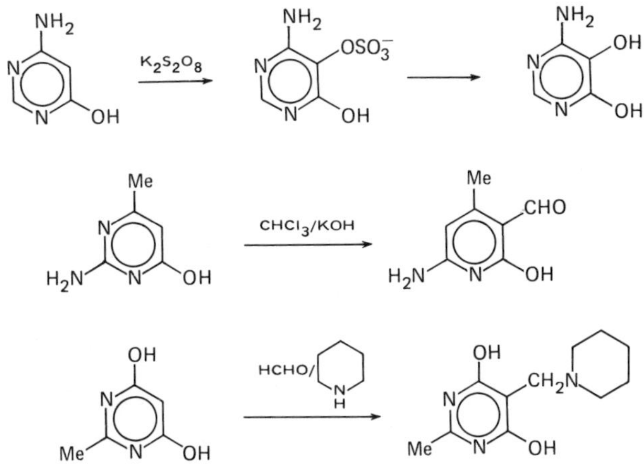

Fig. 2.23. Some electrophilic substitutions of 'activated' pyrimidines

To summarize, it should be noted that for pyrimidines to undergo electrophilic substitution, in general two electron-releasing groups and a free 5-position are necessary. Some pyrimidines having only one electron-releasing group react, and a few cases of 6-attack are known, whilst in some cases reaction at a substituent group will occur in preference to 5-attack. In these respects the general reactions of pyrimidines resemble those of the 1,3-dinitrobenzenes and reflects the deactivating effect of the cyclic nitrogen atoms to electrophilic attack in the ring and the activation of substituents adjacent to the nitrogen atoms to such reactions.

(ii) Nucleophilic Substitution

The 2-,4-, and 6- positions of the pyrimidine ring are 'activated' for nucleophilic attack due to the presence of the adjacent electron-attracting nitrogen atoms, and although there are very few cases known of direct nucleophilic attack on pyrimidines, there are very many recorded examples of nucleophilic displacement of halogens or other labile groups.

Fig. 2.24 indicates the intermediates for nucleophilic attack at the various positions of the pyrimidine ring. The intermediates **(54)** and **(55)** in which the negative charge is held mainly by the nitrogen atoms are more favoured than **(56)**, so nucleophilic attack at the 2- or 4-position is preferred to attack at the 5- position. The symmetry of the anion **(54)** compared to **(55)** should mean that the former intermediate is more stable than the latter; thus nucleophilic attack at position 2 should be more favoured than attack at the 4-position. However, experimental observations do not seem to support this. For example the rates of amination, using piperidine and

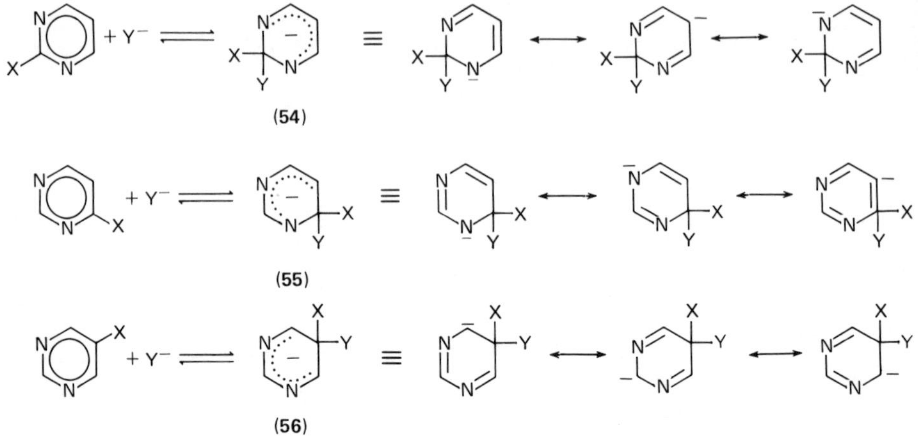

Fig. 2.24. Intermediates in nucleophilic substitution at positions 2, 4, and 5 of the pyrimidine ring

morpholine, of a number of 2- and 4-chloropyrimidine show[60] that the 4-chloro substituent is very much more reactive than than 2 substituent 2,4-Dichloropyrimidine **(57)** reacts with sodium methoxide in methanol at room temperature to give only 2-chloro-4-methoxypyrimidine **(58)**,[61,62] and whereas 2-amino-4-chloropyrimidine **(59)** gives 2-amino-4-hydroxypyrimidine (isocytosine, **60**) on boiling with water for one day, 2-chloro-4-aminopyrimidine **(61)** is unchanged under these conditions and only gives the corresponding hydroxy compound, cytosine **(62)** by heating to 140°C for some time.[63]

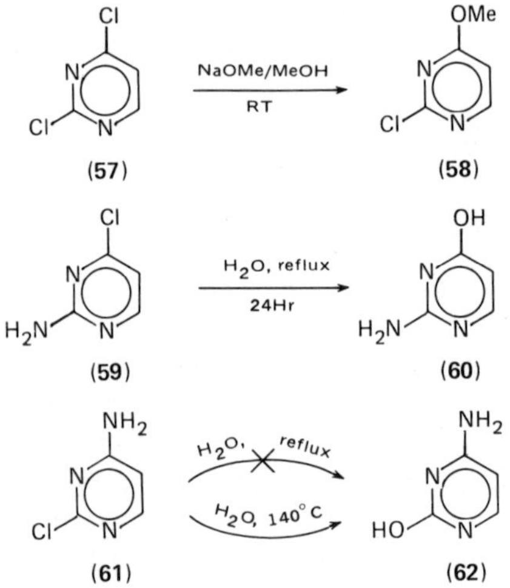

However, on treating 2,4-dichloropyrimidine with alcoholic ammonia at room temperature and allowing it to stand for some time a mixture of

2-amino-4-chloro (60%)- and 4-amino-2-chloropyrimidine (40%) is obtained.[2] In this case the 2-position does seem to be more readily attacked. When 2,4-dichloro-5-nitropyrimidine **(63)** is reacted with alcoholic or aqueous ammonia at 0°C, monoamination takes place in a few minutes and only the 4-amino product seems to be formed, so that the 5-nitro group appears to activate the 4-chloro group more than the 2-chloro group.[43,64–66]

Cl, NO_2, N, N, Cl → (NH_3, 0°C) → NH_2, NO_2, N, N, Cl

(63)

The products of these nucleophilic substitution reactions are under kinetic rather than thermodynamic control and an explanation for the apparent lower activity of the 2-position relative to the 4-position may be that the lone pairs of electrons on the annular nitrogen atoms may exert a greater electrostatic repulsion towards the approach of a nucleophile to the 2- rather than the 4- position.

It is observed that the introduction of electron-attracting groups further enhances nucleophilic attack (note the amination conditions for 2,4-dichloro- and 2,4-dichloro-5-nitropyrimidine above) whilst the introduction of electron-releasing groups causes deactivation. For example, 2-amino-4-chloropyrimidine requires methanolic ammonia at 175°C to produce the diamino product.[67]

Nucleophilic substitution at the 5-position of the pyrimidine ring is comparable to the reaction of aryl compounds and generally requires fairly vigorous reaction conditions, although it is reported that a number of 5-bromopyrimidines can be aminated comparatively easily.[1,2] 5-Bromopyrimidine itself can be reacted with sodium methoxide at 100°C to give 5-methoxypyrimidine,[68] and 5-bromouracil gives isobarbituric acid (2,4,5-trihydroxypyrimidine) merely on boiling with sodium bicarbonate solution.[69] However, comparatively few cases of nucleophilic substitution at the 5-position have been recorded, although a very large number of such reactions are known in the case of 2-, 4-, or 6-attack.

The examples of nucleophilic substitution given above have been those of halogenopyrimidines, which have been the most studied of such reactions. However, frequently the displacement of groups other than halogen provides the most satisfactory synthetic route to a particular pyrimidine. Groups which may often be usefully nucleophilically displaced include MeO, MeS, $MeSO_2$ and $Me_3N^{\oplus}$. Some examples of this type of reaction are included in the list below, which summarises the more common nucleophilic substitution reactions of pyrimidines.

(1) Reactions of 2-, 4-, 6-halogenopyrimidines:

(a) Aminations, reaction with hydrazine, hydroxylamine, etc.

Reference

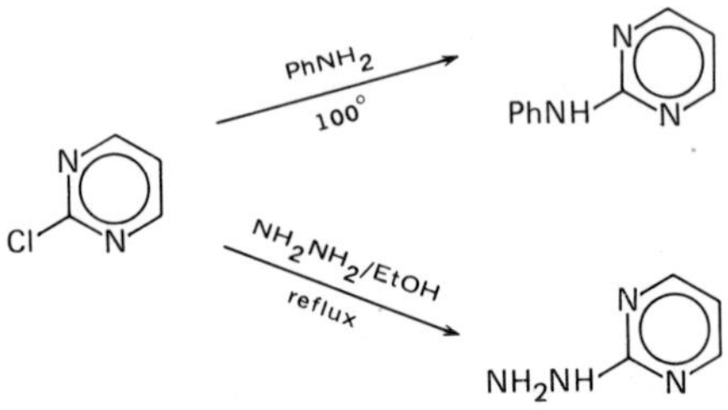

2-Chloropyrimidine $\xrightarrow[100^\circ]{PhNH_2}$ 2-(PhNH)pyrimidine — 70

2-Chloropyrimidine $\xrightarrow[\text{reflux}]{NH_2NH_2/EtOH}$ 2-(NH_2NH)pyrimidine — 71

(b) Hydrolysis (frequently unsuccessful):

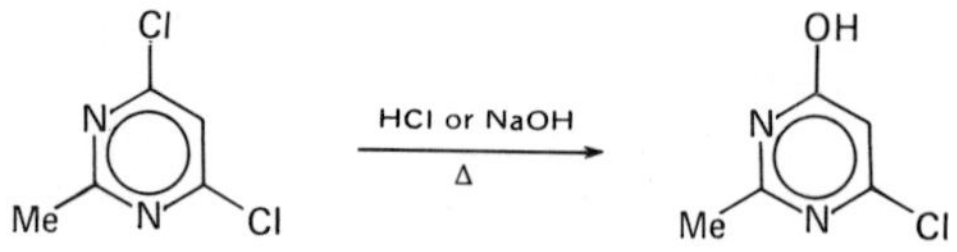

4,6-Dichloro-2-methylpyrimidine $\xrightarrow[\Delta]{\text{HCl or NaOH}}$ 6-chloro-4-hydroxy-2-methylpyrimidine (Me, OH, Cl) — 72

(c) Thiation:

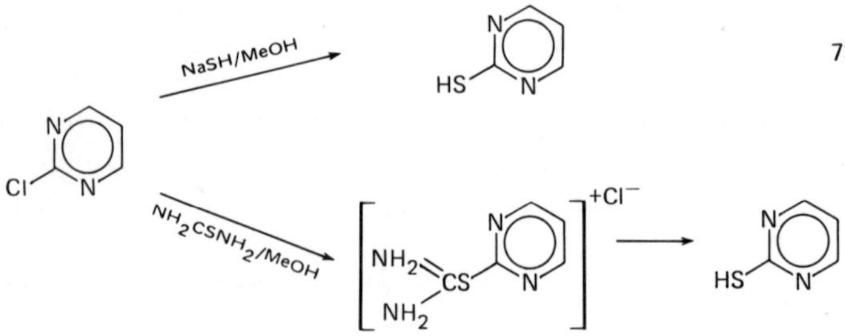

2-Chloropyrimidine $\xrightarrow{NaSH/MeOH}$ 2-HS-pyrimidine — 73

2-Chloropyrimidine $\xrightarrow{NH_2CSNH_2/MeOH}$ $\left[(NH_2)_2CS\text{-pyrimidin-2-yl} \right]^+ Cl^-$ $\longrightarrow$ 2-HS-pyrimidine — 70(a)

(d) Replacement by alkoxy or alkylthio groups:

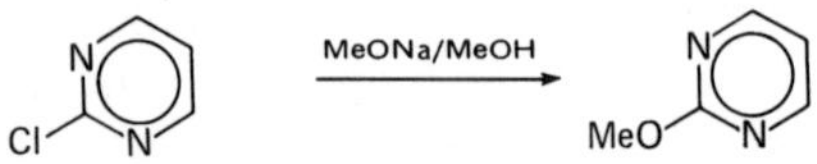

2-Chloropyrimidine $\xrightarrow{MeONa/MeOH}$ 2-MeO-pyrimidine — 74

4-Chloro-6-methylpyrimidine (Me, Cl) $\xrightarrow{MeSNa/MeOH}$ 4-methyl-6-SMe-pyrimidine (Me, SMe) — 75

(e) Halogen exchange:

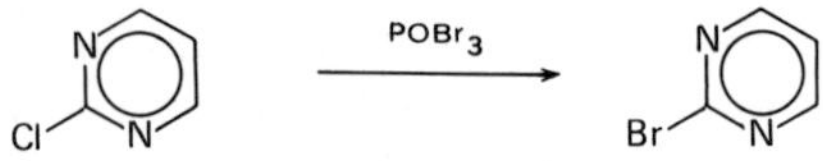

2-Chloropyrimidine $\xrightarrow{POBr_3}$ 2-Br-pyrimidine — 76

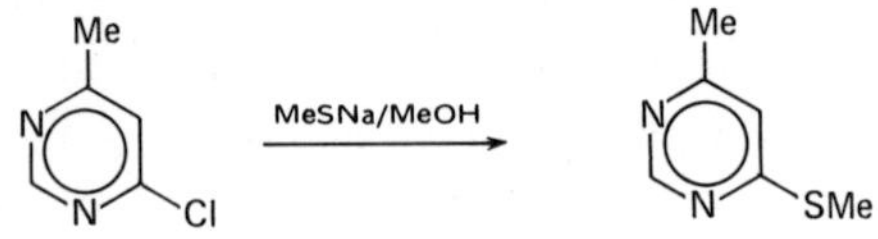

2-Chloro-4,6-dimethylpyrimidine (Cl, Me, Me) $\xrightarrow{HI}$ 2-iodo-4,6-dimethylpyrimidine (I, Me, Me) — 76

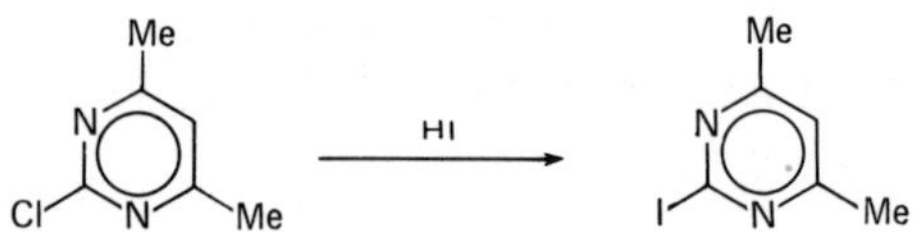

(2) Reactions of 2-, 4-, 6-aminopyrimidines

(a) Hydrolysis:

Reference

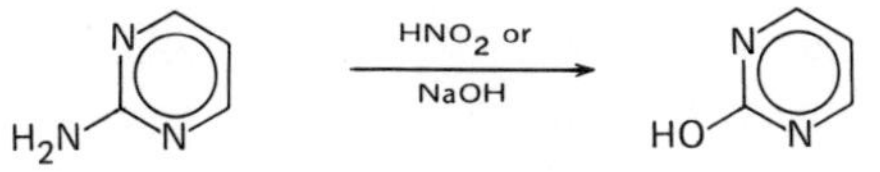

H_2N–pyrimidine $\xrightarrow[\text{NaOH}]{HNO_2\text{ or}}$ HO–pyrimidine 77

(b) Diazotiation followed by halogenation:

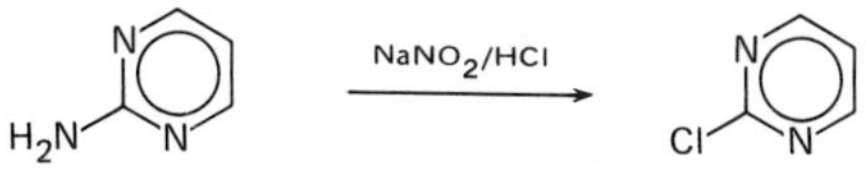

H_2N–pyrimidine $\xrightarrow{NaNO_2/HCl}$ Cl–pyrimidine 78

(c) Displacement of a quaternary ammonium pyrimidine:

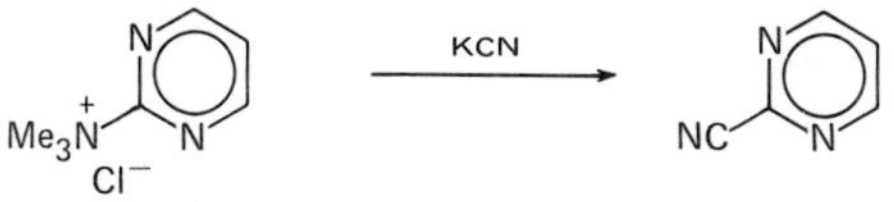

$Me_3\overset{+}{N}$–pyrimidine Cl^- $\xrightarrow{KCN}$ NC–pyrimidine 79

(3) Reactions of 2,4,6-mercaptopyrimidines (and derivatives):

(a) Hydrolysis

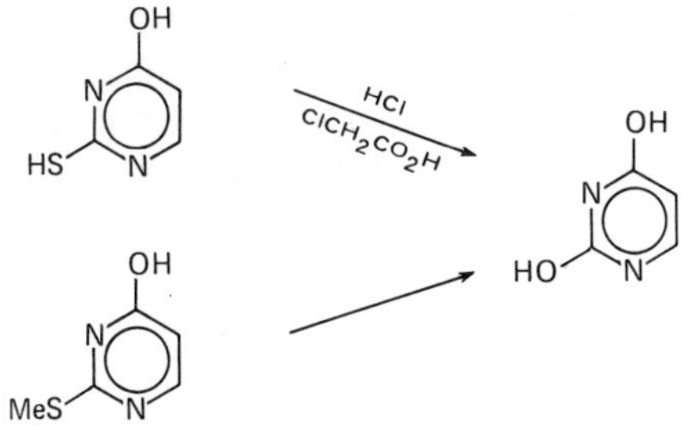

HS, OH-pyrimidine $\xrightarrow[ClCH_2CO_2H]{HCl}$ HO, OH-pyrimidine 80-82

MeS, OH-pyrimidine $\rightarrow$ HO, OH-pyrimidine

$MeSO_2$, OCH_2Ph-pyrimidine $\xrightarrow{NaOH}$ HO, OCH_2Ph-pyrimidine 83

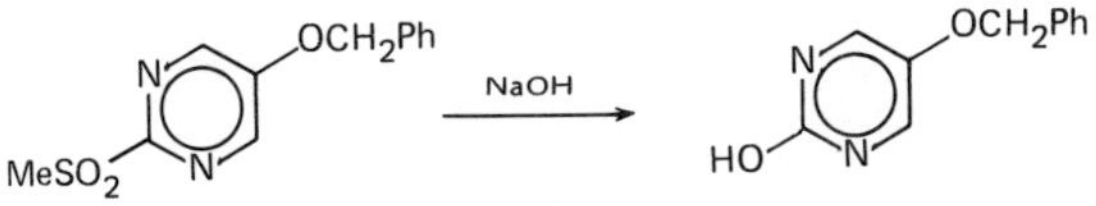

(b) amination

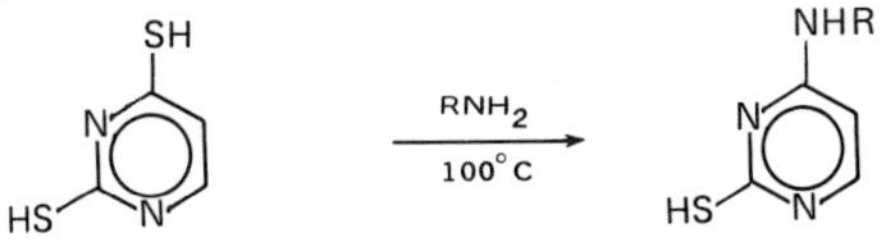

HS, SH-pyrimidine $\xrightarrow[100^\circ C]{RNH_2}$ HS, NHR-pyrimidine 84

Note: only 4-substituent attacked

(c) cyanation

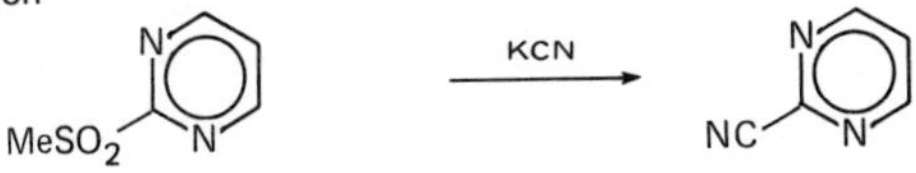

$MeSO_2$–pyrimidine $\xrightarrow{KCN}$ NC–pyrimidine 85

(4) Reactions of alkoxypyrimidines

(a) Reaction with hydrazine

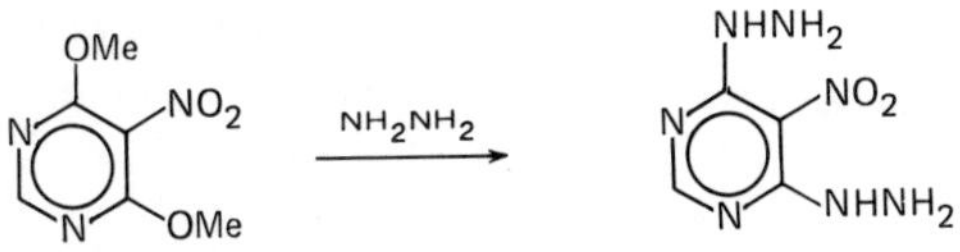

OMe, NO_2, OMe-pyrimidine $\xrightarrow{NH_2NH_2}$ $NHNH_2$, NO_2, $NHNH_2$-pyrimidine 65

(iii) The Reactions of 2,4,6- hydroxypyrimidines

The 2-,4-, and 6-hydroxypyrimidines can be converted to halogens or mercaptopyrimidines by reaction with the phosphorus halides or phosphorus sulphides respectively. However, these are not true nucleophilic reactions of the hydroxypyrimidines since these exist in the lactam tautomeric form but involve intramolecular nucleophilic displacements of pyrimidine–phosphorus intermediates. This is illustrated in Fig. 2.25. The thiation of hydroxypyrimidines occurs via a similar type of reaction.

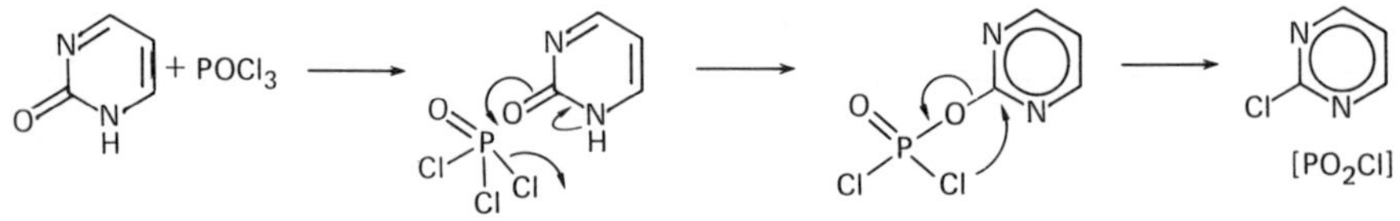

+ $POCl_3$ → … → … → $[PO_2Cl]$

Fig. 2.25. Chlorination of 2-hydroxypyrimidine with $POCl_3$

5-hydroxypyrimidines have not been reported to react with phosphoryl chloride etc., and, for example, isobarbituric acid reacts to form 2,4-dichloro-5-hydroxypyrimidine (**64**),[83] although phenolic and alcoholic hydroxy compounds usually react to form phosphate esters (or similar products), e.g. (**65**).

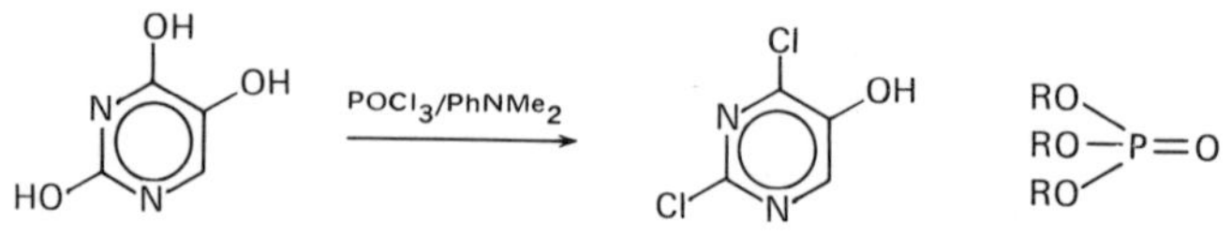

$POCl_3/PhNMe_2$ → (**64**) (**65**)

(iv) Other Reactions of Pyrimidines

(a) Oxidation

Direct oxidative reactions of pyrimidines seem to have been little explored although the introduction of a 5-hydroxy group using the Elb's persulphate oxidation[86(b)] is useful for some pyrimidines having electron-releasing substituents, for example the synthesis of 2,5-dihydroxy-4,6-dimethylpyrimidine (**66**) from 2-hydroxy-4,6-dimethylpyrimidine.[86(a)]

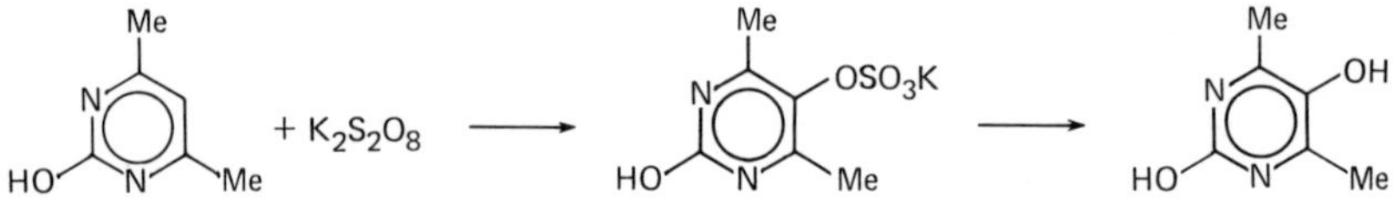

$+ K_2S_2O_8$ → → (**66**)

The oxidation reactions of substituent groups are similar to those in the benzene series of compounds, but generally a 2- or 4-methyl group is oxidized more readily than is a 5-methylpyrimidine (cf. the methylpyridines). For example 4,5-dimethylpyrimidine is oxidized by permanganate to give 5-methylpyrimidine-4-carboxylic acid (**67**)[87]

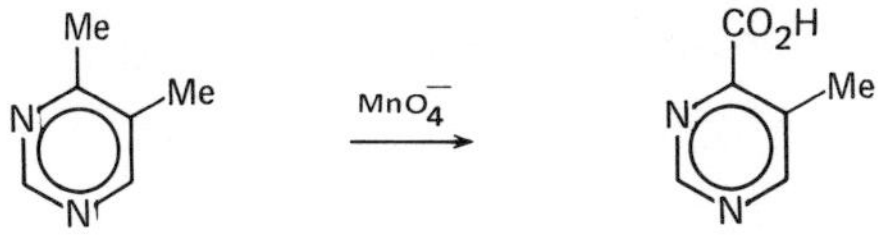

MnO_4^-

CO_2H Me

(**67**)

Some examples of useful oxidative reactions involving substituents are given in Fig. 2.26.

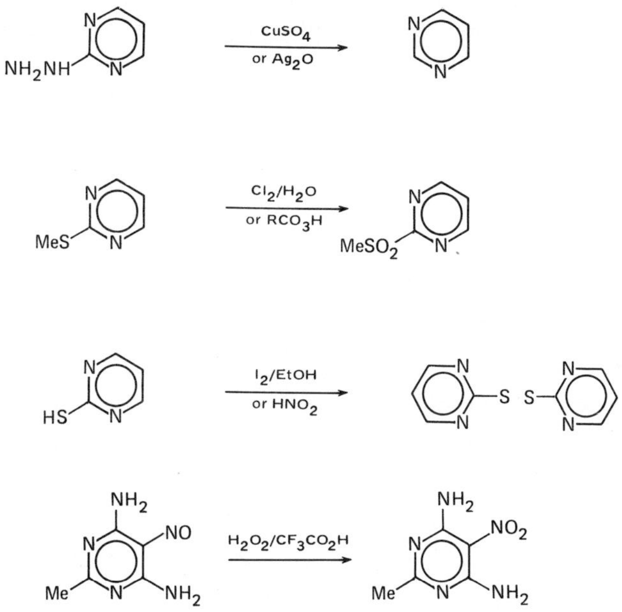

$CuSO_4$ or Ag_2O

Cl_2/H_2O or RCO_3H

$I_2/EtOH$ or HNO_2

H_2O_2/CF_3CO_2H

Fig. 2.26. Oxidative reactions of some pyrimidines

(b) *Reduction*

The field of the reduced pyrimidines is comparatively unknown and no discussion of this subject is intended here, but reductive reactions can give rise to such compounds, for example uracil can be hydrogenated over platinum or palladium to give 4,5-dihydro-2,6-dihydroxypyrimidine (**68**).[88,89]

H_2/Pd

(68)

H_2/Ni

(69)

Boiling with Raney nickel also converts uracil to the dihydro derivative,[88] but under similar conditions 2-hydroxypyrimidine is converted to the tetrahydroderivative (**69**).[90]

The reductive modification of substituents is, however, frequently useful in the pyrimidine field and some examples of its use are given in Fig. 2.27.

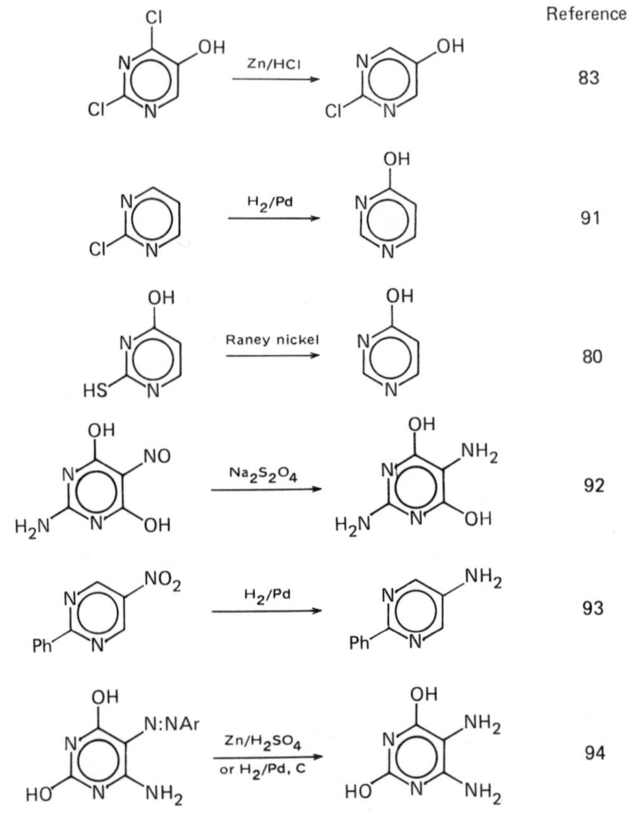

Fig. 2.27. Reduction of some substituted pyrimidines

(c) *Acylation and alkylation of amino, hydroxy and mercaptopyrimidines*

Only 5-hydroxypyrimidines can be *O*-acylated although amino and mercaptopyrimidines acylate on the amino or mercapto group whatever the position of the group. Some hydroxypyrimidines acylate on the ring nitrogen to give unstable N-acyl-oxopyrimidines. For example, uracil gives the compound **(70)**. The 5-amino group acylates in preference to a 2- or 4-group. Mercaptopyrimidines invariably alkylate on the sulphur, but whereas 5-aminopyrimidines may be alkylated at the 5-amino group, with 2- and 4-aminopyrimidines it is usual for alkylation to occur on a ring nitrogen to give compounds of the type **(71)** or **(72)**.

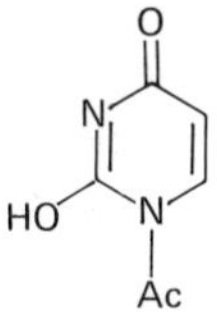

(70) 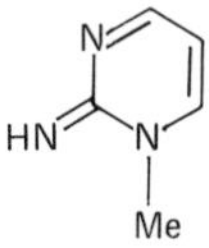 (71) 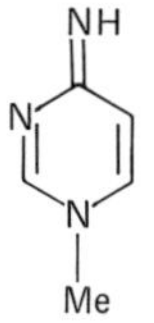 (72)

Alkylation of 5-hydroxypyrimidines gives the 5-alkoxy derivative, but 2- or 4-hydroxypyrimidines usually give a mixture of *O*- and *N*-alkyl products in which the *N* isomer predominates. The 2- or 4-alkoxypyrimidines rearrange to give the *N*-alkyl derivatives on heating to the melting point.

(d) *Reactions of Methylpyrimidines*

Methylpyrimidines in which the methyl groups are situated in the 2- or 4-positions are activated like the 2- and 4-methylpyridines and *o*-nitrotoluene. Such methyl groups are oxidized more readily than 5-methyl groups and give styryl derivatives when heated with benzaldehyde and a Lewis acid catalyst such as zinc chloride. The 2- and 4-methylpyrimidines are also subject to nitrosation to give hydroxyiminomethyl derivatives (the oximes of pyrimidine aldehydes), such a reaction frequently occurring in preference to, and sometimes in addition to, 5-nitrosation of the pyrimidine. Other reactions of 2- and 4-methylpyrimidines include the Claisen and Mannich reactions. Some examples of these reactions of methylpyrimidines are given in Fig. 2.28.

(e) *Diazotization of aminopyrimidines*

Only 5-aminopyrimidines have been diazotized and coupled to activated aromatic compounds, although 2- and 4-aminopyrimidines have been converted to the corresponding 2- or 4-chloropyrimidine by reverse addition diazotization. For example, 2-amino-4,6-dimethyl- and 6-amino-2,4-dimethylpyrimidine have been converted to the chloro derivatives **(73)** and **(74)** by this method.

(A) Nitrosation

Reference

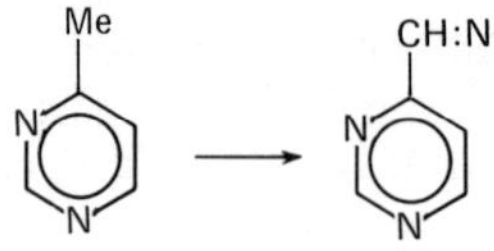

Me → CH:NOH

95

(B) Oxidation

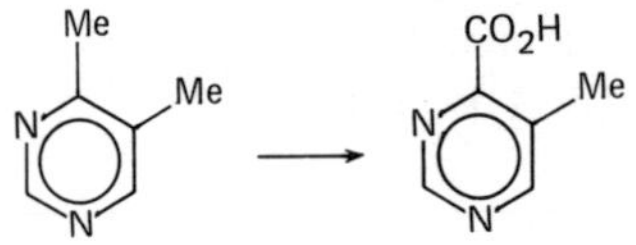

Me, Me → CO_2H, Me

87

(C) Condensation

(a) Me → CH:CHPh

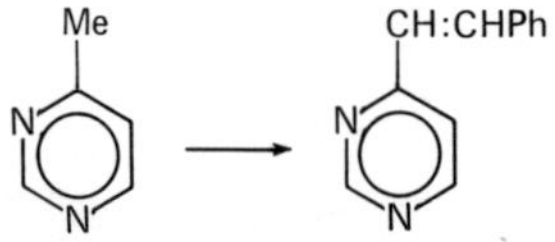

96

(b) Me → $CH_2CHOH.CCl_3$

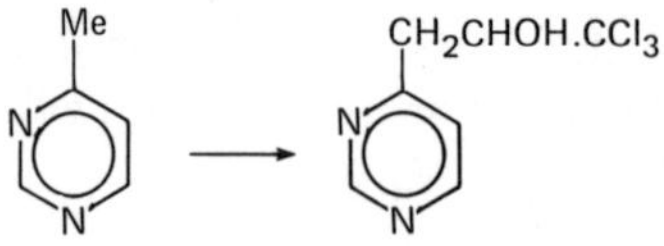

97

(c) Me → CH_2COCO_2Et

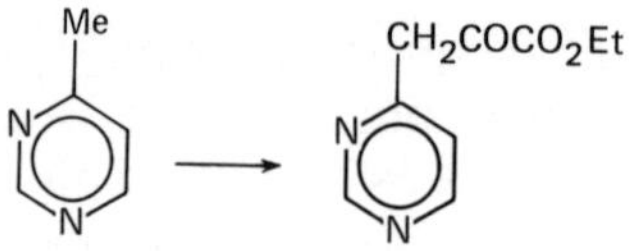

98

(d) Me → $CH_2CH_2NMe_2$

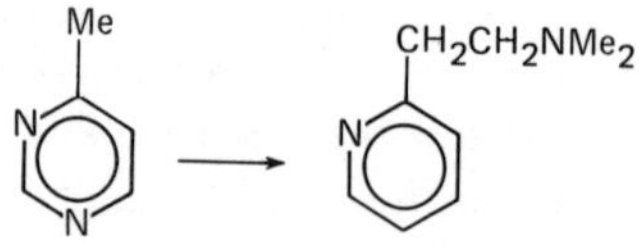

99

(D) Halogenation

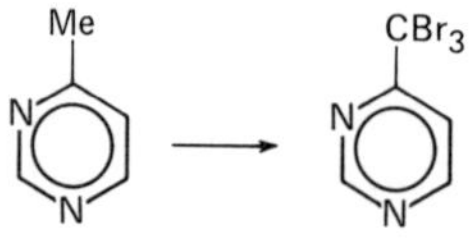

Me → CBr_3

100

Fig. 2.28. Some reactions of 2- or 4-methyl-pyrimidines

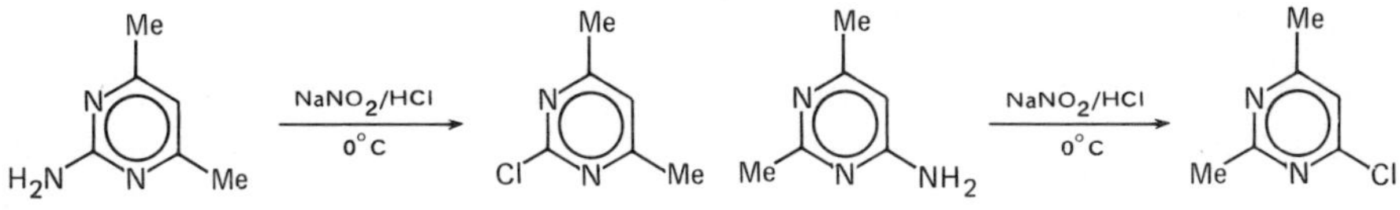

(73) (74)

(f) *Addition Reactions and Covalent Hydration of Pyrimidines*

Although pyrimidine is aromatic and therefore should not be expected to be particularly prone to addition reactions, there is evidence that a number of pyrimidine derivatives undergo some substitution reactions by an addition–elimination mechanism. Also in some cases addition products can be isolated, for example, uracil undergoes bromination in water to give in turn **(75)**, **(76)** and **(77)**.[1]

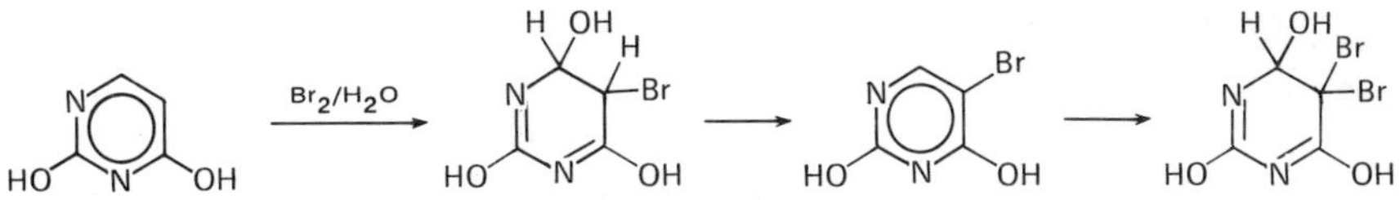

(75) (76) (77)

Some pyrimidines will also add water to give covalent hydrates, although this is not as prevalent as the covalent hydration of pteridines (see Section 3C(iii)). A well-known example is the formation of a hydrate of uracil **(78)** on u.v. irradiation of an aqueous solution. This hydrate has also been synthesized unambiguously.[101]

Spontaneous and complete covalent hydration has been observed with the cations of some pyrimidines having electron-attracting groups at the 5-position, e.g. 5-NO_2, 5-SO_2Me and 5-SOMe, these having the structure **(79)**.[65,102]

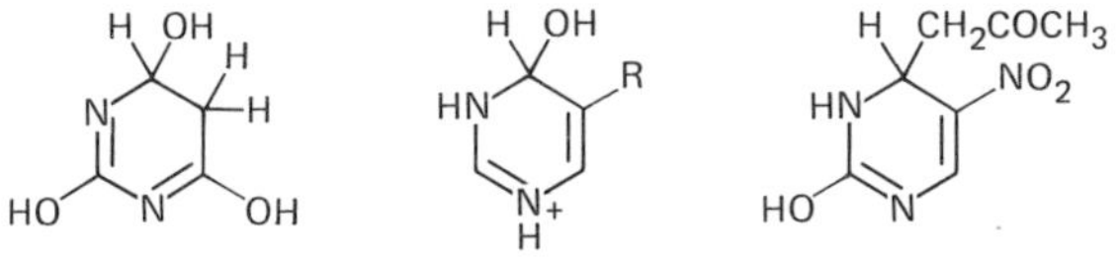

(78) (79) (80)

Such covalent hydration will, of course, profoundly alter the spectral and pK_a properties of the pyrimidines.

Active methyl and methylene compounds may also add to the pyrimidine ring. For example, 2-hydroxy-5-nitropyrimidine and acetone form 4-acetonyl-3,4-dihydro-2-hydroxy-5-nitropyrimidine **(80)**.[2]

(g) *The Dimroth Rearrangement and Ring Transformations*

The Dimroth rearrangement[103] of iminopyrimidines refers to the isomerization proceeding by a ring-fission and subsequent recyclization, whereby a ring nitrogen and its attached substituent exchanges places with an imino group in an adjacent position.

An example is given by the rearrangement of 1,2-dihydro-2-imino-1-methylpyrimidine (**81**) to 2-methylaminopyrimidine (**82**) in aqueous alkali at room temperature.[103]

MeN, HN (**81**) → [HOCH, MeNH, HN, N] → MeNH, N (**82**)

A detailed study of this reaction in pyrimidines, pyridines and related compounds has been made and much information on the mechanism, kinetic studies, etc. has been obtained. This subject has been comprehensively reviewed by D. J. Brown. [2,104]

A variety of ring transformations of pyrimidines is also known. These have also been recently reviewed[105] and it is not intended to give an account here. However, there has been considerable interest in the ring transformations brought about by hydrazine and its derivatives since it was observed that hydrazine causes mutations in microorganisms.

In general pyrimidines are converted to pyrazoles by the action of hydrazine. For example, uracil is converted into (**83**) by the action of hydrazine. A number of examples of this type of transformation and many references are given in ref. 105.

OH, N, HO, N $\xrightarrow{NH_2NH_2}$ HN, N, H, O (**83**)

This chapter has surveyed the chemistry and the physical properties of pyrimidine derivatives. It has not provided an exhaustive and complete record of the reactions and properties, but has attempted to indicate the important attributes of this class of compounds, and to give examples of the general and important reactions of pyrimidines. Although the reactions of pyrimidine derivatives are paralleled in the chemistry of benzene and pyridine, it is the combination of the effects of the two-ring nitrogen atoms, the position of the substituent, and the presence of other groups in the ring which combine to give an uncertainty as to the outcome of a reaction in a particular case and also makes pyrimidine chemistry unique and interesting.

REFERENCES

1. D.J. Brown, *The Pyrimidines*, Wiley-Insterscience, New York and London (1962).
2. D. J. Brown, *The Pyrimidines*, Supplement I, Wiley-Interscience, New York, London, Sydney, Toronto (1970).
3. F. L. Rose and D. J. Brown, *J. Chem. Soc.*, **1956**, 1953.
4. G. A. Jeffrey, S. Ghose, and J. O. Warwicker, *Acta. Cryst.*, **14**, 881 (1961)
5. T. J. Batterham, *NMR Spectra of Simple Heterocycles*, Wiley-Interscience, New York, London, Sydney, Toronto (1973).
6. A. Combes and C. Combes, *Bull. Chim. Soc. (France)*, **7**, 791 (1892).
7. P. N. Evans, *J. Prakt. Chem.*, **46**, 352 (1892); **48**, 489 (1893).
8. D. Davidson and O. Baudisch, *J. Amer. Chem. Soc.*, **48**, 2739 (1926).
9. A. Bowman, *J. Chem. Soc.*, **1937**, 494.
10. P. E. Fanta and E. A. Hedman, *J. Amer. Chem. Soc.*, **78**, 1434 (1956).
11. A. W. Dox and L. Yoder, *J. Amer. Chem. Soc.* **44**, 361 (1922).
12. O. Hromatka, Ger. Pat., 667,990 (1938); US Pat., 2,235,638 (1941); *Chem. Abs.*, **35**, 4041 (1941).
13. F. G. P. Remfry, *J. Chem. Soc.*, **99**, 610 (1911).
14. R. Hull, *J. Chem. Soc.*, **1951**, 2214.
15. H. Bredereck, R. Gompper, and G. Morlock, *Chem. Ber.*, **90**, 942 (1957).
16. J. R. Marshall and J. Walker, *J. Chem. Soc.*, **1951**, 1004.
17. D. J. Brown, E. Hoerger and S. F. Mason, *J. Chem. Soc.*, **1955**, 211.
18. D. J. Brown, P. W. Ford, and K. H. Tratt, *J. Chem. Soc.*, (C), **1967**,1445.
19. R. Mason, *J. Chem. Soc.*, **1959**, 1253.
20. W. W. Zorbach and R. S. Tipson (eds.), *Synthetic Procedures in Nucleic Acid Chemistry*, Vol. 2, Wiley—Interscience, New York, London, Sydney, Toronto (1973).
21. A. R. Katritzky (ed), *Physical Methods in Heterocyclic Chemistry*, Vols. 2 and 4, Academic Press, New York and London (1963, 1971).
22. D. J. Brown and L. N. Short, *J. Chem. Soc.*, **1953**, 331.
23. R. Mason, *J. Chem. Soc.*, **1957**, 4874.
24. R. Mason, *J. Chem. Soc.*, **1958**, 3619; **1959**, 1281.
24. D. J. Brown, E. Hoerger and S. F. Mason, *J. Chem. Soc.*, **1955**, 4035.
26. A. J. Boulton, D. T. Hurst, J. F. W. McOmie, and M. S. Tute, *J. Chem. Soc.* (C), **1967**, 1204.
27. G. D. Davies, D. E. O'Brien, L. R. Lewis, and C. C. Cheng in *Synthetic Procedures in Nucleic Acid Chemistry*, Vol. 1, W. W. Zorbach and R. S. Tipson (eds.), Wiley-Interscience, New York, London, Sydney, Toronto, (1968) p. 80.
28. J. Elguero, C. Marzin, A. R. Katritzky, and P. Linda, The tautomerism of heterocycles, in *Advances in Heterocyclic Chemistry*, Supplement I, Academic Press, New York, San Francisco, London (1976).
29. Y. Inone, N. Turntachi and K. Nakanishi, *J. Org. Chem.*, **31**, 175 (1966).
30. D. J. Brown and S. Teitei, *Austral. J. Chem.*, **17**, 567 (1964).
31. A. R. Katritzky, F. D. Popp and A. J. Waring, *J. Chem. Soc.*, (B), **1966**, 565.
32. G. M. Kheifets, N. V. Khromov-Borisov, A. I. Kol'tsov, and M. V. Volkenstein, *Tetrahedron*, **23**, 1197 (1967) (and earlier refs.)
33. T. Nishiwaki, *Tetrahedron*, **22**, 3117 (1966); **23**, 1153 (1967).

34. Q. N. Porter and J. Baldas, *Mass Spectrometry of Heterocyclic Compounds*, Wiley-Interscience, New York, London, Sydney, Toronto (1971), p. 54.
35. D. C. DeJongh in, *Synthetic Procedures in Nucleic Acid Chemistry*, Vol. 2. W. W. Zorbach and R. S. Tipson (eds), Wiley-Interscience, New York, London, Sydney, Toronto (1973).
36. D. Davidson and M. T. Bogert, *Proc. Natl. Acad. Sci. U.S.*, **18**, 490 (1932).
37. J. H. Chesterfield, D. T. Hurst, J. F. W. McOmie and M. S. Tute, *J. Chem. Soc.*, **1964**, 1001.
38. J. H. Chesterfield, J. F. W. McOmie and E. R. Sayer, *J. Chem. Soc.*, **1955**, 3478.
39. (a) H. Schroeder, *J. Amer. Chem. Soc.*, **82**, 4115 (1960); (b) H. Schroeder, E. Kober, H. Ulrich, R. Rätz, H. Agahigian and C. Grundmann, *J. Org. Chem.*, **27**, 2580 (1962).
40. M. J. Robins and S. R. Naik, *J. Amer. Chem. Soc.*, **93**, 5277 (1971).
41. D. H. R. Barton, W. A. Bubb, R. H. Hesse and M. M. Pechet, *J. Chem. Soc. (Perk I)*, **1974** 2095 and refs. therein.
42. T. B. Johnson and I. Matsuo, *J. Amer. Chem Soc.*, **41**, 782 (1919).
43. D. J. Brown, *J. Appl. Chem.* **2**, 239 (1952).
44. S. Gabriel and J. Colman, *Ber.*, **34**, 1234 (1901).
45. D. J. Brown, E. Hoerger, and S. F. Mason, *J. Chem. Soc.*, **1954**, 3832.
46. W. R. Boon, W. G. M. Jones and G. R. Ramage, *J. Chem. Soc.*, **1951**, 96.
47. J. W. Daly and B. E. Christensen, *J. Org. Chem.*, **21**, 177 (1956).
48. I. Wempen, H. V. Blank, and J. J. Fox, *J. Heterocyclic Chem.*, **6**, 593 (1969).
49. D. E. O'Brien, C. C. Cheng, and W. Pfleiderer, *J. Medicin Chem.*, **9**, 573, (1966).
50. B. Lythogoe, A. R. Todd, and A. Topham, *J. Chem. Soc.*, **1944**, 315.
51. M. Polonovski and M. Pesson, *Bull. Soc. Chim., France,* **15**, 688 (1948).
52. E. A. Falco, G. H. Hitchings and P. Russell, *J. Amer. Chem. Soc.*, **71**, 362 (1949).
53. T. B. Johnson and J. Clapp, *J. Biol. Chem.*, **5**, 163 (1908).
54. E. D. Day and W. A. Mosher, *J. Biol. Chem.*, **197**, 227 (1952).
55. T. J. Bardos, R. R. Herr, and T. Enkoji, *J. Amer. Chem. Soc.*, **77**, 960 (1955).
56. G. R. Barker, N. G. Lutly and M. M. Dhar, *J. Chem. Soc.*, **1954**, 4206.
57. N. V. Khromov-Borisov and R. S. Karlinskaya, *J. Gen. Chem. (USSR)*, **27**, 2518 (1957).
58. N. V. Khromov-Borisov and R. S. Karlinskaya, *J. Gen. Chem., (USSR)*, **24**, 2212 (1954).
59. W. T. Caldwell and G. E. Jaffe, *J. Amer. Chem. Soc.*, **81**, 5166 (1959).
60. N. B. Chapman and C. W. Rees, *J. Chem. Soc.*, **1954**, 1190.
61. G. W. Kenner, C. B. Reese, and A. R. Todd, *J. Chem. Soc.*, **1955**, 855.
62. H. Yamanaka, *Chem. Pharm. Bull. (Japan)*, **7**, 158 (1959).
63. G. E. Hilbert and T. B. Johnson, *J. Amer. Chem. Soc.*, **52**, 1152 (1930).
64. O. Isay, *Ber.*, **39**, 250 (1906).
65. M. E. C. Biffin, D. J. Brown, and T. C. Lee, *J. Chem. Soc. (C),* **1967**, 573.
66. A. Albert, D. J. Brown, and G. H. Cheeseman, *J. Chem. Soc.*, **1951**, 474.
67. J. P. English and J. W. Clapp, USP 2, 416, 617 (1947); *Chem. Abs.*, **41**, 3493 (1947).
68. H. Bredereck, R. Gompper and H. Herlinger, *Chem. Ber.*, **91**, 2832 (1958).
69. S. Y. Wang, *J. Amer. Chem. Soc.*, **81**, 3786 (1959).
70. (a) M. P. V. Boarland and J. F. W. McOmie, *J. Chem. Soc.,* **1951**, 1218; (b) T. Matsukawa and K. Sirakawa, *J. Pharm. Soc., Japan,* **71**, 993 (1951); *Chem. Abs.,* **46**, 10212 (1952).

71. (a) K. Sirakawa, S. Ban and M. Yoneda, *J. Pharm. Soc., Japan,* **73**, 598 (1953); Chem. Abs., **48**, 9362 (1954); (b) H. Vanderhaeghe and M. Claesen, *Bull. Soc. Chim. Belges*, **68**, 30 (1959).
72. (a) H. R. Henze, W. J. Clegg, and C. W. Smart, *J. Org. Chem*., **17**, 1320 (1952); (b) F. R. Basford, F. H. S. Curd, and F. L. Rose, *J.Chem. Soc.,* **1946**, 713.
73. R. O. Roblin and J. W. Clapp, *J. Amer. Chem. Soc*., **72**, 4890 (1950).
74. M. P. V. Boarland and J. F. W. McOmie, *J. Chem. Soc*., **1952**, 3716.
75. J. R. Marshall and J. Walker, *J. Chem. Soc*., **1951**, 1004.
76. M. P. L. Caton, D. T. Hurst, J. F. W. McOmie, and R. R. Hunt, *J. Chem. Soc*. (C), **1967**, 1204.
77. D. J. Brown, *Nature*, **165**, 1010 (1950).
78. (a) C. G. Overberger and I. C. Kogon, *J. Amer. Chem. Soc*., **76**, 1065 (1954); (b) K. L. Howard, USP 2, 477, 409 (1949).
79. F. H. Case and E. Koft, *J. Amer. Chem. Soc*., **81**, 905 (1959).
80. D. J. Brown, *J. Appl. Chem*., **2**, 239 (1952); *J. Soc. Chem. Ind*., **69**, 353 (1950).
81. H. L. Wheeler and L. M. Liddle, *Am. Chem. J*., **40**, 547 (1908).
82. H. L. Wheeler and H. F. Merriam, *Am. Chem. J*., **29**, 478 (1903).
83. D. T. Hurst, J. F. W. McOmie, and J. B. Searle, *J. Chem. Soc*., **1965**, 7116.
84. (a) D. J. Brown and L. N. Short, *J. Chem. Soc.* **1954**, 331. (b) P. Russell, G. Elion, E. A. Falco, and G. H. Hitchings, *J. Amer. Chem. Soc.,* **71**, 2279 (1949).
85. D. J. Brown and P. W. Ford, *J. Chem. Soc.* (C), **1967**, 568.
86. (a) R. Hull, *J. Chem. Soc.,* **1956**, 2033; (b) K. Elbs, *J. prakt. Chem.,* **42**, 179 (1893).
87. J. Schlenker, *Ber.,* **34**, 2812 (1901).
88. E. B. Brown and T. B. Johnson, *J. Amer. Chem. Soc*., **45**, 2702 (1923); *ibid*, **46**, 702 (1924).
89. F. J. DiCarlo, A. S. Schlutz, and A. M. Kent, *J. Biol. Chem*., **199**, 333 (1952).
90 J. J. Fox and D. Van Praag, *J. Amer. Chem. Soc*., **82**, 486, (1960).
91. M. P. V. Boarland, J. F. W. McOmie, and R. N. Timms, *J. Chem. Soc*., **1952**, 4691.
91. P. D. Landor and H. N. Rydon, *J. Chem. Soc*., **1953**, 3721.
93. P. E. Fanta and E. A. Hedman, *J. Amer. Chem. Soc*., **78**, 1434 (1956).
94. M. Ishidate, M. Sekiya, Y. Daki, I. Kurita, and M. Harada, *Chem. Pharm. Bull (Tokyo),* **3**, 224 (1955).
95. H. Bredereck, G. Simchen, and P. Speh, *Annalen*, **737**, 29 (1970).
96. T. D. Heyes and J. C. Roberts, *J. Chem. Soc.,* **1951**, 328.
97. R. G. Jones, E. C. Kornfeld, and K. C. McLaughlin, *J. Amer. Chem. Soc.,* **72**, 3539 (1950).
98. W. Pfleiderer and H. Mosthaf, *Chem. Ber*., **90**, 728 (1957).
99. G. G. Overberger and I. C. Kogon, *J. Amer. Chem. Soc*., **76**, 1879 (1954).
100 H. Bredereck, W. Jentzsch and G. Morlock, *Chem. Ber*., **93**, 2405 (1960).
101. H. Gattnner and E. Fahr, *Annalen,* **670**, 84 (1963).
102. D. J. Brown, P. W. Ford, and M. N. Paddon-Row, *J. Chem. Soc.* (C), **1968**, 1452.
103. D. J. Brown and J. S. Harper, *J. Chem. Soc*., **1963**, 1276.
104. D. J. Brown in B. S. Thyagarajan (ed.) *Mechanisms of Molecular Migrations*, Wiley, New York (1968).
105. H. C. Van der Plas, *Ring Transformations of Heterocycles*, Vol. 2, Academic Press, London and New York, 1973.

Chapter 3

The Chemistry of the Purines and Pteridines

(A) INTRODUCTION

The purine uric acid (1) was probably the first N heterocyclic compound to be isolated in a pure form (Scheele 1776), although elucidation of the structure did not come until about 100 years later. There was also a long interval between the first isolation of a pteridine derivative (Hopkins 1889) and the recognition of the structure of such compounds (Purrmann 1940). Purines and pteridines have now been isolated from a variety of natural sources and derivatives of both ring systems play important roles in the biochemistry of all cells. A number of purines and pteridines having biological activity have now been synthesized and examples of such compounds and the structures and roles of the naturally occurring compounds will be discussed later in the book.

Purine is the compound **(2)** having the numbering system shown, and pteridine **(3)** is shown with the numbering system which will be used in this account.

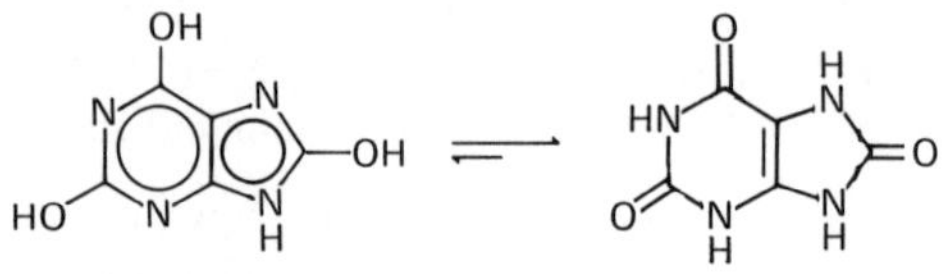

(1)(a) (b)

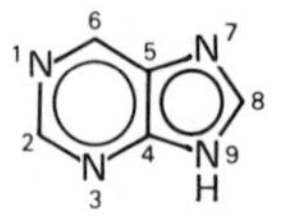

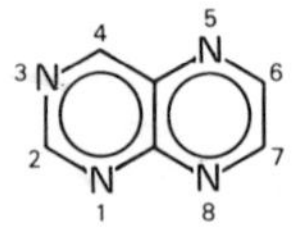

(2) (3)

Like pyrimidine, both purine an pteridine are π deficient ring systems and in general the usual principles of nitrogen heterocyclic chemistry apply to them, but in these two cases there is no position equivalent to the

Table 3.1 π Electron densities in purine and pteridine

Position	1	2	3	4	5	6	7	8	9
Purine	1.195	0.902	1.216	—	—	0.907	1.308	0.895	1.592
Pteridine	1.182	0.862	1.156	0.864	1.100	0.924	0.897	1.133	—

5-position of the pyrimidine ring, so that all of the carbon sites are electron deficient. Thus all sites in both the purine and pteridine rings are activated towards nucleophilic attack and deactivated towards electrophilic attack. Some values of the π electron densities at the various sites in purine and pteridine are listed in Table 3.1.

However, the 8-position of purine, being equivalent to the 2-position of imidazole, and having an electron-releasing −NH− group adjacent in addition to the electron-attracting =N−, would be expected to be less π electron deficient than the sites in the pyrimidine ring. Although the π electron density seems to be about the same as that of the carbon sites in the pyrimidine ring of purine, the 8-position is the most susceptible to electrophilic attack, but the presence of electron-releasing groups in the pyrimidine ring is necessary for such a reaction to occur. Because of the pronounced electron deficiency of the carbon sites in pteridine, no electrophilic reactions of pteridine or its simple derivatives have been observed. But pteridines quite readily undergo addition reactions—across the 6,7 bond—and form covalent hydrates even more readily than pyrimidines (see Section 2E).

As in the case of the pyrimidines, the purines and pteridines also exist in oxo, thione, and amino tautomeric forms but, for the sake of simplicity, the hydroxy, mercapto and amino purines and pteridines will be represented in the aromatic form unless there is a particular need to use the alternative form. Also the imidazole H in purines will normally be represented as a 9-substituent (although certain purines seem to favour 7-positioning) except where it is necessary to use such an alternative form.

(B) THE PURINES

(i) Properties

Many of the reactions of purine can be explained by a consideration of the properties of the N-heterocyclic ring systems in a similar way to that described for the pyrimidines in Section 2B. However, in the case of the purines the effect of the pyrrole-type NH in the imadazole ring has also to be considered. Lister has written a comprehensive review[1] of the purines.

Because of the electron-attracting character of the cyclic nitrogen atoms, =N−, the 2-, 6-, and 8-sites in purines are active, and substituents at such positions are susceptible to the reactions described for pyrimidine derivatives in Section 2E. Nucleophilic displacement of appropriate substituents usually occurs readily, the order of reactivity of the sites being governed by the ionization of the molecule at the time of attack. If the molecule is un-ionized, the 8-position seems to be the most electron deficient and substituents at this position are displaced first, the order usually being C-8, C-6, C-2. If the imidazole −NH− is ionized, the proton readily being removed in alkaline conditions, then nucleophilic attack occurs less readily and is directed to the pyrimidine ring, the order of displacement now usually being C-6, C-2, C-8. In acid media, when purines may be protonated, the order usually followed for nucleophilic displacement is 8, 6, 2.

Electrophilic attack is possible in the case of purines but it occurs only at position 8, and only when at least one, and usually two, strong electron-releasing substituents are present in the pyrimidine part of the molecule.

Purine and its simple alkyl derivatives have low melting points and are readily soluble in water. The introduction of substituents such as hydroxy, amino, thio, etc., generally considerably decreases the water solubility and increases the melting point due to the increased possibility of interpurine hydrogen bonding. Many hydroxy and mercaptopurines show no well-defined melting points but decompose on heating to high temperatures.

Owing to the presence of the imidazole −NH−, purine has both acidic and basic properties. Purine itself is an acid, comparable in strength to phenol, and has pK_a 8.9, whilst the basic ionization constant (pK_a 2.4) shows it to be a weaker base than aniline. The alkyl purines are little different in their ionization properties from purine, but the hydroxy purines become more acidic in character with the progressive introduction of the hydroxy groups, 2,6,8-trihydroxypurine (uric acid) having pK_a 5.8. The mercaptopurines are more acidic than the hydroxy analogues, but the introduction of amino groups leads to a progressive increase in basic strength 2,6,8-triaminopurine having pK_a 6.2 N alkylation at positions 7 or 9 gives purines which do not have acidic properties but generally alkylation of purines causes weak acid-weakening and base-strengthening properties.

(ii) Synthesis

The most common synthesis of the purine ring system utilizes preformed 4,5-diaminopyrimidines for condensation with an active one carbon unit to provide C-8. This sythesis is generally referred to as the Traube synthesis.[2] As 4,5-diaminopyrimidines are comparatively readily available, this synthesis can lead to a great number of substituted purines.

Several different active one-carbon units are also available, for example (**4**), X = OH, Y = O (carboxylic acids); X = NH_2, Y = O (amides; X = Cl, Y = O acid chlorides); X = OC_2H_5, Y = Cl (chloroformic

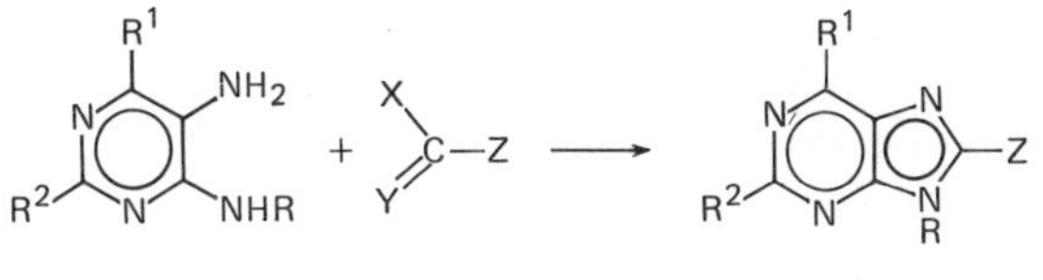

(4) (5)

esters); and $X = NH_2$, $Y = NH$ (ureas, thioureas, and amidines). By an appropriate choice of the pyrimidine, the carbon unit, and reaction conditions, a wide variety of purines (**5**) can be obtained. However, not all combinations react, and because of the fairly vigorous conditions needed to effect ring closure in many instances, frequently reactions of the pyrimidine substituents R^1, R^2 occur (e.g. hydrolysis, attack by the one-carbon unit, cyclization to the 6-substituents instead of a 4-substituent, etc.). For example, the following types of reaction may occur:

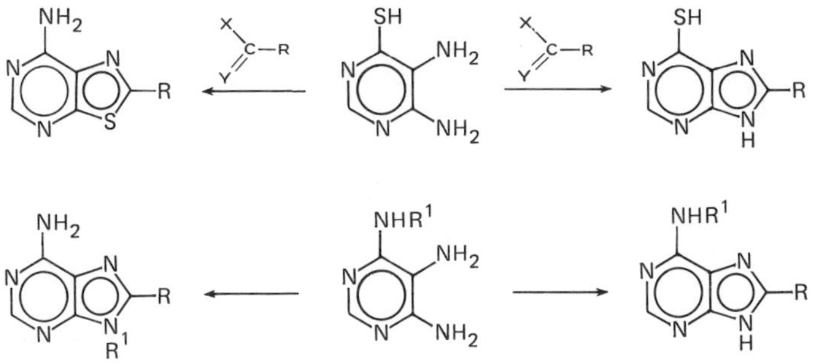

In these condensations of 5,6-diaminopyrimidines with one-carbon fragments the course of the reaction is usually attack at the 5-amino group to give an isolable product of the type (**6**) in many cases which cyclizes to give the purine on further reaction.

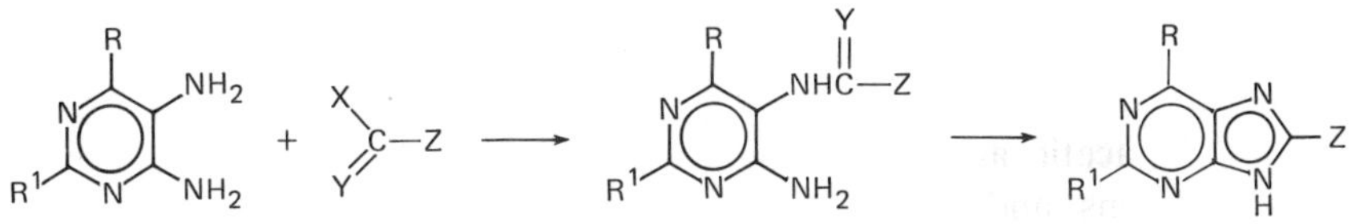

(6)

The use of formic acid as the one-carbon unit results in the production of purines having no substituent at the 8-position. The ease of formation of the 5-formamidopyrimidine varies with the pyrimidine, but generally formylation is carried out with formic acid under reflux. It is only in a minority of cases that direct ring closure to the purine occurs under these conditions, for example, guanine (**8**) is obtained in good yield on refluxing 6-hydroxy-2,4,5-triaminopyrimidine (**7**), as the sulphate, with formic acid for several hours.[2,3] However, generally conversion of the 4-amino-5-formamidopyrimidine to the purine requires more vigorous reaction conditions, for example fusion, heating in aqueous alkali, the use of a high-boiling inert organic solvent, or the use of higher reflux temperatures by the addition of sodium formate, of the use of formic acid–acetic anhydride mixtures. 5,6-Diamino-2-methoxypyrimidine (**9**) reacts[4] with formic acid–acetic anhydride (probably via formic acetic anhydride formed *in situ*) to give 2-methoxypurine (**10**).

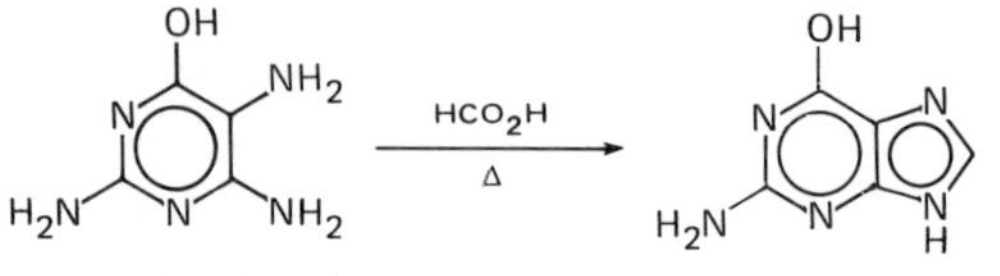

(**7**) $\xrightarrow[\Delta]{HCO_2H}$ (**8**)

(**9**) $\xrightarrow[\Delta]{HCO_2H/Ac_2O}$ (**10**)

Other reagents which can be used to synthesize purines which are unsubstituted at position 8 are formamide (which gives better yields of product under acid conditions), triethyl orthoformate (which reacts faster in the presence of acetic anhydride), and dithioformic acid, which reacts under mild conditions and which has been used for the synthesis of purine glycosides since the cyclization can be effected under conditions in which the glycosyl group remains intact. Examples of these reactions are given in Fig. 3.1.

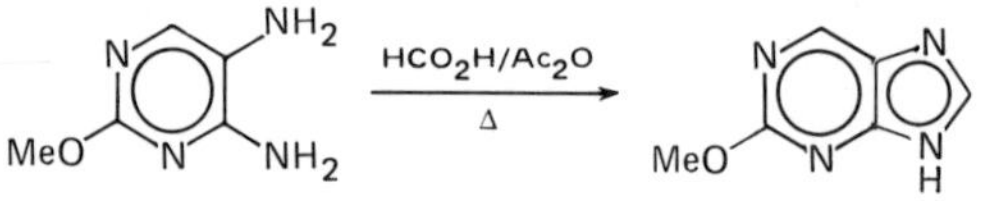

The use of carboxylic acids other than formic acid, acid anhydrides, and acid chlorides results in the production of 8-substituted purines. The range of such compounds available enables a wide variety of 8-substituted purines to be sythesized. Generally the 5-acyl (or aryl) aminopyrimidine is formed first, which requires further reaction, e.g. fusion or refluxing in alkaline

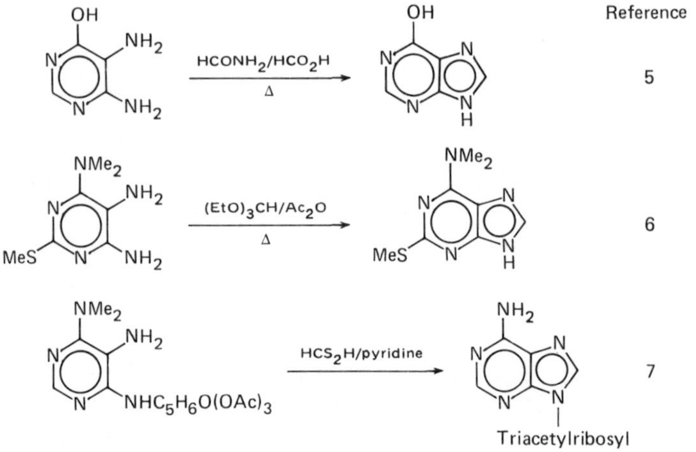

	Reference
$\xrightarrow[\Delta]{HCONH_2/HCO_2H}$	5
$\xrightarrow[\Delta]{(EtO)_3CH/Ac_2O}$	6
$NHC_5H_6O(OAc)_3$ $\xrightarrow{HCS_2H/pyridine}$ Triacetylribosyl	7

Fig. 3.1.

Fig. 3.2.

solution, to effect ring closure. Examples of the use of these reagents are given in Fig. 3.2.

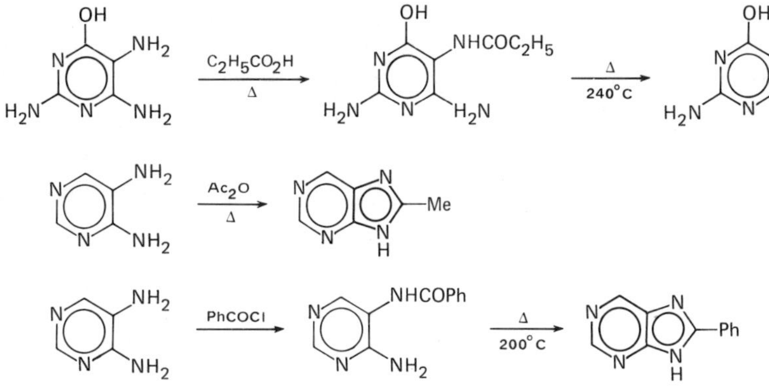

The reaction of 5,6-diaminopyrimidines with urea, ethyl chloroformate, or carbonyl chloride, results in the production of 8-hydroxypurines whilst the use of thiourea or carbon disulphide affords the 8-mercapto purines. However, the use of amidines

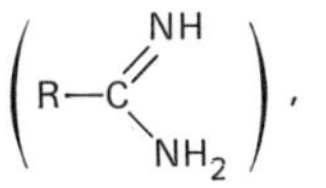

$$\left(R{-}C{\overset{\displaystyle NH}{\diagup\!\!\diagup}}\!\!\!\!\!\!\!\!\underset{\displaystyle NH_2}{\diagdown}\right),$$

guanidine, and aldehydes and ketones, does not seem to have been extensively studied, and although products have been obtained in some cases, frequently the product which is obtained is better synthesized by another route. Some examples of these other purine syntheses are given in Fig. 3.3.

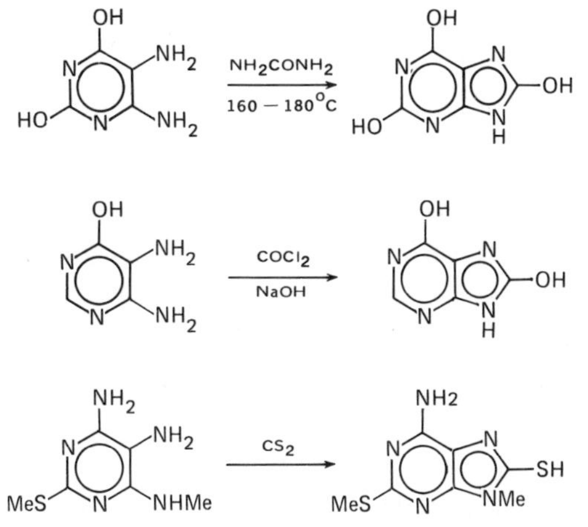

Fig. 3.3.

In addition to the Traube synthesis of purines, another method available involves the use of 6-amino-5-nitroso- or 6-amino-5-nitropyrimidines. Two techniques are possible using such compounds, one of which involves an *in situ* reduction of the 5-nitroso or nitro group and is, in effect, a modified Traube synthesis, whilst the other involves an intramolecular cyclodehydration reaction of the 5-nitroso or nitro group and an alkylamino substituent at position 6.

An example of this 'modified Traube synthesis' is the preparation of guanine by heating 2,6-diamino-4-hydroxy-5-nitrosopyrimidine (**11**) in 90% formic acid in the presence of zinc dust of Raney nickel.[13] This method, or minor modification of it, has resulted in the preparation of a number of purines.

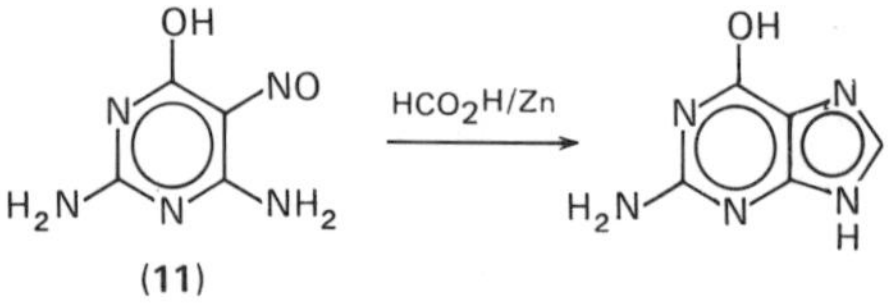

(11)

The cyclodehydration reaction has been particularly useful in the synthesis of purines alkylated in the imidazole part of the molecule. The reaction involves loss of elements of water between a 5-nitrosopyrimidine and a C-4-alkylaminogroup and it seems to be general for derivatives of such a type, the thermal cyclizations occurring on heating in a solvent such as dimethylformamide or on fusion. An example of this reaction is the synthesis of theophylline (**13**) on heating the methylated pyrimidine (**12**) in butanol or xylene.[14]

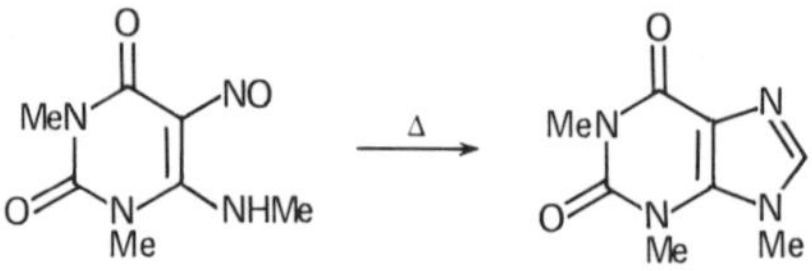

(12) (13)

Some other methods of synthesis of purines from pyrimidine precursors such as 5-amino-4-hydroxy,4,5-dihydroxy, and 4-amino-5-carbamoylpyrimidines are useful in some cases, but generally this type of reaction is much less useful than those indicated above. Examples of such reactions are given by Lister.[1] However, although the use of pyrimidine precursors has been the most common route to purines to date, the use of alternative intermediates, particularly imidazole derivatives, is becoming more widely used for purine syntheses. It is of interest that in the biosynthesis of purines it is the imidazole ring which is formed first, although chemically the appropriately substituted pyrimidines are usually more readily available than suitable imidazole precursors. Certain purines have also been obtained from acyclic precursors, much interest in this approach being centred upon the synthesis of vital biological chemicals

$3HCN \longrightarrow H_2NCH(CN)_2 \xrightarrow{+2NH_3} H_2N{-}C(=NH){-}C(NH_2){=}C(NH_2){-}NH_2$

(14) (15)

$HCN + NH_3 \longrightarrow CH(NH_2){=}NH$

(16)

(15) + (16) ⟶ (17) $\xrightarrow{+(16)}$ (18)

Fig. 3.4. Abiotic synthesis of adenine

from simple organic molecules when the earth was in an early state in the evolution of biological species.

It has been shown[15] that adenine (**18**) is produced on passing HCN into aqueous ammonia, the sequence of reactions which occurs[16,17] being that indicated in Fig. 3.4. The intermediates aminomalonitrile (**14**), aminomalondiamidine (**15**), and 4-amino-5-amidino-imidazole (**17**) have been isolated[17] from such a reaction mixture.

A number of combinations are possible for the condensation of substituted imidazoles with other fragments to lead to the synthesis of purines. Some examples of the more important types of condensation are shown in Fig. 3.5.

+ Z ⟶

In practice the syntheses of the 4- or 5-CN substituted imidazoles frequently start from the 4- or 5-aminoimidazole and one carbon unit, so that

Table 3.2

Type	X	Y	Z	example of imidazole
I	NC	CN	—	5-Formylamino-4-carbamoylimidazole (**19**)
II	N	CN	C	5-Amino-4-carbamoylimidazole (**20**)
III	NCN	C	—	5-Ureidoimidazole-4-carboxylic acid (**21**)
IV	NC	C	CN	5-Amino-4-methoxycarbonylimidazole (**22**)

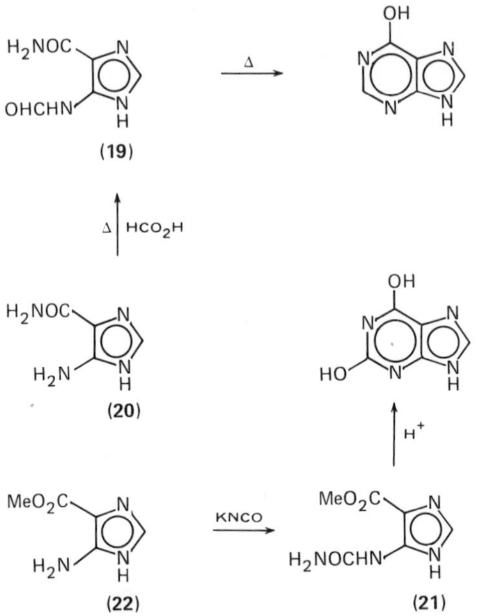

Fig. 3.5. Some examples of purine synthesis from imidazoles

types I and II, and III and IV, in the classification of Table 3.2 may be considered to be essentially the same processes in such cases.

(iii) Reactions

Substituted purines undergo essentially the same reactions as pyrimidines, having the same substituent in the 2-,4-, or 6-position. Purine has no position equivalent to the 5-position of pyrimidine and none of its derivatives show the same 'aromatic' properties which are shown by 5-substituted pyrimidines. As a consequence, the majority of the reactions which purines undergo involve nucleophilic substitution. However, some purines having electron-releasing groups to counteract the electron-attracting property of the cyclic =N− groups, will undergo electrophilic attack to give the 8-substituted product in all cases, with the exceptions of alkylation which occurs at nitrogen, and free radical reactions which occur at C-6 if this position is free, then at C-8.

(a) 8-Halogenation of purines

Direct chlorination of purines seems to require two electron-releasing groups in the pyrimidine ring; however, bromination seems to be a more widely applicable reaction and even simple monosubstituted purines can be directly brominated. Adenine **(23)**, for example, does not seem to chlorinate directly, but with bromine first gives a bromine adduct[18] which then breaks down to give 8-bromoadenine **(24)**. The naturally occurring methylated derivatives of xanthine, theobromine **(25)**, theophylline **(26)**, and caffeine **(27)** readily brominate or chlorinate at position 8 under a wide variety of conditions (see ref. 1, pp. 146, 147).

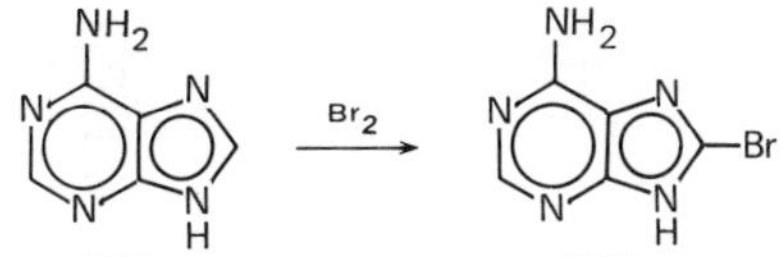

(23) $\xrightarrow{Br_2}$ **(24)**

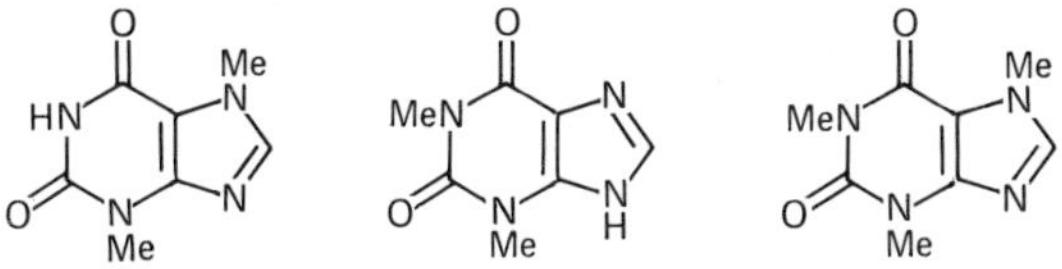

(25) **(26)** **(27)**

Very few examples of the direct iodination of purines have been recorded, but caffeine has been converted to the 8-iodo derivative by heating with iodine in a sealed tube at 150°C for 6 h.[19] No examples of the direct fluorination of purines have been reported to date.

(b) Nitration, nitrosation and diazo coupling of purines

Direct nitration is successful only with polysubstituted purines having two strong electron-releasing groups. The nitration may be carried out using nitric acid without solvent, but in most cases the reaction is best carried out in acetic acid. Theophylline gives good yields of the 8-nitro derivative using this method.[20] Whereas a number of pyrimidines have been shown to nitrosate at position 5, and a few to nitrosate at position 6, no direct nitrosation of purines has been shown to occur, although there was a claim,[21] now disproved, that nitrosation does occur. The only known example of a nitrosopurine is 6-nitrosopurine **(28)** formed from the 6-hydroxyamino derivative by oxidation.[22,23]

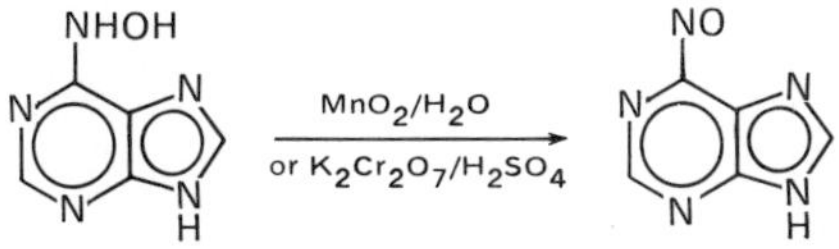

$\xrightarrow[\text{or } K_2Cr_2O_7/H_2SO_4]{MnO_2/H_2O}$ **(28)**

The presence of two electron-releasing groups in the pyrimidine ring also seems to be necessary for purines to undergo diazo-coupling reactions at position 8, but the reaction does seem to be more widely applicable than the nitration of purines. Some claims for the coupling of diazotized aromatic amines with adenine, 2-aminopurine, and hypoxanthine do not seem to have been substantiated.[4,11,24,25] However, with compounds such as xanthine, guanine and theophylline, 8-diazo coupled products are readily obtained, although theobromine and caffeine do not couple,[24] presumably because these compounds are prevented from forming an anion since they are alkylated on the imidazole acidic position.[26]

(c) Free radical reactions

Purines react with free radicals such as hydroxyl radicals generated by Fenton's reagent or by irradiation of an aqueous solution of a purine to give hydroxypurines. If the 6-position is vacant, this is attacked preferentially followed by attack at the 8-position, but attack at the 2-position has not been observed. Using such a method guanine is obtained from 2-aminopurine,[27] whilst adenine gives the 8-hydroxy product.[27,28]

(d) N-alkylation of purines

The site of N-alkylation in purines is dependent on the substitution in the purine, the alkylating reagent, the conditions under which the alkylation is being carried out, and also the presence and position of other alkyl groups. For example, purine itself reacts with dimethyl sulphate in aqueous solution to give only 9-methyl purine.[29] The use of an excess of methyl iodide in dimethylformamide also gives[30] 7,9-dimethylpurinium iodide (**29**), also obtained by heating 7-methylpurine with methyl iodide in methanol.[31] However, 6-methylpurine reacts with dimethyl sulphate in methanolic potassium hydroxide to give two products: 6,9-dimethylpurine (**30**) as the major product, and 3,6-dimethylpurine (**31**) as the minor product.[32]

MeI/DMF

(**29**)

MeSO4/MeOH, KOH

(**30**) (**31**)

The alkylation of hydroxypurines tends to give dialkyl products rather than monosubstituted products even if the amount of alkylating agent is limited. Hypoxanthine reacts with methyl iodide in alkaline solution at 80°C to give the 1,7-dimethyl derivative (**32**).[8,33] However, in a non-basic reaction mixture (dimethyl sulphate in dimethylacetamide at 125°C) hypoxanthine is converted to the 7,9-dimethylhypoxanthinium salt (**33**).[34]

Xanthine dimethylates on treatment with dimethyl sulphate in potassium hydroxide at 60°C to give theobromine (3,7-dimethylxanthine, **34**), whilst a

MeI/OH^-

(32)

Me_2SO_4/DMA

NH_3/MeOH

(33)

detailed study[35] of the methylation of guanine using methyl chloride in sodium hydroxide at 70°C indicated the presence of the following products: 9-methyl- (33%), 7-methyl- (18%), 3-methyl- (11%), and only a trace of 1-methylguanine.

With 1-alkylguanines further alkylation gives the 1,7-dialkyl derivatives, whilst either 7- or 9-alkylguanines react to give 7,9-dialkyl guanines. The 7-position of guanine residues present in nucleic acids seems to be the most favoured site for attack by alkylating agents, the 7,9-disubstituted guanine which is formed being labile even under neutral conditions, with the result that guanine residues are lost from the nucleic acid chain. This reaction is believed to be the prime effect responsible for the mutagenic action of alkylating agents on nucleic acids (see Section 9B(*ii*)).

Me_2SO_4/KOH 60°C

(34)

caffeine

In alkaline solution adenine gives the 9-alkyl product with small quantities of the 3-isomer, but in neutral solution the 3-alkyl product predominates with small amounts of the 9 and 1-alkyl isomers. However, in acid media adenine does not normally alkylate.

Further alkylation of 3-alkyladenines occurs at N-7, whilst further alkylation of a 9-alkyladenine occurs at N-1. The general scheme for the methylation of adenine is given in Fig. 3.6.

Alkylation of chloro- and other halogenopurines tends to give mixtures of 7- and 9- substituted products, with the 9-isomer predominating, whilst with the mercapto purines both N and S methylation can occur but in many cases only the N-alkylated thiopurine is obtained. This is in contrast to pyrimidines when the S-alkylated products are obtained. Studies of this reaction[36] have indicated that an S-alkylated purine is probably first formed which is subsequently reconverted to a thiopurine during the subsequent N-alkylation.

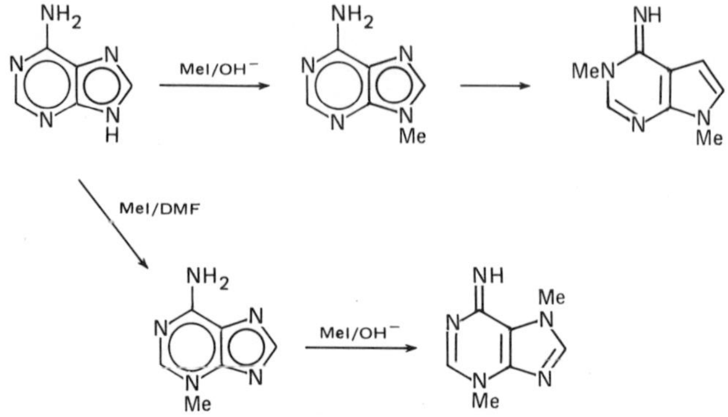

Fig. 3.6. Methylation of adenine (see ref. 1, pp. 342–348)

Thus in considering the alkylation of purines the products obtained depend on the substituents on the purine ring (and possibly their orientation), the alkylating agent, the reaction conditions, the site of the first alkylation, and the severity of the conditions which will affect the extent of alkylation.

(e) Nucleophilic Substitutions

The majority of the reactions of purines are nucleophilic reactions and a wide variety of these are known, the general mechanisms being similar to the nucleophilic metatheses of the 2-, 4-, and 6-substituted pyrimidines. However, there are three different sites in purines and consideration should be given to the relative reactivity of each site.

If we consider, for example, 2,6,8-trichloropurine (**35**), in carrying out a nucleophilic reaction in basic solution the anion of the purine (**36**) is the species which reacts. The order of ease of displacement of the halogens is $6 > 2 > 8$. However, in the case of 2,6,8-trichloro-9-methylpurine (**37**), where such ionization is prevented, the order of ease of displacement of the halogens is $8 > 6 > 2$. These effects are explained by considering that in the unionized species, attack at position 8 only removes the aromaticity of the pyrrole ring whilst attack at positions 2 or 6 would disrupt the pyrimidine ring. In the ionized form the negative charge is held mainly by the imidazole portion of the ring, leaving the pyrimidine ring relatively unaltered, giving a greater reactivity to sites 6 and 2.

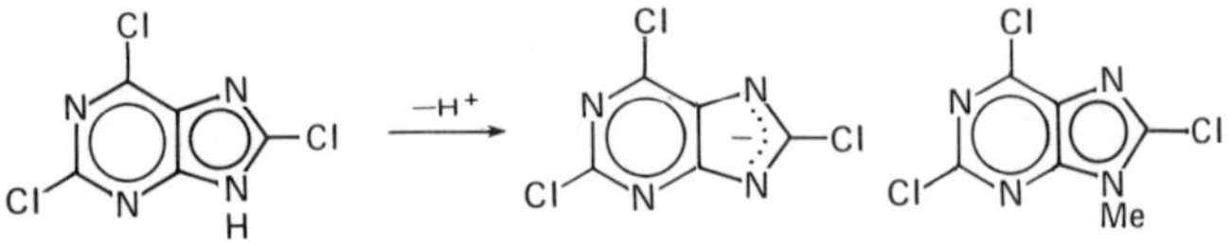

(**35**) (**36**) (**37**)

Although they are basically similar to the corresponding pyrimidine reactions, some points of interest and some examples of interesting variations of purines are given below:

(1) Amination of halogenopurines: The halogenopurines are less reactive than comparable halogenopyrimidines, but amination can usually be achieved quite readily with primary or secondary amines or hydrazines. However, reaction with ammonia usually requires heating to well above 100°C in a sealed tube. For example adenine **(38)** is obtained[37] by heating 6-chloropurine with butanol saturated with ammonia in a sealed tube at 150°C.

Once one halogen of a polyhalogenopurine has been replaced by an amine or ammonia, the second halogen is more difficult to displace due to the deactivating effect of the electron-releasing −NHR group. The 2-position of the purine ring is particularly difficult to react,[1] but in the case of 6-chloro-2-fluoropurine riboside **(39)** the order of reactivity is reversed and methanolic ammonia caused the preferential removal of the 2-fluorine atom, giving 2-amino-6-chloropurine riboside **(40)**.[38] This reflects the ease of displacement in aromatic nucleophilic substitution of fluorine relative to chlorine.

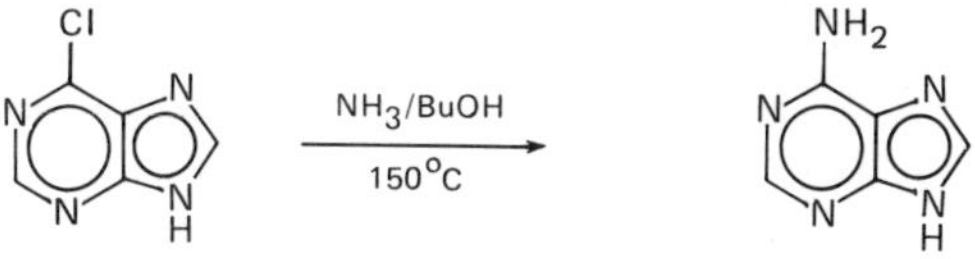

Cl N N N N H —NH_3/BuOH, 150°C→ NH_2 N N N N H **(38)**

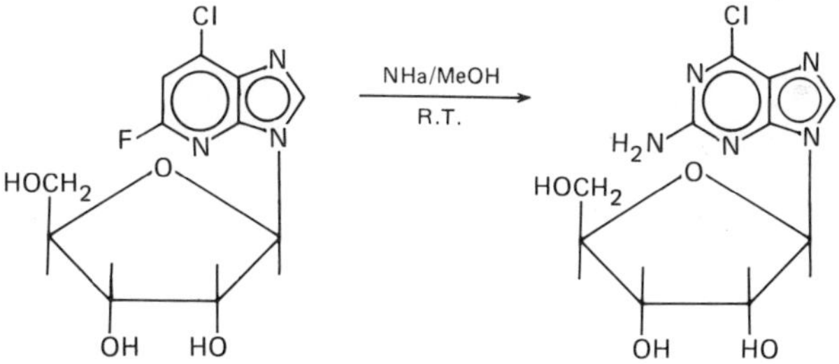

Cl F N N N N $HOCH_2$ O OH HO **(39)** —NHa/MeOH, R.T.→ Cl H_2N N N N N $HOCH_2$ O OH HO **(40)**

(2) Reaction of halogenopurines with alkoxide, hydroxide, and alkylthio anions: The highly basic nature of the alkoxide, hydroxide and alkylthio anions means that the favoured reaction with purines is removal of the 9-H atom to form the anion, consequently subsequent nucleophilic displacements of halogens etc. require rather vigorous conditions. Again the 2-position is the one which is most resistant to attack—for example 2-chloropurine is converted to the ethoxy derivative **(41)** in a sealed tube at 150°C, whilst the 9-methyl compound **(42)** is obtained under reflux conditions;[39] 8-chloropurine reacts with sodium ethoxide at 150°C (1 h) to give the 8-ethoxy

derivative (**43**) whilst 8-chloro-9-methylpurine gives the 8-ethoxy compound (**44**) on leaving at room temperature for some hours.[39]

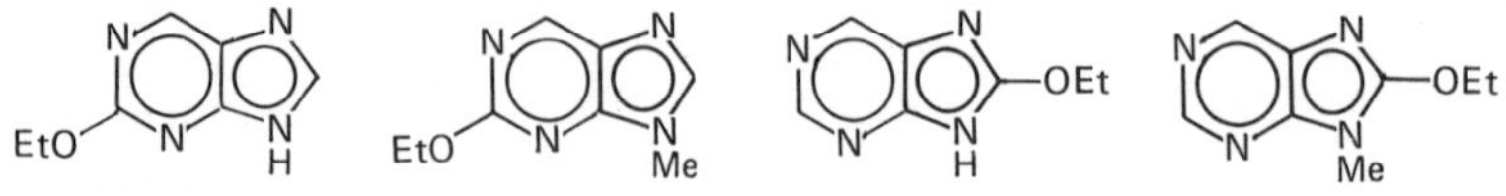

(41) (42) (43) (44)

An exception to the general rule concerning the lack of reactivity of a 2-halogen and the greater difficulty in displacing a second halogen (owing to the stabilizing electron-releasing effect of the alkoxy group first introduced) is the case of 2,8-dichloropurine (**45**) which is converted to the dimethoxy derivative using sodium methoxide in refluxing methanol.[40] A similar treatment of 6,8-dichloropurine gives only 8-chloro-methoxypurine (**46**).[41]

The hydrolysis of halogenopurines with hydroxide ion seems to occur more readily than nucleophilic attacks by other anions, 6-chloropurine, for example, being converted to hypoxanthine by heating with 0.1 M sodium hydroxide for 4 h.[37] However, a side reaction which occurs in some cases is cleavage of the imidazole ring to give a diaminopyrimidine. It has been shown[42] that alkali instability is characteristic of purines which have no strong electron-releasing groups attached.

The purine ring system is more stable to hydrolysis in acid media, and in some cases hydrolysis of a halogen is easier than in alkaline media. For example, 6-chloropurine gives hypoxanthine in comparable yield in a quarter of the time when 0.1 M HCl is used rather than 0.1 M NaOH.[37]

Generally, the reactivity of the halogenopurines towards alkaline hydrolysis is similar to their reactivity towards other nucleophiles, the ease of replacement being $6 > 2 > 8$, but in the case of acid hydrolysis the 8-halogen seems to be the more labile. For example, in 50% aqueous acetic acid, 6,8-dichloropurine reacts at 100°C to give 6-chloro-8-hydroxypurine (**47**).

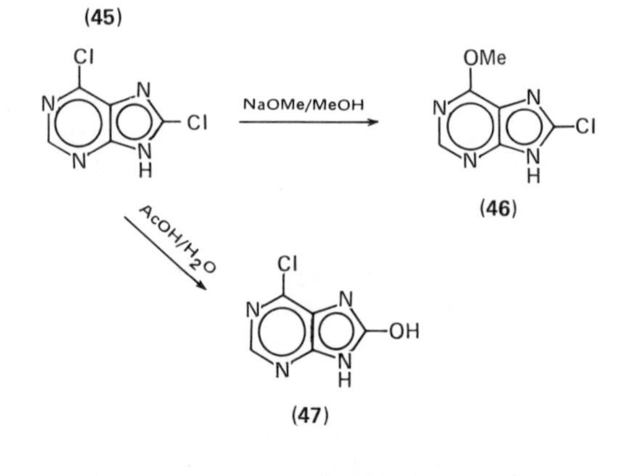

NaOMe/MeOH

(45)

NaOMe/MeOH

(46)

AcOH/H_2O

(47)

The reactions of purines with alkylthio anions seems to follow the same pattern as for alkoxy anions, and although the simple alkylthiopurines may be made by alkylation of mercaptopurines in many cases, direct replacement of halogen by the alkylthio group is of importance in a number of cases.

(3) Other nucleophilic reactions of purines. Halogen exchange reactions can be carried out, although only a few examples seem to have been reported. A 2-chloro group seems to be resistant, but a 6- or 8-chlorine can be replaced by iodine by the action of hydriodic acid at 0°C. For example, 2,6,8-trichloropurine can be converted to 2-chloro-6,8-di-iodopurine (**48**).[43] Although it is not a general reaction, some fluoropurines have been obtained from chloropurines by halogen exchange using silver fluoride in boiling toluene, but sometimes 2-replacement is unsuccessful. This generally requires a higher-boiling solvent (xylene for example).[43]

The diazotization of an aminopurine using a Modified Schiemann method (sodium nitrite followed by 48% fluorboric acid) has been used to obtain some fluoropurines—for example 2-fluoropurine.[44,45] However, other reactions of a similar nature seem to be mostly unsuccessful, but many aminopurines can be converted to the corresponding hydroxy derivatives by nitrous acid.

Chloropurines are probably the most versatile and widely used derivatives for nucleophilic substitution reactions, but in a number of cases other starting products can be most useful. For example, alkoxy, alkylthio and alkylsulphonylpurines can be hydrolysed to hydroxypurines with acid and can be converted to aminopurines by reaction with appropriate amino compounds. Such compounds are also capable of undergoing other typical nucleophilic metatheses, although the reactions have been less well studied than those of the halogenopurines.

(f) Other Reactions of Purines

(1) Formation of N-oxides: Hydrogen peroxide in acid converts a number of purines to the N-1 oxide, yet other purines give the N-3 oxide. For example, 6-methylpurine gives the 1-oxide (**49**) on treatment with hydrogen peroxide in acetic acid at 80°C for 12 h,[46] but 6-chloropurine gives the 3-oxide (**50**) using either perphthalic acid or acetic acid—hydrogen peroxide.[47] However, the formation of the 1-oxide seems to be the more widely occurring reaction.

(2) Diazotization of aminopurines: Although the diazotization of 8-aminopurines to give stable 8-diazopurines (**51**) is well known, in general the 2- and 6-aminopurines do not show this same ability, although some cases are known. For example, 2-aminopurine has been converted to the 2-chloro compound (**52**) using sodium nitrite in hydrochloric acid[44,38] and the only known 2-nitropurine has been obtained[48] by a diazotization reaction of

guanine when the compound 2-nitro-6-hydroxypurine (**53**) was obtained. The diazotization reactions of purines seem to have been comparatively little investigated.

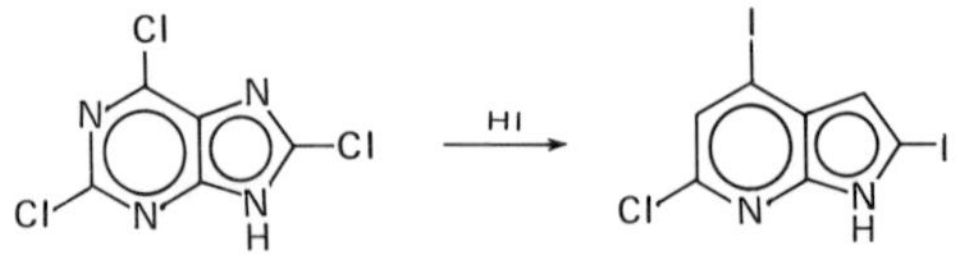

(48)

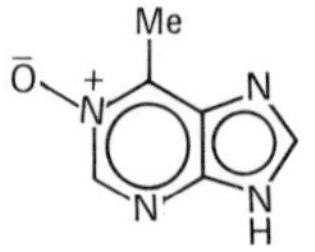

(49) 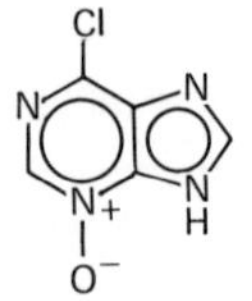 (50)

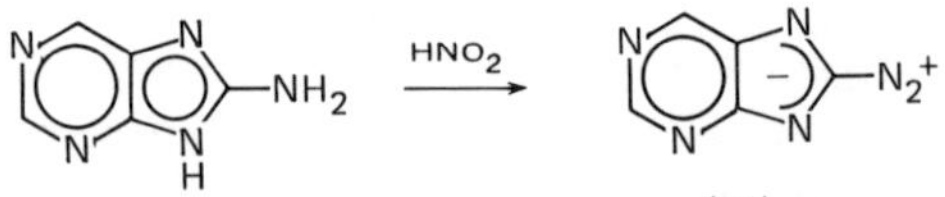

(51)

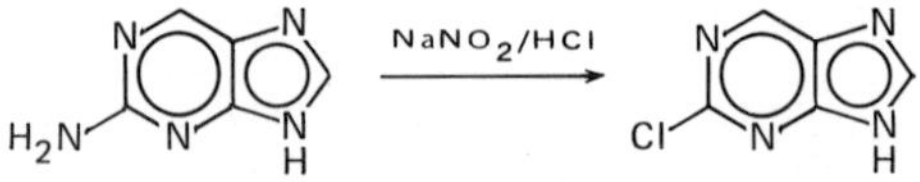

(52)

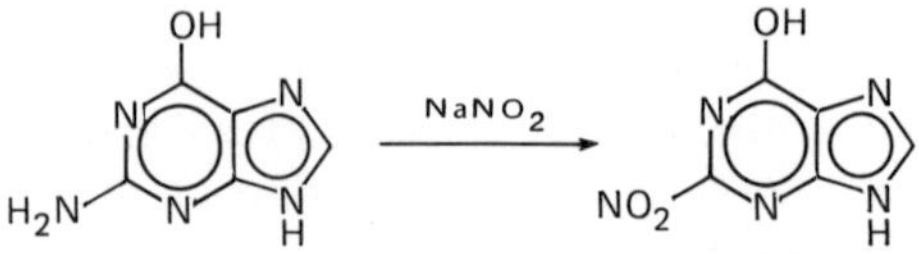

(53)

(3) Formation of covalent hydrates: Although pyrimidines and pteridines show well-defined hydrates by addition of water across the 5,6- and 3,4-bond respectively, no positive evidence for the formation of this type of compound seems to be obtained in the case of the purines.[49] However, it has been observed that isolatable adducts are obtained by u.v. irradiation of anhydrous, deoxygenated solutions of purine in some alcohols.[50] Methanol gives, for example, the adduct (**54**).

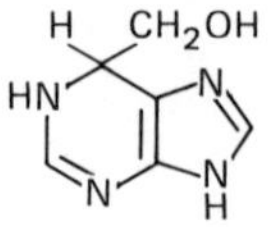

(54)

However, the addition of stronger nucleophiles has been observed.[51] For example, bisulphite adds across the 1,6-bond and also the addition of barbituric acids to purines has been observed.

Of other reactions of substituted purines, it should be noted that the hydroxy derivatives do not acylate or alkylate (except some derivatives of uric acid), indicating their oxo rather than hydroxy form, whilst the mercaptopurines will undergo such reactions, although they exist in the thione rather than the thiol form.

In general the methylpurines do not undergo oxidation to give aldehyde, carboxy-, or hydroxyiminomethylpurines as readily as the pyrimidines, and such reactions have been little investigated. But they do undergo reaction at the side-chain methyl with halogens.

(iv) Spectra of Purines

(a) Ultraviolet spectra

The u.v. absorption spectra of purines generally gives four groups of bands, although only two are usually observed. In some cases there are weak n–π* bands at about 300 nm and above. These arise from 'forbidden' transitions and are of weak intensity and also show bathochromic shifts in passing from polar to less polar solvents. In the region 230–300 nm, purines show absorbances attributed to π–π* transitions. These bands have relatively high extinction coefficients and show hypsochromic shifts in passing from more to less polar solvents.

A third set of absorbance, also due to π–π* transitions, may be observed below 230 nm having an extinction greater than the 230–300 nm band. However, these last bands sometimes occur below 210 nm, which is beyond the range of u.v. spectrometers in normal use.

The u.v. spectra of purines have been used for investigating the tautomerism of these compounds and are also used for structure elucidation. However, the situation is not as simple as in the case of the pyrimidines and evidence from u.v. spectra alone is sometimes ambiguous.

Like the pyrimidines, the purines have u.v. spectra which are dependent on the pH of the solution and this can be used to determine the pK_a of the purines.

The effects of substituents on the u.v. spectrum of purine are generally to increase the intensity of absorption and to cause bathochromic shifts. 8-Substitution seems to cause the greatest intensity increase, followed by 6-substitution, while 2-substitution gives a very much lower increase. However, a substituent in a 2-position has a greater bathochromic effect than such a substituent in the 8-position, which is also greater than the effect of 6-substitution. The usual 'additive' nature of the effects of polysubstitution does not seem to hold for purines in the same way as for pyrimidine derivatives.

For a detailed account of the u.v. spectra of purines reference should be made to ref. 1, Chap XIII, ref. 52, Chap. II, and ref. 53, Vol. 2.

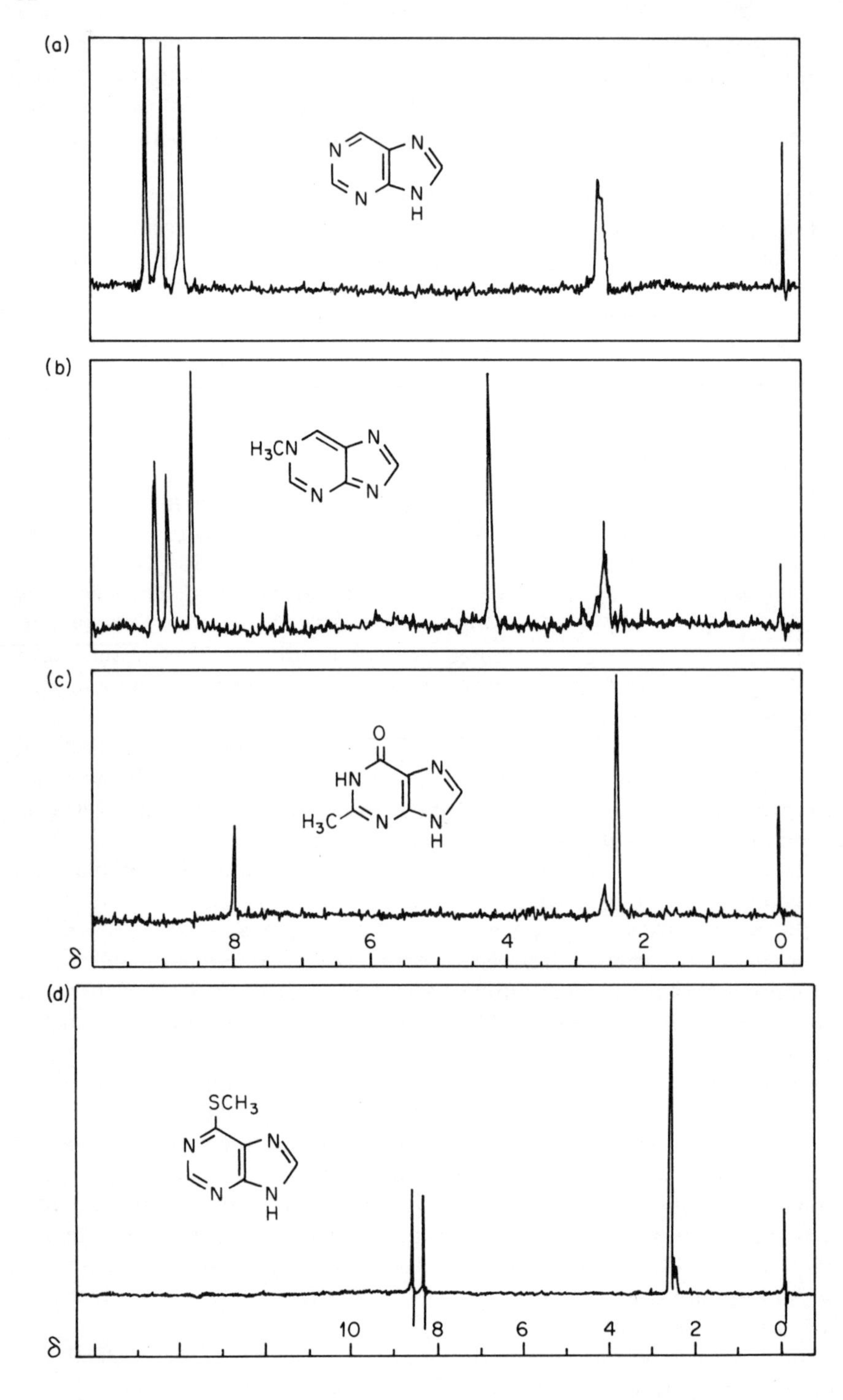
(a)
N
N
N
N
H
(b)
H_3CN
N
N
N
(c)
O
HN
N
H_3C
N
N
H
8
6
4
2
0
δ
(d)
SCH_3
N
N
N
N
H
10
8
6
4
2
0
δ

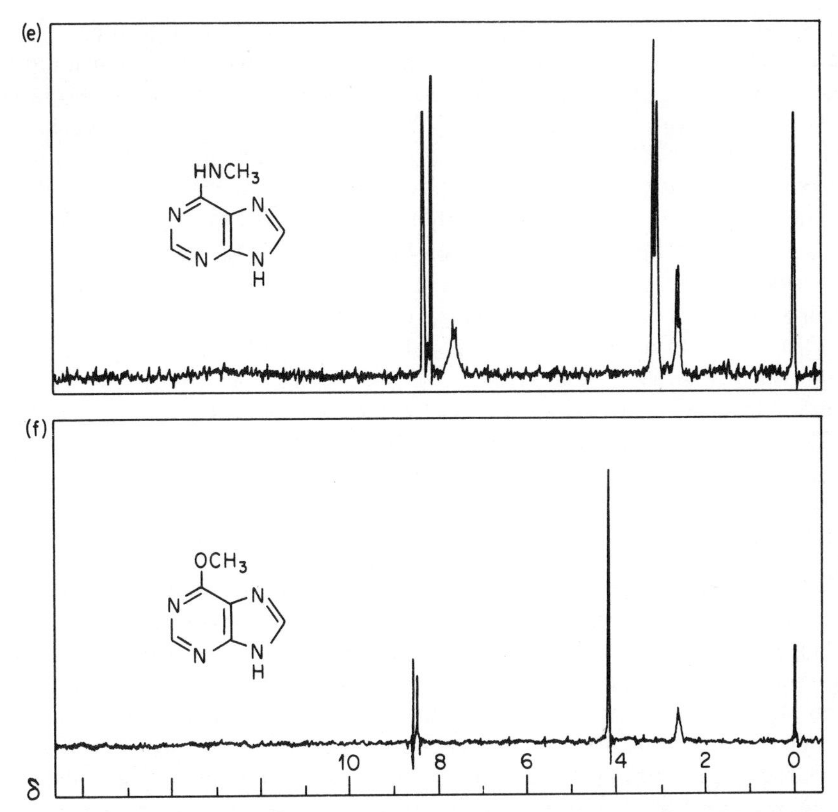

Fig. 3.7. ^{1}H-n.m.r. spectra of some simple purines. Reproduced by permission of John Wiley & Sons Inc.

(b) Infrared spectra

I.r. spectra of purines have been used for studies of the tautomeric nature and for structure-elucidation purposes in relation to purines. The spectra of purines tend to be complex and no discussion will be given here, but reference may be made to ref. 1, Chap. XIII, ref. 52, Chap. VI, and ref. 53, Vols. 2 and 4.

(c) Nuclear magnetic resonance

The ^{1}H n.m.r. spectrum of purine itself in D_2O, or d_6-DMSO, shows three singlet peaks in the aromatic region of the spectrum. The acidic N–H proton of the imidazole ring is a broad, low-intensity band, at even lower field. The peaks due to C-2, C-6, and C-8 show a noticeable broadening which has been ascribed to proton interaction with the nitrogen nuclei of adjacent molecules. By the use of specifically deuterated derivatives it has been established that the assignments of the peaks, in d_6-DMSO, are: C-8 (δ, 8.73), C-2 (δ, 9.03), C-6 (δ, 9.27). Thus the C-6 proton appears at a lower field than the C-2 proton, in contrast to the case of the pyrimidines. Some interesting effects are observed in substituted purines: for example, with 8-methylpurine it is H-6 rather than H-2 which appears at the higher field,[54] whilst in the cases of 6-iodopurine and adenine, the proton at position 2 appears at a higher field than that due to H-8.[55]

The ^{1}H n.m.r. spectrum of 1-methylpurine in d_6-DMSO and of purine in an acid solvent has shown that spin–spin coupling between the H-2 and H-6 does occur in cases where the nuclear quadrupole effect of the intervening nitrogen is inhibited. Thus the spectrum of 1-methylpurine in d_6-DMSO shows doublets for the H-2 (δ, 9.07) and H-6 (δ, 9.20) the coupling constant being ~2Hz (Fig. 3.7). The n.m.r. spectra of 3-, 7-, and 9-methylpurine showed only singlets for the two protons at C-2 and C-6.[53]

The ^{1}H n.m.r. spectra of some simple substituted purines are shown in Fig. 3.7.

(d) Mass spectra

The purine ring system is inherently stable, and, in the majority of cases, the base peak in the mass spectrum is that due to the molecular ion. The pyrimidine ring is usually the site of initial fragmentation, with further disruption being dependendent on the substitution of the purine. The fragmentation patterns of monosubstituted purines are usually complex, indicating that fragmentation is proceeding via multiple pathways with no particular one predominating. However, polysubstituted purines usually have more regular fragmentation patterns, having high intensity fragment ions from which major fragmentation pathways can frequently be inferred. The mass spectra of some typical purines are shown in Fig. 3.8.

The mass spectrum of the parent purine (Fig. 3.8a) has the molecular ion M^+ at *m/e* 120. Two molecules of HCN are lost from scission of the C-2–N-3 and C-6–N-1 bonds, metastable peaks being observed for each process.[56,57] With the 2- and 6-methylpurines, losses due to HCN and CH_3CN are observed.[57]

The three main fragmentation peaks given by adenine are due to the successive loss of HCN. N-6-alkylated adenines usually show patterns resulting from fragmentation of the side chain and then loss of HCN.[58]

The fragmentation patterns of guanine and xanthine and their derivatives are essentially similar and involve the loss of NH_2CN or HNCO (or similar species) respectively followed by further fragmentation.

The mass spectra of a number of purines are given in refs. 1, 52, 53, Vol. 3 and 59, and recourse to the original literature is also recommended.

(e) Fluorescence of Purines

Börressen[60] found that purine exhibits a slight fluorescence, this fluorescence increasing markedly in acid or base solution. A number of other

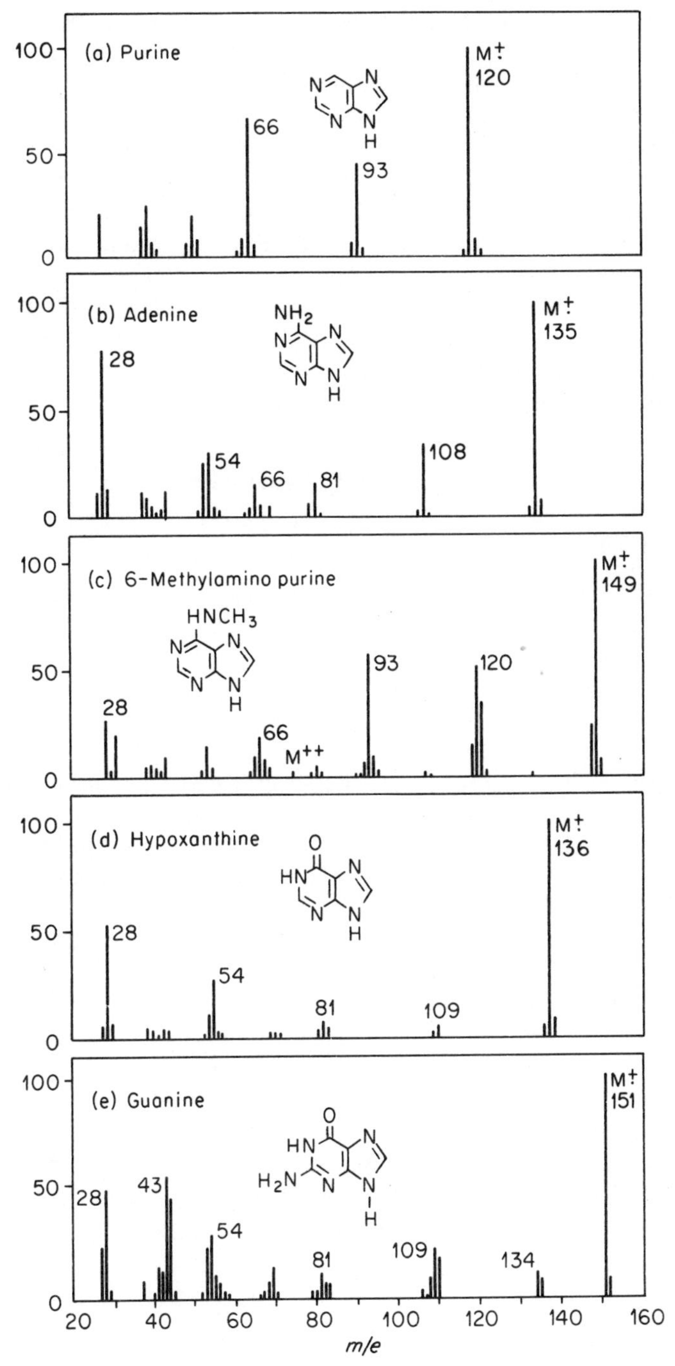

Fig. 3.8. Mass spectra of purine, adenine, 6-methylamino-purine, hypoxanthine, and guanine

purines also show this property, adenine and guanine being important examples. This marked fluoresence of guanine has been used for a fluorimetric assay for guanine deaminase.[61]

The fluoresence spectra of purines are pH dependent, since the quantum efficiencies of emission for the anionic, neutral and cationic species are different. For example, neutral adenine lacks fluorescence, but appreciable fluorescence can be observed for the cation. Similarly, guanine in neutral solution is non-fluorescent, but is appreciably fluorescent as the cation or anion. In the case of 2,6-diaminopurine the neutral molecule is more fluorescent than either of the charged species. Solvent and temperature effects have also been observed for the fluorescence of purines, but for a more detailed account of the phenomenon of fluorescence the reader is directed to ref. 1, p. 472 and the references cited therein.

(C) THE PTERIDINES

(i) Properties

As a class of compounds the pteridines seem to have been less thoroughly investigated than the pyrimidines and purines and their chemical and physical properties are less well documented. However, some reviews of pteridines are available (refs. 62–66). Due to the presence of four nitrogen atoms in the molecule, each being present in a six-membered aromatic ring, pteridine is a compound which has a very marked electron deficiency at each carbon site (Section 3(A)). As a result pteridine and its simple alkyl derivatives lack any appreciable aromatic stability and undergo ring opening or addition reactions very easily. Very few pteridines having strong electron-attracting substituents are known, whilst those having electron-releasing substituents, e.g. hydroxy and amino, are stable to ring-opening reactions but are very insoluble in organic solvents and rather insoluble in polar solvents. They tend therefore to be difficult to crystallize and also tend to decompose rather than melt on heating.

The ring structure of pteridine, with the numbering system used in this account, is shown in Fig. 3.9.

Pteridine is basic having a pK_a of 4.12,[63] the order of the basic strengths of the nitrogen atoms being[67] N-3 > N-1 > N-5 > N-8. The introduction of alkyl or amino groups increases the basicity, some values being given in Table 3.2. The introduction of hydroxy groups results in acidic compounds which are stronger acids than analogous compounds with fewer cyclic nitrogen atoms.

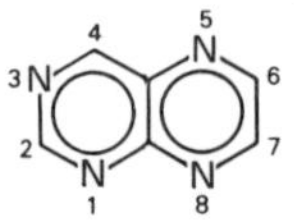

4 5 3N N 6 2 7 N N 1 8

Fig. 3.9. Pteridine

Table 3.2 pK_a values of some pteridines (Data from ref. 63)

Pteridine	pK_a
Pteridine	4.12
2-Amino	4.29
4-Amino	3.56
6-Amino	4.15
2-Methoxy	2.13
4-Methoxy	<1.5
6-Methoxy	3.60
2-Hydroxy	<2, 11.13
4-Hydroxy	<1.3, 7.89
6-Hydroxy	3.67, 6.7
7-Hydroxy	1.2, −2.0, 6.41
2,4-Dihydroxy	<1.3, 7.91
6,7-Dihydroxy	<2.7, 6.87, 10.0
6-Hydroxy-7,8-dihydro	4.78, 10.54
7-Hydroxy-5,6-dihydro	3.36, 9.94
2-Amino-4-hydroxy	2.31, 7.92

The unexpectedly high basicity of pteridine is due to covalent hydration of the 3,4-bond which results in N-3 becoming sp^3 hybridized. Such covalent hydration is very prevalent in the case pteridines (see Section 3C(*iii*)f) and may lead to anomalous pK_a values and spectral data.

Like the analogous derivatives of pyrimidines and purines, the amino derivatives of pteridine have the amino tautomer as the predominant form,

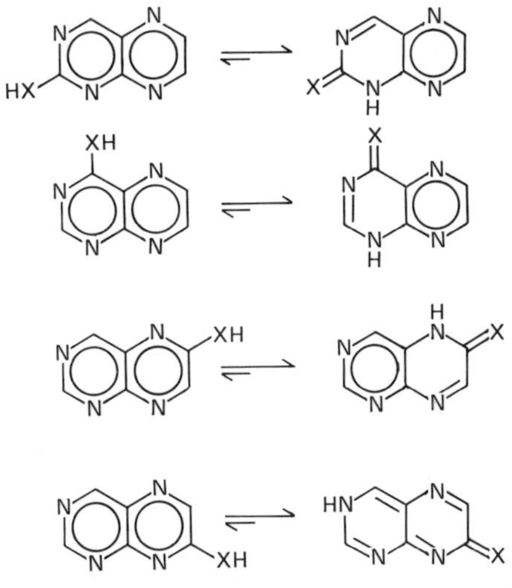

Fig. 3.10. Predominant tautomers of substituted pteridines, X = O or S

whilst the hydroxy and mercapto derivatives exist in the oxo forms respectively. The possibility of tautomerism in the hydroxy and mercapto pteridines is more complex than for pyrimidines and purines. The forms for 2-, 4-, 6-, and 7-substituted pteridines which seem to be most favoured[63,68] are shown in Fig. 3.10.

(ii) Synthesis of Pteridines

The most widely used method for the synthesis of pteridines starts with a 5,6-diaminopyrimidine. This reaction is commonly called the Isay reaction and was first used[9] for the synthesis of 6,7-diphenylpteridine (**55**). Most pyrimidines having 5,6-diamino groups seem to undergo this reaction with appropriate doubly activated two-carbon fragments. The types of such compound which may be used include 1,2-diketones, aldehydoketones, glyoxal

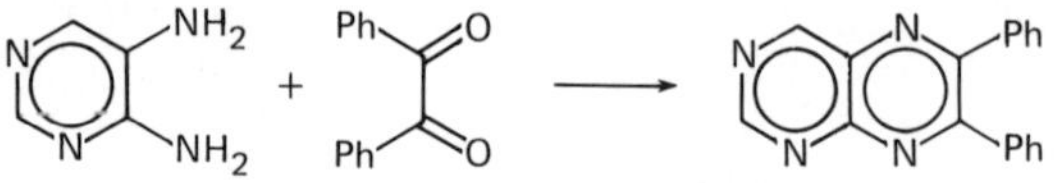

(**55**)

(a dialdehyde), α-keto acids (or esters), α-aldehydoacids, 1,2,-dicarboxylic acids (or esters). A few examples of such reactions are given in Fig. 3.11.

In the case of unsymmetrical bifunctional 2-carbon units two products may be obtained, e.g.:

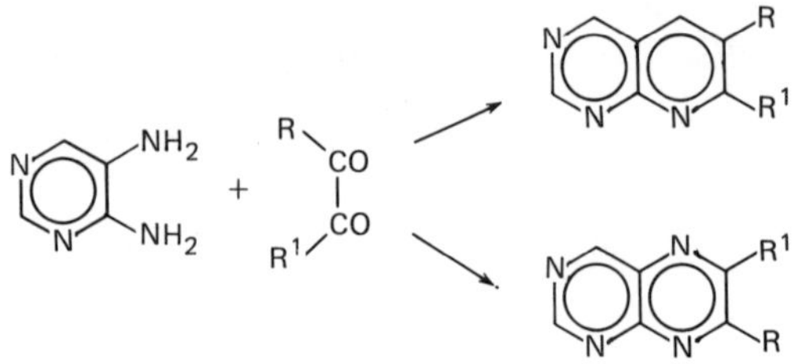

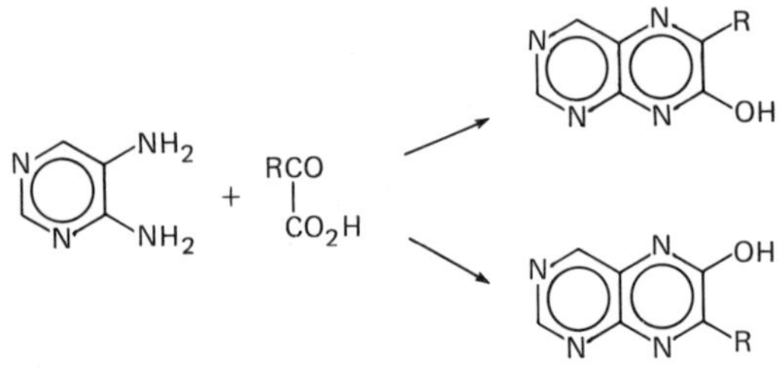

This can lead to difficulties in the separation and identification of the products. Generally it is observed that if the dicarbonyl compound is an aldehydo or keto acid or ester highly acid reaction conditions tend to favour the formation of the 6-hydroxypteridine product whilst mildly acid or neutral conditions tend to favour the production of 7-hydroxypteridines. Some examples of this effect are given in Table 3.3.

Reference

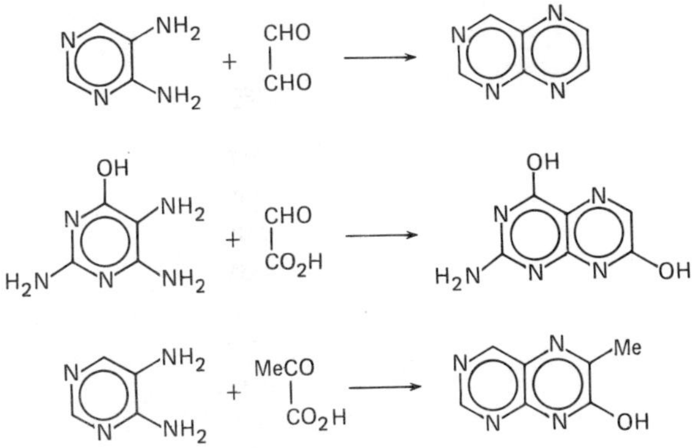

69

70

71

Fig. 3.11. Synthesis of pteridines from 5,6-diaminopyrimidines

The reason for the orientating effect of pH has been attributed[72] to differences in the basic strength of the 5 and 6- amino groups and to the assumption that the 5-amino group and aldehyde or keto carbonyl react preferentially in neutral conditions but not in strong acid. However, a comprehensive explanation of these orientation effects does not seem to have been proposed as yet, although the empirical observations indicate that, in general, in neutral or mildly acid solution, the aldehyde or keto carbonyl tends to react preferentially with the pyrimidine 5-amino group and in strong acid with the 6-amino group.

When the dicarbonyl compound is a diketone or ketoaldehyde changes in pH do not seem to be important in controlling the orientation of the

Table 3.3 Product distribution in pteridine synthesis at different pH.[72]

		Yield (%)			
		pH 5 (acetate buffer)		pH 0.25 (2N sulphuric acid)	
Pyrimidine	Reagent	6-OH	7-OH	6-OH	7-OH
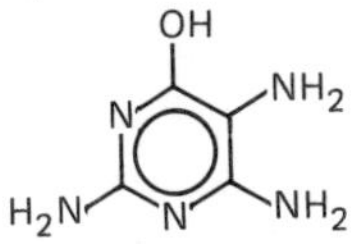 2-amino-4-hydroxy-5,6-diaminopyrimidine (OH, NH_2, H_2N, NH_2)	CH_3COCO_2H	42	10	76	0
	$C_2H_5O_2CCOCH_2CO_2C_2H_5$	trace	40	69	11
	$CO(CO_2C_2H_5)_2$	0	85	42	29
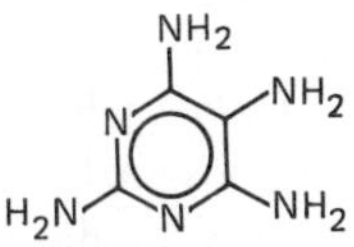 2,4,5,6-tetraaminopyrimidine (NH_2, NH_2, H_2N, NH_2)	CH_3COCO_2H	13	37	92	3
	$C_2H_5O_2CCOCH_2CO_2C_2H_5$	0	66	49	22
	$CO(CO_2C_2H_5)_2$	0	87	0	90

reaction. However, the presence of reagents which react with carbonyl groups does. For example 2,4,5,6-tetra-aminopyrimidine and methyl glyoxal give 2,4-diamino-7-methylpteridine (**56**) in 0.25 M hydrochloric acid in 65% yield, this being the only product isolated, but in aqueous sodium sulphite (a 'carbonyl-binding agent') only 5% of this isomer was obtained but 65% of the 6-methyl isomer (**57**) resulted.[73]

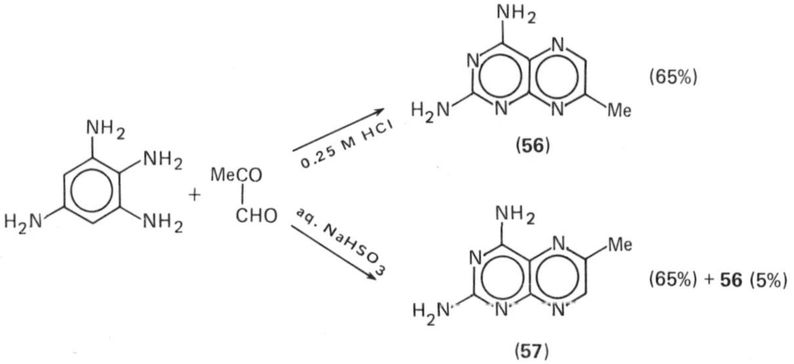

(65%)

(**56**)

(**57**)

(65%) + **56** (5%)

Other examples of this effect have been recorded using hydrazine instead of sodium bisulphite as the carbonyl-binding agent.

The synthesis of the pteridine ring system first described by Timmis[74] provides a convenient synthesis for some compounds. It involves condensation of a 6-amino-5-nitrosopyrimidine with an α-ketomethylene compound. For example, 2,6,-diamino-4-hydroxy-5-nitrosopyrimidine (**58**) reacts with ethyl phenylketone to give the product (**59**).[74] whilst phenylacetonitrile reacts with (**58**) in a similar way to give (**60**).[75]

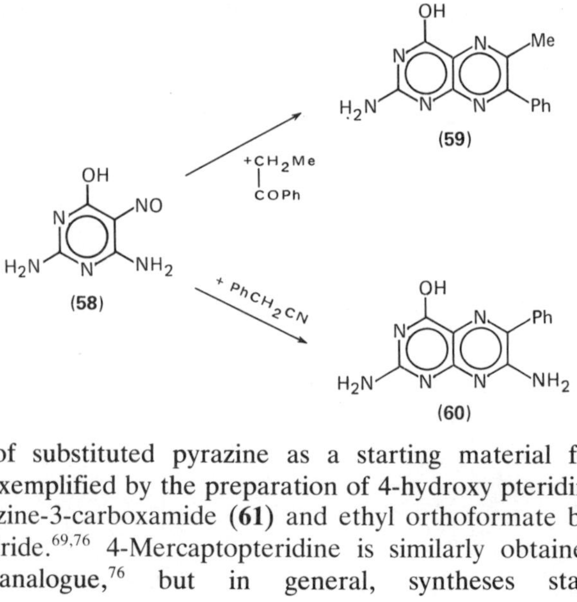

(**59**)

(**58**)

(**60**)

The use of substituted pyrazine as a starting material for pteridine synthesis is exemplified by the preparation of 4-hydroxy pteridine (**62**) from 2-aminopyrazine-3-carboxamide (**61**) and ethyl orthoformate by heating in acetic anhydride.[69,76] 4-Mercaptopteridine is similarly obtained from the thioamide analogue,[76] but in general, syntheses starting from pyrazine precursors are less frequently used than syntheses starting from pyrimidine precursors.

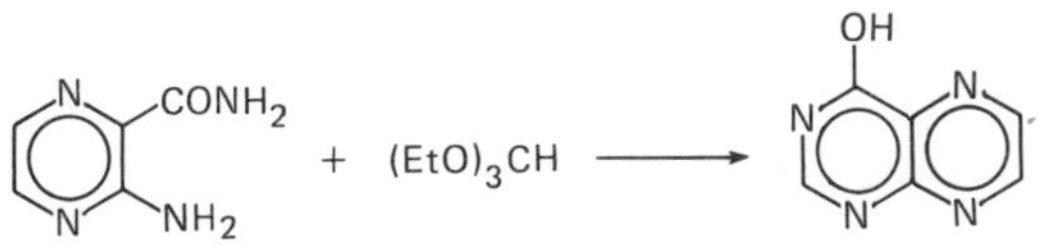

(61) + $(EtO)_3CH$ ⟶ (62)

However E. C. Taylor has recently described (Chemical Society, Perkin Division Meeting, UWIST, March 1979) a new approach to pteridine synthesis starting from oximinoketones and N–C–N fragments such as aminomalontrile to give pyrazine *N*-oxides. Further condensations of these compounds lead to pteridine *N*-oxides which may be subsequently deoxygenated to give the pteridine.

Such an approach has lead to synthesis of L-*erythro*biopterin. An example of such a reaction is given below:

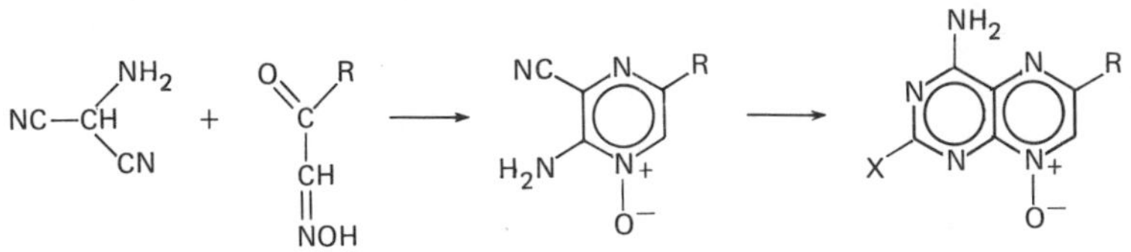

It is of interest to note that pteridines may also be obtained by reactions involving purines. For example, when 2-hydroxypurine is incubated in water with glyoxal 2-hydroxypteridine is formed[77] by the route shown in Fig. 3.12. Some other pteridines syntheses are given by E. C. Taylor (ref. 62, p. 543).

The biosynthesis of pteridines may proceed from purines by an analogous route and it has been shown that in the toad (*Xenopus*) the administration of 2-C^{14}-guanine resulted in the production of radioactive 2-amino-4-hydroxypteridine-6-carboxylic acid.[78] Further discussion of the biochemistry of pteridines will be considered in Chapter 6.

Other examples of the chemical synthesis of pteridines are given in the review by Pfleiderer.[79]

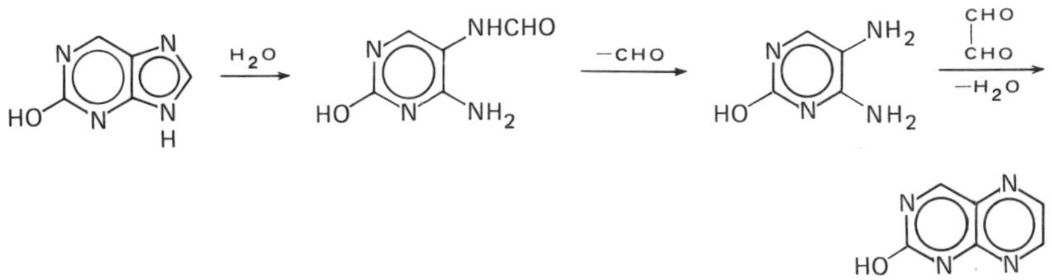

Fig. 3.12. Conversion of 2-hydroxypurine to 2-hydroxypteridine

(iii) Reactions of Pteridines

(a) Ring-opening reactions

Pteridine has a delocalized 10π electron system similar to naphthalene but there is considerable localization of the electrons in certain bonds and the π electron density calculations (see Section 1C) show that the C atoms are all electron deficient. A result of these effects is that pteridine, and many of its derivatives, undergo ready hydrolytic cleavage reactions catalysed by both acids and bases. The cleavage usually occurs in the pyrimidine ring to result in the formation of substituted pyrazines, this mode of cleavage being in agreement with calculations based on the electron-density distribution which show that the pyrimidine ring has the greater polarity of the two.[63] The delocalization energies of a number of nitrogen heterocycles have been recorded[80] but that of pteridine is not quoted.

Pteridine itself is converted to 2-aminopyrazine-3-aldehyde (**63**) by acid[81] and is also decomposed by alkali.

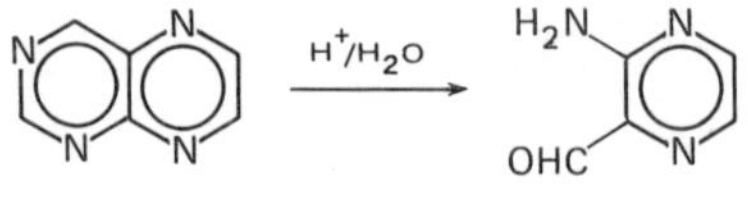

H^+/H_2O

(**63**) 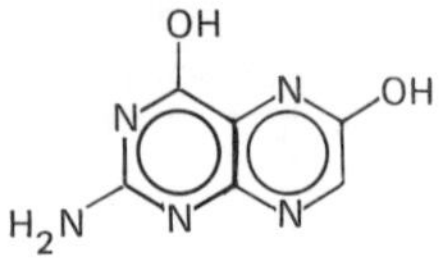 (**64**)

The introduction of electron-releasing groups increases the stability of the pteridine ring system: for example, xanthopterin (2-amino-4,6-dihydroxypteridine, **64**) is stable to boiling 7 M hydrochloric acid[82] and is little affected by being boiled with 0.75 M barium hydroxide for 20 h.[83] However, 4-hydroxypteridine (**65**) is degraded on refluxing with 1 M sul-

Table 3.4 Comparative stabilities of the hydroxy pteridines (Data from Ref. 63).

	Decomposed, %, in 1 h by	
	0.5 M H_2SO_4	10 M NaOH
2,4,6,7-Tetrahydroxy	*	6
4,6,7-Trihydroxy	*	4
6,7-Dihydroxy	7	12
2,4-Dihydroxy	6	4
2-Hydroxy	55	89
4-Hydroxy	60	94
6-Hydroxy	2	100
7-Hydroxy	52	76
Pteridine	74	>57

*Apparently unaffected, but insolubility prevents useful comparison.

phuric acid to give 2-aminopyrazine-3-carboxamide (**66**) and the corresponding acid, and is also degraded to give this same acid in quantitative yield on refluxing with 10 M sodium hydroxide.[84] The comparative stabilities of hydroxypteridines is given in Table 3.4.

OH

IM H_2SO_4

H_2NCO

H_2N

(65) (66)

(b) *Electrophilic reactions*

So far electrophilic substitutions have not been reported for the pteridine series.

(c) *Nucleophilic reactions*

Because of the presence of the cyclic nitrogen atoms and the resulting electron deficiency at carbon sites, substituents present in the pteridines are very susceptible to nucleophilic displacement.

The hydroxy pteridines can be converted into the corresponding chloro compound using $PCl_5/POCl_3$ or PCl_5/PCl_3 in an analogous way to the reaction of 2,4- and 6-hydroxypyrimidines but 7-chloropteridine does not seem to have been synthesized.[63] These chloropteridines may then be readily reacted with reagents such as ammonia, amines, hydroxide ion, or alkoxides to give the corresponding amino, alkylamino, hydroxy or alkoxy pteridine. Hydrogen iodide (*d* 1.7) has been used for the reductive removal of chlorine from some pteridines.

The ease of hydrolysis of the simple 2-, 4-, and 6- chloro pteridines is in the order $4 > 6 > 2$. Some typical reactions of chloropteridines are shown in Fig. 3.13.

Many of the reactions typical of substituted pyrimidines and purines are not generally applicable in the pteridine series. For example, the removal of thiol groups using Raney nickel is unsuccessful; the diazotization and subsequent conversion of aminopteridines to halogenopteridines is unsuccessful, and some hydrolysis reactions, e.g. the acid hydrolysis of 2-aminopteridine, results in hydrolytic cleavage of the pyrimidine ring, whilst nitrous acid also causes the destruction of the pteridine ring system of 2-aminopteridine and some other aminopteridines. However, alkaline hydrolysis of aminopteridines to the corresponding hydroxy compound is usually a facile reaction.

Certain mercaptopteridines can be oxidized to hydroxy compounds using cold, alkaline, hydrogen peroxide and can be converted into amines by heating with the appropriate amino compound.

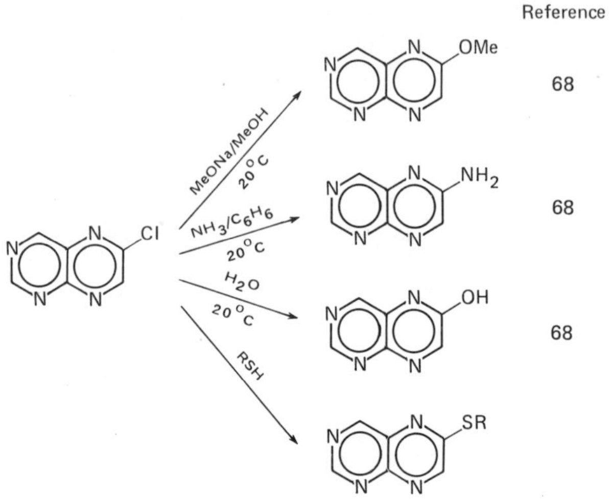
Reference
68
68
68
MeONa/MeOH
20 °C
NH3/C6H6
20 °C
H2O
20 °C
RSH

Fig. 3.13.

(d) Alkylation and Acylation

Alkylation reactions of pteridines result in complications owing to the number of potential sites of alkylation and the products which are observed depend on the number, type and orientation of the substituents, the alkylating agent, and the conditions under which the reaction is carried out, Alkylation of mercaptopteridines usually results in reaction at the sulphur atom, but in the case of hydroxy and aminopteridines the results are frequently unpredictable and O- and N-alkylation may be observed.

For example, leucopterin (2-amino-4,6,7-trihydroxypteridine, **67**) is methylated by diazomethane in methanol to give a mixture of **(68)** and **(69)**[85,81] 4-Hydroxypteridine reacts with dimethyl sulphate in dilute alkali to give a mixture of the N-alkyl products **(70)** and **(71)**,[81] and whilst 6-methyl-2,4,7-trihydroxy pteridine **(72)** also undergoes N-methylation, under similar conditions the 1,3-dimethyl derivative **(73)** undergoes O-methylation to give **(74)**[86]

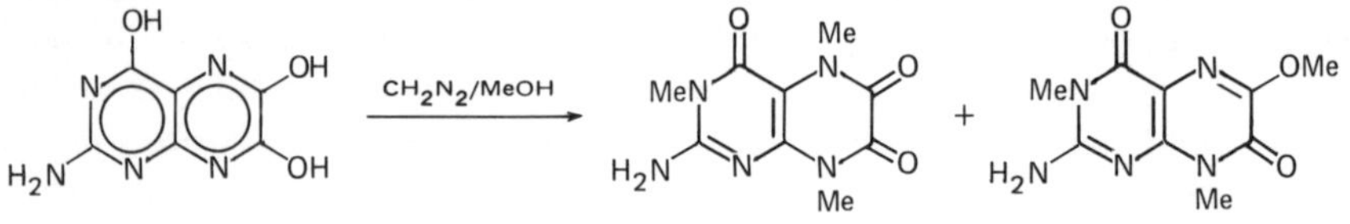
CH2N2/MeOH

(67) (68) (69)

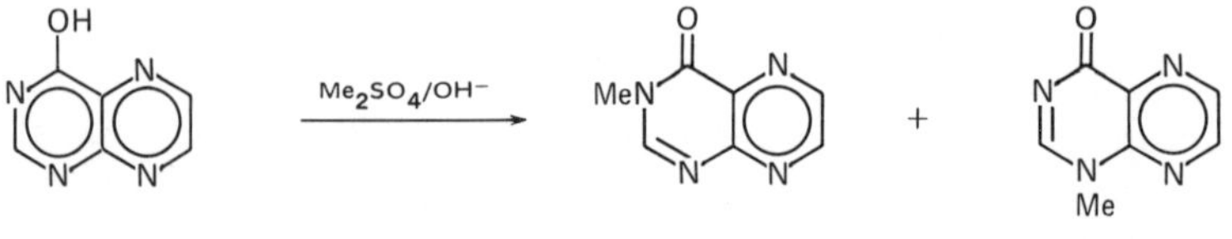
Me2SO4/OH−

(70) (71)

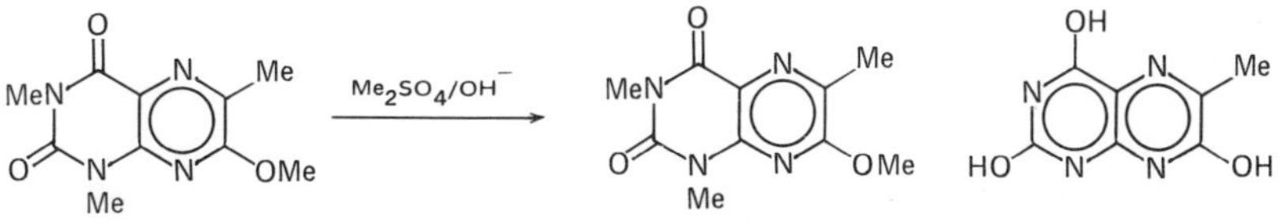

(73) $\xrightarrow{Me_2SO_4/OH^-}$ (74) (72)

It seems that steric factors affect the site of alkylation of pteridines and these may be such that alkylation may occur at a site other than the most basic site which would be the predicted point of attachment.

The conversion of some 6-hydroxy pteridines which have tautomeric groups in the pyrimidine part of the molecule into 6-alkoxypteridines by reaction by an alcohol in acid conditions is a reaction not undergone by the pyrimidines or purines. The reaction is probably an acid-catalysed addition of the alcohol to the lactam with subsequent dehydration.[87]

The hydroxypteridines cannot be acylated but the aminopteridines can. The simple substituted aminopteridines normally acetylate on refluxing with acetic anhydride, but some aminopteridines only acetylate satisfactorily in the presence of sulphuric acid, Leucopterin, for example, does not acetylate with acetic anhydride alone, but does so if sulphuric acid is added.[88]

An unusual property of many acetamidopteridines is that they are lower melting and more soluble in all solvents than the amines from which they are obtained, in contrast to simple aliphatic and aromatic amides. The effect of lowering the possibility of intermolecular H-bonding of the acetamidopteridines relative to the aminopteridines, must be of greater importance than the effect of introducing a polar acetamido group and increasing the molecular weight.

(e) Some Reactions of Pteridine Substituents

Alkoxypteridines can be hydrolysed to hydroxypteridines by alkali but this is frequently accompanied by degradation of the pteridine ring system.

All of the C sites in pteridine are adjacent to nitrogen atoms and so methylpteridines are activated in the same way as the 2-, 4-, and 6-methylpyrimidines. Methylpteridines can thus be oxidised to the corresponding pteridine carboxylic acids. Nitrosation of methylpteridines to give hydroxyiminomethyl derivatives does not seem to have been carried out, but selenium dioxide oxidation of a methylpteridine to a pteridine aldehyde has been observed.[89]

The pteridine carboxylic acids can be esterified but the formation of amides does not seem to have been carried out. The pteridine carboxylic acids readily decarboxylate on heating except for pteridine-7-carboxylic acids which only undergo degradation. Pteridine-6,7-dicarboxylic acids give the corresponding 7-monocarboxylic acid when heated in boiling quinoline.[90]

Many other reactions of pteridine are similar to the reactions of analogous pyrimidines.

(f) Formation of Covalent Hydrates

A feature of pteridine chemistry is the ease with which many of these compounds add water across a C=N bond to give stable hydrates. Thus 6-hydroxypteridine (**75**) exists as the hydrate (**76**) in aqueous solution.[91,68,65]

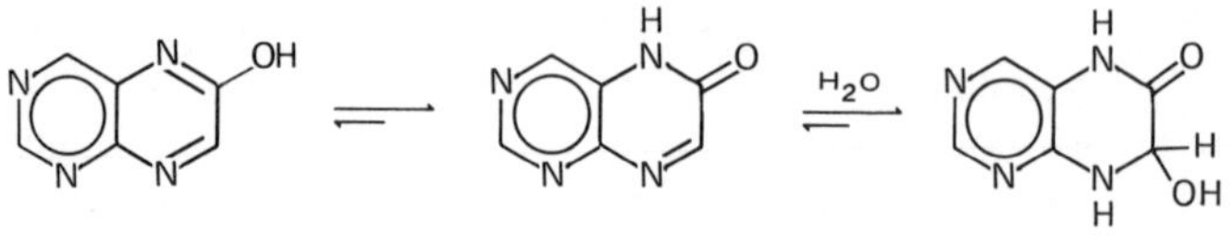

(75) (76)

Pteridine itself adds water across the 3,4-bond as does 2-hydroxy- and 2-aminopteridine[92] (Fig. 3.14).

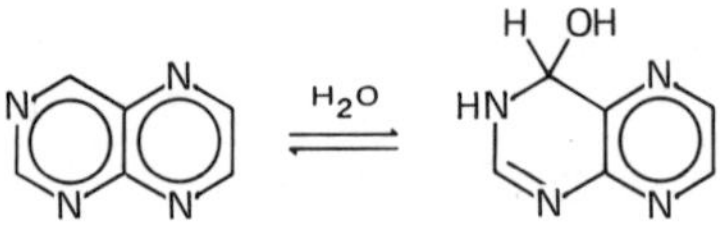

Fig. 3.14.

A wide variety of such pteridine hydrates is known and further hydration has also been observed. For example, in acid solution, the cation of 4-methylpteridine gives the dihydrate (**77**).[93] An extensive review of covalent hydration in nitrogen heterocycles has been written by Albert.[94]

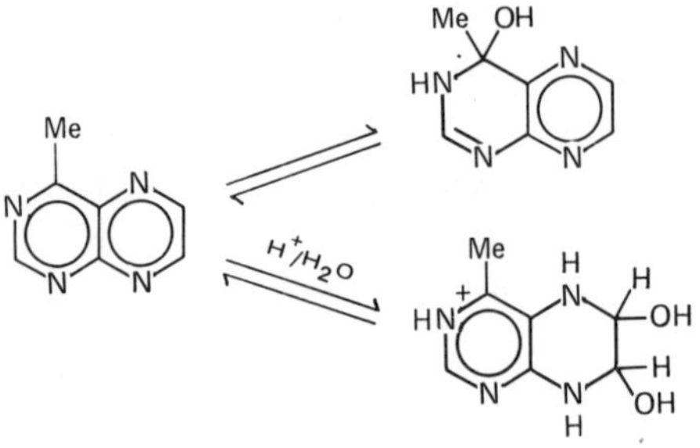

(77)

In addition to the covalent addition of water, pteridines will also add alcohols,[95] amines and ammonia,[96] and active methyl and methylene compounds.[97]

For example, diethylmalomate gives (**78**) and ethyl acetoacetate gives (**79**).

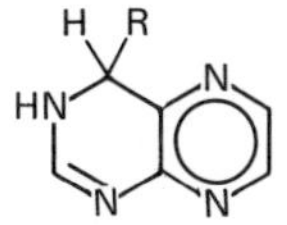

R = $CH(CO_2Et)_2$

(78)

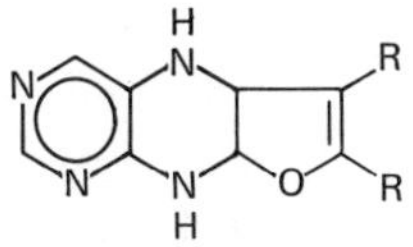

R = CO_2Et

(79)

Table 3.5 Ultraviolet spectra of some pteridines (data from ref. 98)

Pteridine	λ_{max}(log ϵ)	pH of aqueous solution
Parent	233sh (3.47) 298 (3.88) 308 (3.82)	7.4
(3,4-hydrate)	228 (3.66) 269 (3.69) 318 (3.84)	7.4
2-OH	230 (3.88) 269sh (3.58) 307 (3.83)	7.1
	260 (3.85) 375 (3.78)	13.0
4-OH	230 (3.88) 265 (3.54) 310 (3.82)	5.6
	242 (4.23) 333 (3.79)	10.0
6-OH	266sh (3.85) 289 (4.00)	5.2
	224 (4.29) 256 (3.97) 356 (3.84)	
7-OH	227 (3.79) 248 (3.44) 256 (3.45) 303 (4.00)	4.0
	226 (4.27) 260 (3.76) 326 (4.04)	9.0
2-NH_2	225 (4.39) 259 (3.81) 370 (3.82)	7.0
	232 (3.92) 302 (3.87)	2.0
4-NH_2	244 (4.20) 335 (3.82)	7.3
	229 (4.10) 324 (3.99)	1.1
6-NH_2	223 (4.30) 258 (4.01) 362 (3.75)	7.0
7-NH_2	228 (4.26) 262 (3.80) 334 (4.03)	5.1
	217 (4.25) 326 (4.00)	0.6
2-Me	227 (3.70) 269 (3.74) 319 (3.89)	7.4
	247 (3.59) 284 (3.88) 300 (3.90)	2.0
4-Me	258 (3.61) 298 (3.86)	7.4
	305 (4.00)	2.0
6-Me	303 (3.94) 315 (3.90)	6.0
	306 (3.96)	2.0
7-Me	232 (3.63) 270sh (3.57) 322 (3.92)	8.4
	251sh (3.39) 303 (3.98)	

(iv) Spectra of Pteridines

(a) Ultraviolet Spectra

The pteridines usually have a well-defined u.v. spectrum having two or three major bands. The u.v. spectrum of the parent pteridine has λ_{max} at 210, 233 (sh), 298 (π–π^*) and 308 nm (n–π^*) which compares with the spectrum of naphthalene 220, 275, 311 nm (π–π^*) which also has a delocalized 10π -electron system.

A large collection of u.v. data on the pteridines and other heterocyclic compounds is given in a review[98] and elsewhere.[63] The data for some selected pteridines are given in Table 3.5.

Generally the spectra of monosubstituted pteridines (as neutral molecules) resemble that of pteridine, but the bands are shifted to longer wavelength.

In the case of the aminopteridines (as neutral molecules) the spectra resemble that of pteridine, but shifted to longer wavelength, and also resemble the spectra of the anions of the corresponding hydroxy compounds. Cation formation from the aminopteridines causes the bands to shift to much shorter wavelengths, which is the reverse of what usually happens with heteroaromatic bases.[69,68,63]

The difference between the spectra of the hydroxypteridines and their methoxy anologues and the similarities to the *N*-methylpteridones, e.g. 7-hydroxypteridine and *N*-methyl-7-pteridone (Fig. 3.15), establishes the existence of the hydroxypteridines as pteridones,[63,68] and u.v. spectroscopy has been extensively used to investigate the tautomerism of substituted pteridines in the same way as for other nitrogen heterocycles.

The introduction of further substituents into the pteridine ring usually results in the expected spectral changes, although this is not always the case. For example, the aminohydroxypteridines absorb at shorter

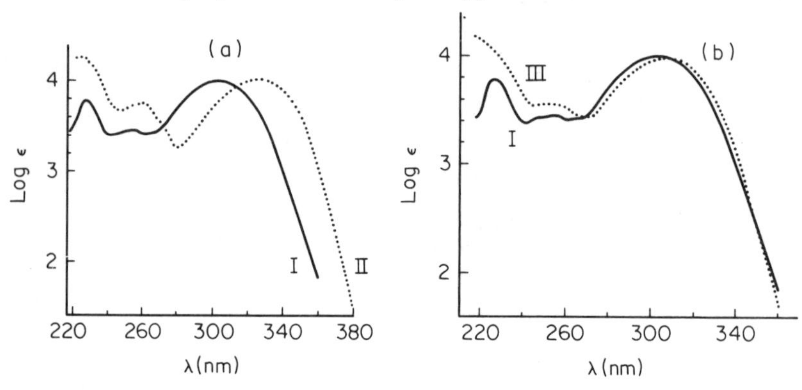

Fig. 3.15. Ultraviolet spectra of I, [structure]; II, [structure]; III, [structure]

(a) I(pH4) and II(pH9); (b) I and III at pH 4. (Data from ref. 63.) Reproduced by permission of The Chemical Society

wavelength than would be expected, presumably due to the electronic distribution in these molecules.[63] However, in general, the points made concerning the u.v. spectra of pyrimidines (Section 2D(ii)a) and purines (Section 3B(iv)) also apply to the pteridines.

(b) Infrared spectra

Like those of the purine, the i.r. spectra of purines tend to be complex. No discussion will be given here, but the reader is directed to a standard text on i.r. spectroscopy for a general account of the subject and to ref. 53, Vols. 2 and 4, for details of the spectra of pteridines.

(c) Nuclear Magnetic Resonance Spectra

The ^{1}H n.m.r. spectrum of pteridine itself in[99] $CDCl_3$ shows singlets for H-2 at 9.65δ and H-4 (9.80δ) with doublets for H-6 at 9.15δ and H-7 (9.33δ). The spectra of simple pteridines are as expected but ^{1}H n.m.r. spectroscopy has been useful in the study of covalent hydration and in the study of the tautomerism of substituents. A further discussion of n.m.r. is not intended here but further details of the ^{1}H n.m.r. spectra of both pteridines and purines may be found in refs. 53 and 100.

(d) Mass Spectra[59,56]

Pteridine itself, like the other aromatic nitrogen heterocycles, shows loss of HCN, this HCN being eliminated from the pyrimidine ring in preference to the pyrazine ring:

$M^{+\cdot}$, *m/e* 132 (100%) —HCN→ *m/e* 105 —HCN→ *m/e* 78

$M^{+\cdot}$, *m/e* 132 (100%) —HCN→ *m/e* 105 —HCN→ *m/e* 78

The methylpteridines which have the methyl group attached to the pyrimidine ring undergo similar fragmentation processes of the pyrimidine ring. However, 7- and 8-methylpteridines undergo initial fragmentation in the pyrazine ring:

$M^{+\cdot}$, *m/e* 146 —HCN→ *m/e* 119 —MeCN→ *m/e* 78

In the case of the hydroxypteridines and the 6- and 7-hydroxy isomers fragment initially with loss of CO to give a species which subsequently undergoes purine fragmentation:

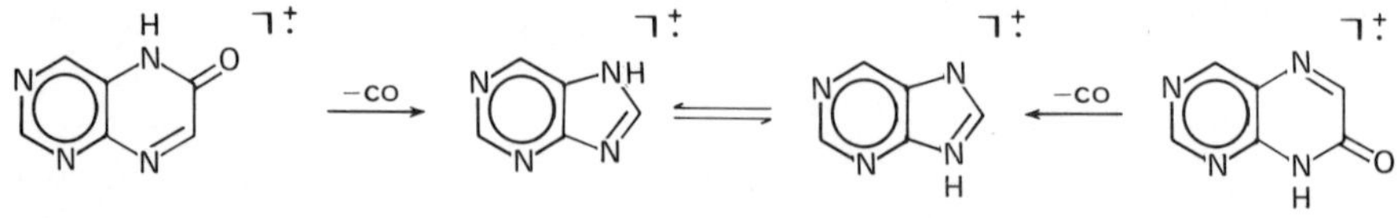

M$^{+\cdot}$; m/e 148 —(−CO)→ m/e 120 ←(−CO)— M$^{+\cdot}$; m/e 148

Other types of fragmetation are shown by the 2- and 4-hydroxy compounds which show an initial loss of HCN followed by CO and by 2,4-dihydroxypteridine which shows an initial loss of HNCO.

The initial fragmentations of these pteridines are shown in Fig. 3.16.

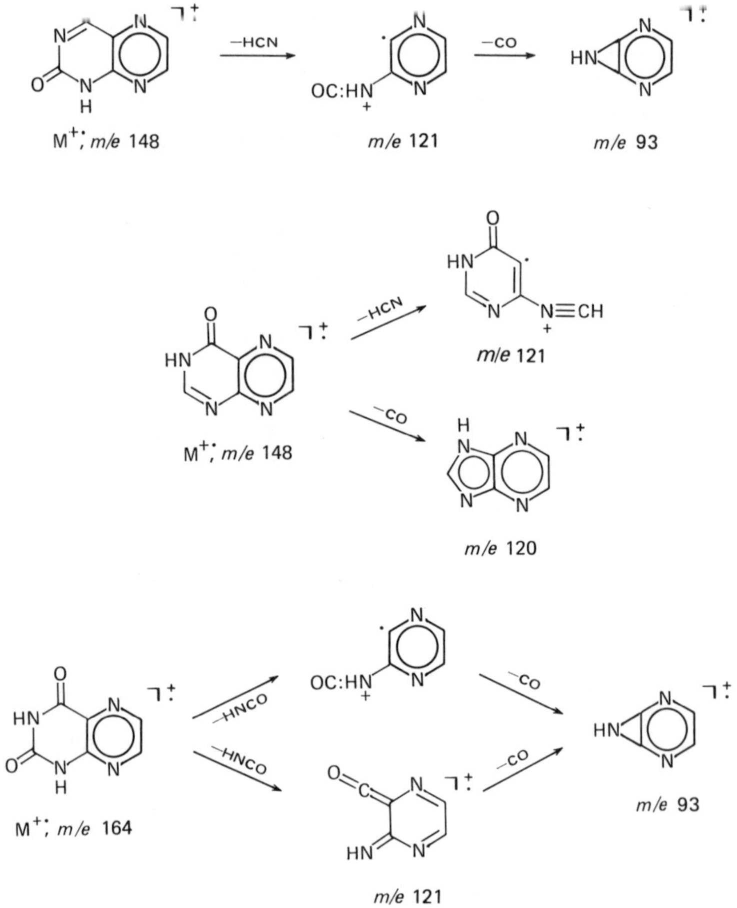

M$^{+\cdot}$; m/e 148 —(−HCN)→ OC:HN^{+} m/e 121 —(−CO)→ HN m/e 93

M$^{+\cdot}$; m/e 148 —(−HCN)→ HN, N≡CH^{+} m/e 121; —(−CO)→ H N m/e 120

M$^{+\cdot}$; m/e 164 —(−HNCO)→ OC:HN^{+}; —(−HNCO)→ O=C, HN m/e 121; —(−CO)→ HN m/e 93

Fig. 3.16.

REFERENCES

1. J. H. Lister, *Fused Pyrimidines, Part II, The Purines*, Wiley-Interscience New York, London, Sydney, Toronto (1971).
2. W. Traube, *Ber.*, **33**, 1371, 3035 (1900).
3. R. K. Robins, K. J. Dille, C. H. Willits, and B. E. Christensen, *J. Amer. Chem. Soc.*, **75**, 263 (1953).
4. A. Albert and D. J. Brown, *J. Chem. Soc.*, **1954**, 2060.
5. H. Getler, P. M. Role, J. F. Tinker, and G. B. Brown, *J. Biol. Chem.*, **178**, 259 (1949).
6. L. Goldman, J. W. Marsico, and A. L. Gazzola, *J. Org. Chem.*, **21**, 599 (1956).
7. G. W. Kenner, C. W. Taylor and A. R. Todd, *J. Chem. Soc.*, **1949**, 1620.
8. W. Traube, *Annalen,* **432**, 266 (1923).
9. O. Isay, *Ber.*, **39**, 250 (1906).
10. S. Gabriel and J. Colman, *Ber.*, **34**, 1234 (1901).
11. F. Cavalieri and A. Bendich, *J. Amer. Chem. Soc.*, **72**, 2587 (1950).
12. A. H. Cook and E. Smith, *J. Chem. Soc.*, **1949**, 3002.
13. C. E. Liau, K. Yamashita, and M. Matsui, *Agric, and Biol. Chem.,* (Japan), **26**, 624 (1962).
14. H. Goldner, G. Dietz, and E. Carstens, *Naturwissenschaft,* **51**, 137 (1964).
15. J. Oro and A. P. Kimball, *Arch. Biochem. Biophys.*, **94**, 217 (1961).
16. J. Oro, *Nature*, **191**, 1193 (1961).
17. C. U. Lowe, M. W. Rees, and R. Markham, *Nature*, **199**, 219 (1963).
18. G. Bruhns, *Ber*, **23**, 225 (1890).
19. Y. Yoshitomi, *J. Pharm. Soc., (Japan)*, **508**, 460 (1924).
20. B. F. Duesel, H. Bermann, and R. J. Schachter, *J. Amer. Pharm. Assoc. Sci. Edn.*, **43**, 619 (1954).
21. H. Biltz and J. Sauer, *Ber.,* **64**, 752 (1931).
22. A. Giner-Sorolla, *J. Heterocyclic Chem.*, **7**, 75 (1970).
23. A. Giner-Sorolla, *Galenica Acta*, **19**, 97 (1966).
24. R. Burian, *Z. Physiol. Chem.*, **51**, 425 (1907).
25. R. Burian, *Ber.*, **37**, 696 (1904).
26. E. Y. Sutcliffe and R. K. Robins, *J. Org. Chem.*, **28**, 1662 (1963).
27. C. Nofre, A. Lefier, and A. Cier, *Compt. Rend.*, **253**, 687 (1961).
28. C. Ponnamperuma, R. M. Lemmon, and M. Calvin, *Radiation Res.*, **18**, 540 (1963).
29. H. Bredereck, H. Ulmer, and H. Waldmann, *Chem. Ber.,* **89**, 12 (1956).
30. E. C. Taylor, Y. Maki, and A. McKillop, *J. Org. Chem.*, **34**, 1170 (1969).
31. E. Fischer, *Ber.*, **31**, 2550 (1898).
32. A. Vincze and S. Cohen, *Israel J. Chem.*, **4**, 23 (1966).
33. M. Kruger, *Ber.*, **26**, 1914 (1893).
34. J. W. Jones and R. K. Robins, *J. Amer. Chem. Soc.*, **84**, 1914 (1962).
35. M. D. Litwack and B. Weissmann, *Biochemistry,* **5**, 3007 (1966).
36. Z. Nieman and F. Bergmann, *Israel J. Chem.*, **3**, 161 (1966).
37. A. Bendich, P. J. Russell, and J. J. Fox, *J. Amer. Chem. Soc.*, **76**, 6073 (1954).
38. J. F. Gerster and R. K. Robins, *J. Org. Chem.*, **31**, 3258 (1966).
39. G. B. Barlin, *J. Chem. Soc.* (B), **1967**, 954.
40. A. F. Lewis, A. G. Beaman, and R. K. Robins, *Can. J. Chem.*, **41**, 1807 (1963).
41. R. K. Robins, *J. Amer. Chem. Soc.*, **80**, 6671 (1958).
42. A. S. Jones, A. M. Mian, and R. T. Walker, *J. Chem. Soc.* (C), **1966**, 692.
43. A. G. Beaman and R. K. Robins, *J. Org. Chem.*, **28**, 2310 (1963).

44. J. A. Montgomery and K. Hewson, *J. Amer. Chem. Soc.*, **82**, 463 (1960).
45. J. A. Montgomery and K. Hewson, *J. Amer. Chem. Soc.*, **79**, 4559 (1957).
46. M. A. Stevens, A. Giner-Sorolla, H. W. Smith, and G. B. Brown, *J. Org. Chem.*, **27**, 567 (1962).
47. G. B. Brown, and G. Levin, and S. Murphy, *Biochemistry*, **3**, 880 (1964).
48. (a) R. Shapiro, *J. Amer. Chem. Soc.*, **86**, 2948 (1964); (b) R. Shapiro and S. H. Pohl, *Biochemistry,* **7**, 448 (1968).
49. (a) A. Albert and W. L. F. Armarego in A. R. Katritzky (ed.), *Advances in Heterocyclic Chemistry*, Vol. 4, Academic Press, New York (1965), p. 1; (b) *A. Albert, J. Chem. Soc.* (B), **1966**, 438.
50. J. S. Connolly and H. Linschitz, *Photochem. and Photobiol.*, **7**, 791 (1968).
51. W. Prendergast, *J. Chem. Soc. Perk. 1,* **1973**, 2759; **1975**, 2240.
52. W. Zorbach and R. Tipson, (eds.), *Synthetic Procedures in Nucleic Acid Chemistry*, Wiley-Interscience, New York, London, Sydney, Toronto (1973).
53. A. R. Katritzky (ed.), *Physical Methods in Heterocyclic Chemistry*, Vols. 1–4 Academic Press, New York and London (1962–1971).
54. S. Matsuura and T. Goto, *J. Chem. Soc.,* **1965**, 623.
55. W. C. Coburn, M. C. Thorpe, J. A. Montgomery, and K. Hewson, *J. Org. Chem.*, **30**, 1114 (1965).
56. T. Goto, A. Tatematsu, and S. Matsuura, *J. Org. Chem.*, **30**, 1844 (1965).
57. T. Goto, A. Tatematsu, and S. Matsuura, *Nippon Kagaku Zasshi*, **87**, 71 (1966).
58. J. M. Rice and G. O. Dudek, *J. Amer. Chem. Soc.*, **89**, 2719 (1967).
59. Q. N. Porter and J. Baldas, *Mass Spectrometry of Heterocyclic Compounds*, Wiley-Interscience, New York, London, Sydney, Toronto, (1971).
60. H. C. Börressen, *Acta Chem. Scand.*, **17**, 921 (1963).
61. A. L. Bieber, *Anal. Biochem.*, **59**, 354 (1974).
62. W. Pfleiderer (ed.), *Chemistry and Biology of Pteridines*, de Gruyter, Berlin and New York, 1975.
63. A. Albert, *Quart. Rev.*, **6**, 197 (1952).
64. A. Albert, *Fortschr. Chem. Org. Naturst.*, **11**, 350 (1954).
65. *Ciba Symposium on Chemistry and Biology of Pteridines*, Churchill, London, 1954.
66. G. R. Ramage and T. S. Stevens in E. H. Rodd (ed.) *Chemistry of Carbon Compounds,* Vol. IVc, Elsevier, Amsterdam, London, New York, Princeton (1960).
67. O. Chalvet and C. Sandorfy, *Compt. rend.*, **228**, 566 (1949).
68. A. Albert, D. J, Brown, and G. H. Cheeseman, *J. Chem. Soc.*, **1952**, 1620.
69. A. Albert, D. J. Brown, and G. H. Cheeseman, *J. Chem. Soc.,* **1951**, 474.
70. W. Koschara, *Z. Physiol. Chem.*, **277**, 159 (1943).
71. G. B. Elion and G. H. Hitchings, *J. Amer. Chem. Soc.*, **69**, 2553 (1947).
72. G. B. Elion, G. H. Hitchings, and P. B. Russell, *J. Amer. Chem. Soc.*, **72**, 78 (1950).
73. D. R. Seeger, D. B. Cosulich, J. M. Smith, and M. E. Hultquist, *J. Amer. Chem. Soc.*, **71**, 1753 (1949).
74. G. M.Timmis, *Nature*, **164**, 139 (1949).
75. R. G. W. Spickett and G. M. Timmis, *J. Chem. Soc.*, **1954**, 2887.
76. O. Vogl and E. C. Taylor, *J. Amer. Chem. Soc.*, **81**, 2472 (1959).
77. A. Albert, *Biochem. J.*, **65**, 124 (1957).
78. I. Ziegler-Günder, H. Simon, and A. Wacher, *Z. Naturforsch*, **11B**, 82 (1956).
79. W. Pfleiderer, *Angew. Chem. Int. Edn.*, **3**, 114 (1964).
80. K. Pihlaja and E. Taskinen in A. R. Katritzky (ed.), *Physical Methods in Heterocyclic Chemistry*, Academic Press, New York and London, Vol. 4. (1974) p. 199.

81. A. Albert, D. J. Brown, and H. C. S. Wood, *J. Chem. Soc.*, **1956**, 2066.
82. H. Wieland and C. Schöpf, *Chem. Ber.*, **58**, 2178 (1925).
83. H. Wieland and R. Purrmann, *Annalen*, **544**, 163 (1940).
84. A. Albert, D. J. Brown, and G. H. Cheeseman, *J. Chem. Soc.*, **1952**, 4219.
85. W. Pfleiderer and M. Rukwied, *Chem. Ber.*, **94**, 118 (1961).
86. W. Pfleiderer, *Chem. Ber.*, **90**, 2588 (1957).
87. W. Pfeiderer, E. Liedek, and M. Rukwied, *Chem. Ber.*, **95**, 755 (1962).
88. H. Wieland, H. Metzger, C. Schöpf, and H. Bülow, *Annalen*, **507**, 226 (1933).
89. R. Tschesche and F. Korte, *Chem. Ber.*, **84**, 641, 801 (1951).
90. C. K. Cain, M. F. Mallette, and E. C. Taylor, *J. Amer. Chem. Soc.*, **70**, 3026 (1948).
91. Third International Symposium on Pteridines, Stuttgart, 1962, *Reports and Discussions*, Pergamon Press, London, 1963.
92. D. D. Perrin, *J. Chem. Soc.*, **1962**, 645.
93. A. Albert, D. J. Brown, and J. J. McCormack, *J. Chem. Soc.* (B), **1966**, 1105.
94. A. Albert in A. R. Katritzky and A. J. Boulton (eds.) *Advances in Heterocyclic Chemistry*, Vol. 20, Academic Press, New York and London, (1977).
95. (a) A. Albert and H. Mizuno, *J. Chem. Soc.* (B), **1971**, 2423; (b) A. Albert and J. J. McCormack, *J. Chem. Soc.*, **1965**, 6930; (c) A. Albert and C. F. Howell, *J. Chem. Soc.*, **1962**, 1591.
96. B. E. Evans, *J. Chem. Soc. Perk. I*, **1974**, 357.
97. A. Albert and H. Mizuno, *J. Chem. Soc. Perk. I*, **1973**, 1615, 1974.
98. W. L. F. Armarego in A. R. Katritzky (ed.) *Physical Methods in Heterocyclic Chemistry*, Vol. 3, Academic Press, New York and London, (1971).
99. S. Matsuura and T. Goto, *J. Chem. Soc.*, **1963**, 1773.
100. T. J. Batterham, *N.M.R. Spectra of Simple Heterocycles*, Wiley-Interscience, New York, London, Sydney, Toronto (1973).

Chapter 4

Nucleosides and Nucleotides

(A) INTRODUCTION: THE STRUCTURE AND NOMENCLATURE OF NUCLEOSIDES AND NUCLEOTIDES

The term *nucleoside* was originally applied to the purine-ribose derivatives obtained by Levene and Jacob[1] from alkaline hydrolysates of ribonucleic acid isolated from yeast, these compounds being adenosine (**1**) and guanosine (**2**).

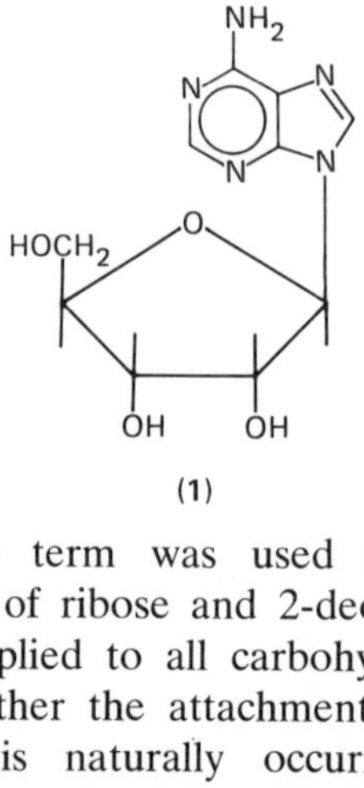

(1) 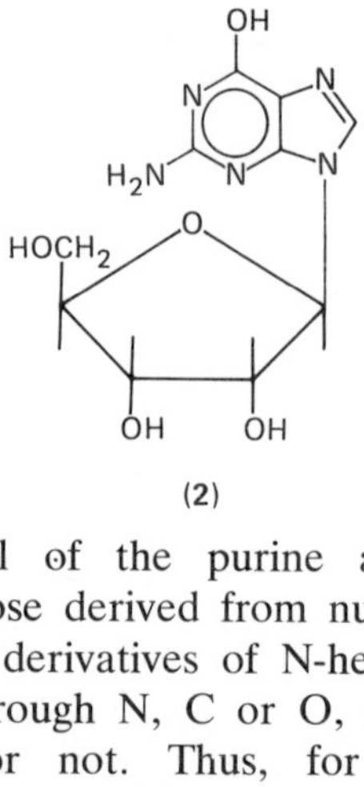 (2)

Later, the term was used for all of the purine and pyrimidine *N*-glycosides of ribose and 2-deoxyribose derived from nucleic acids, but now it is applied to all carbohydrate derivatives of N-heterocyclic compounds, whether the attachment is through N, C or O, or whether the heterocycle is naturally occurring or not. Thus, for example, the C-glycoside pseudouridine (**3**) isolated from RNA from certain sources, cytosine arabinoside (**4**), isolated from certain marine sponges, and the synthetic antiviral and antineoplastic agent 5-fluorodeoxyuridine (**5**) are all termed nucleosides. Many such compounds are known, involving a variety of heterocyclic compounds and a variety of carbohydrates.

The term *Nucleotide* was also introduced by Levene[2] to denote the products isolated from acid digests of nucleic acid isolated from calf thymus and which were shown to be phosphate esters of nucleosides. The first nucleotide to be isolated, inosinic acid (**6**), was obtained from beef extract by Leibig[3] as early as 1847, but it was not until the development in the 1950s of enzymic techniques for nucleic acid degradation that considerable

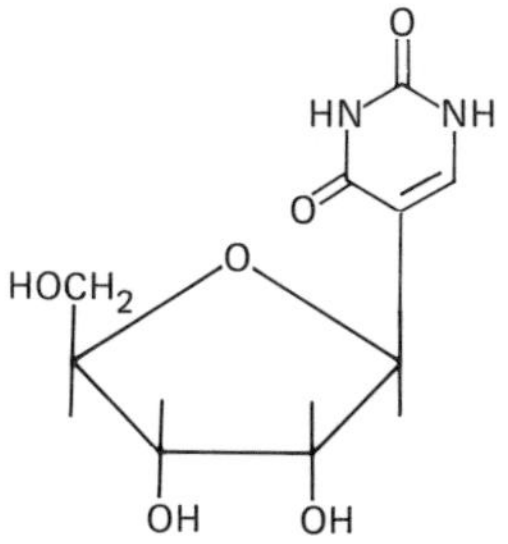

(3) 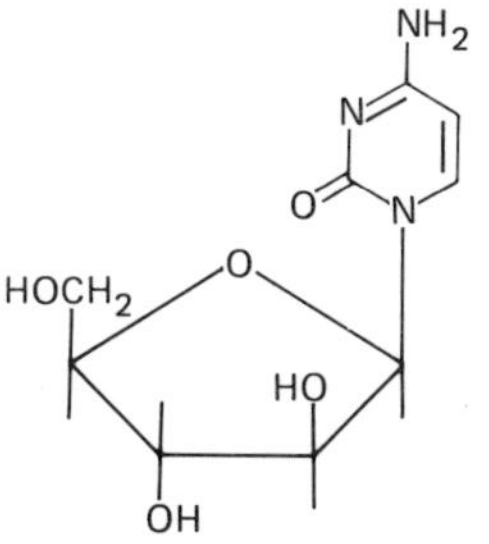 (4) 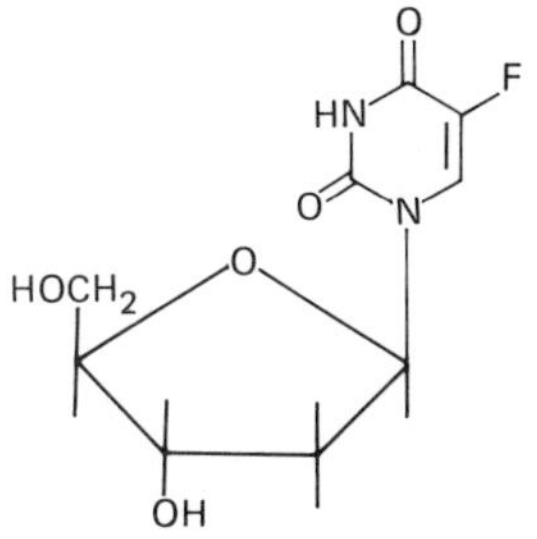 (5)

advances were made in nucleotide chemistry. In the early days of nucleotide chemistry the phosphate esters of nucleosides were named according to their source, e.g. 'yeast adenylic acid' and 'muscle adenylic acid', having the structures (**7**) and (**8**) respectively, but such nomenclature is not used now.

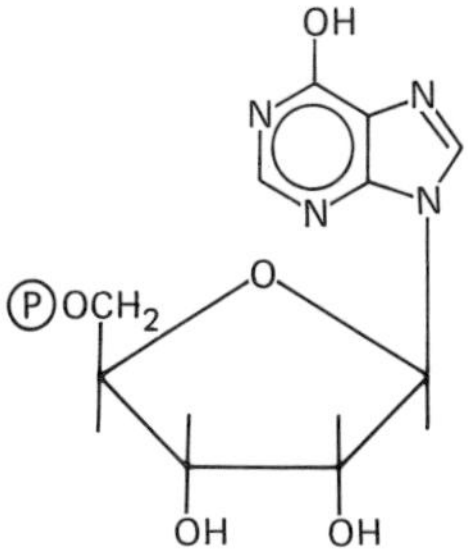

(6) 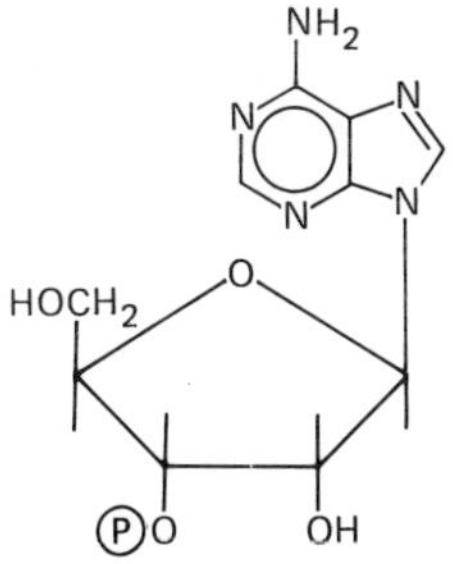 (7) 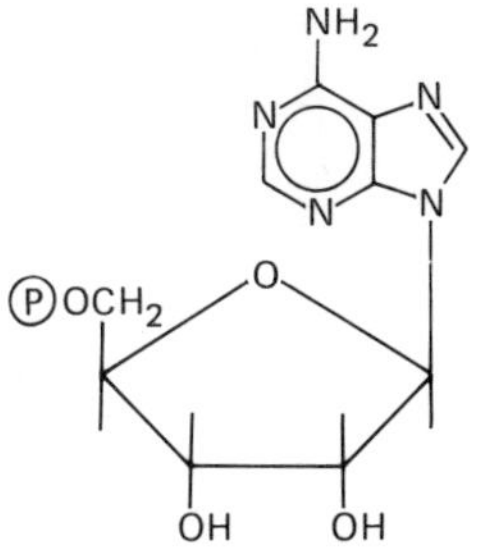 (8)

In order to describe a nucleoside or nucleotide completely the information needed is:

(i) the structure of the heterocyclic compound;
(ii) the structure of the carbohydrate;
(iii) the site of attachment of the carbohydrate to the heterocycle;
(iv) the ring structure of the carbohydrate;
(v) the configuration of the glycosidic bond;
(vi) the site (or sites) of attachment of the phosphate group (or groups); and
(vii) the number of attached phosphate groups.

Table 4.1 lists the structures, trivial names and systematic names of some of the common nucleosides, and Table 4.2 gives this information for some nucleotides.

A study of the chemistry and biochemistry of nucleosides and nucleotides has occupied an important role for many years, not only because of the relevance to nucleic acid chemistry and molecular biology, but also because a number of nucleosides have been found to have useful antibiotic

Table 4.1

Structure	Common name	Heterocycle (common name)	Carbohydrate and nature of glycosidic link	Systematic name
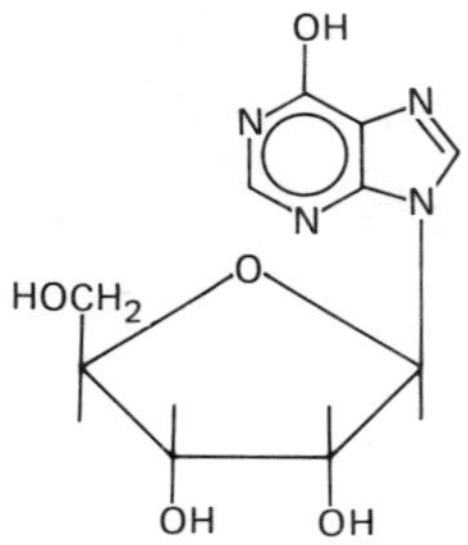	Inosine	Hypoxanthine	D-Ribofuranose (β)	9-(β-D-Ribofuranosyl) hypoxanthine
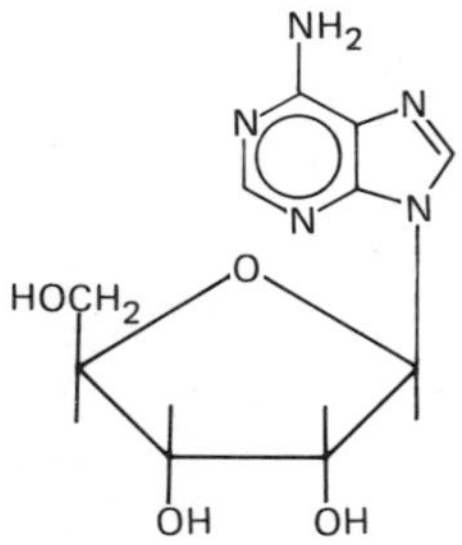	Adenosine	Adenine	D-Ribofuranose (β)	9-(β-D-Ribofuranosyl) adenine
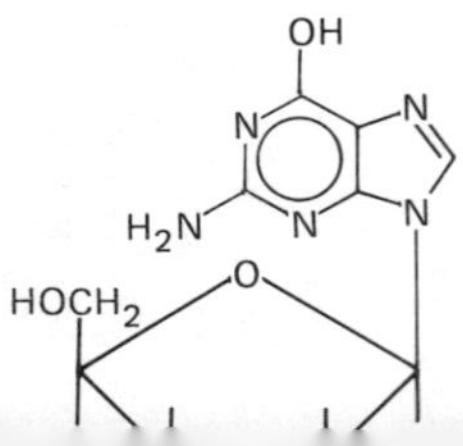	Guanosine	Guanine	D-Ribofuranose (β)	9-(β-D-Ribofuranosyl) guanine

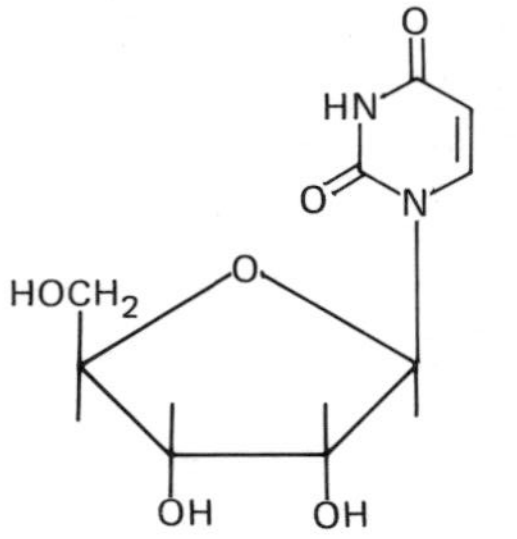	Uridine	Uracil	D-Ribofuranose (β)	1-(β-D-Ribofuranosyl) uracil
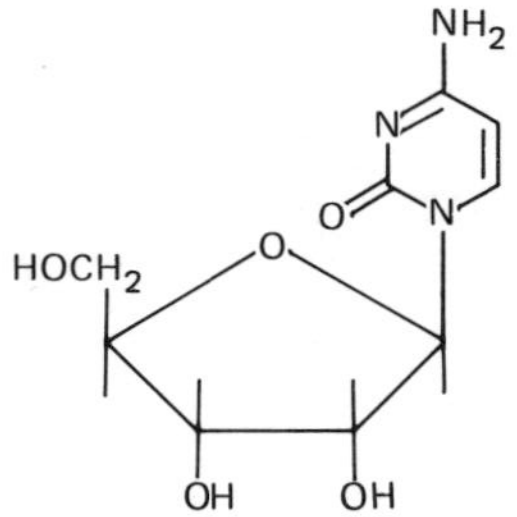	Cytidine	Cytosine	D-Ribofuranose (β)	1-(β-D-Ribofuranosyl) cytosine
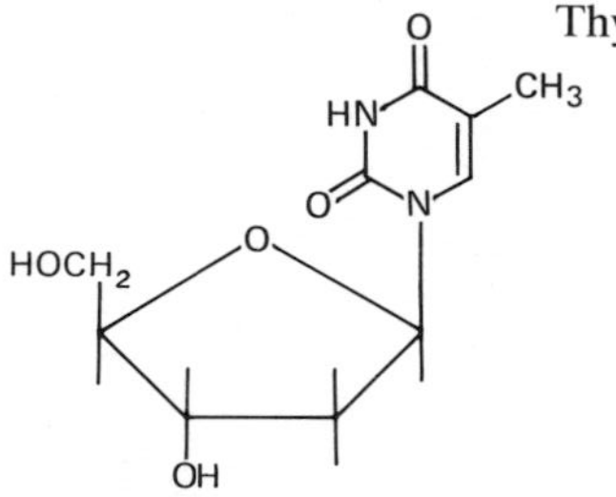	Thymidine	Thymine	2-Deoxy-D-ribofuranose (β)	1-(β-2′-Deoxy-D-ribofuranosyl) thymine

Table 4.2

Structure	Common name	Systematic name	Abbreviation
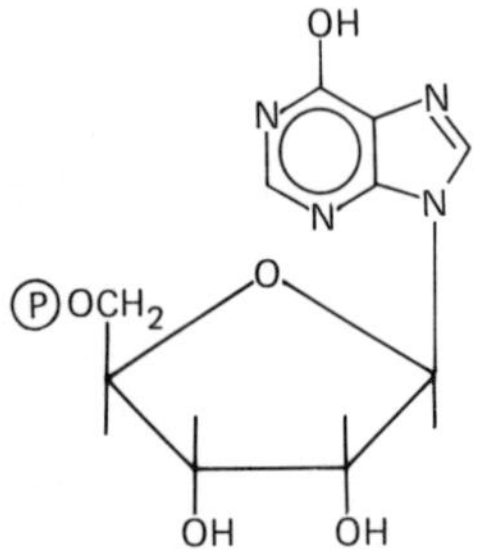	Inosinic acid	Inosine-5′-monophosphate	IMP
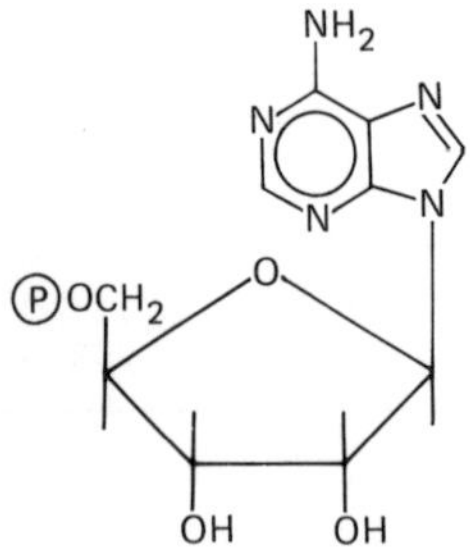	Adenylic acid	Adenosine-5-′-monophosphate	AMP
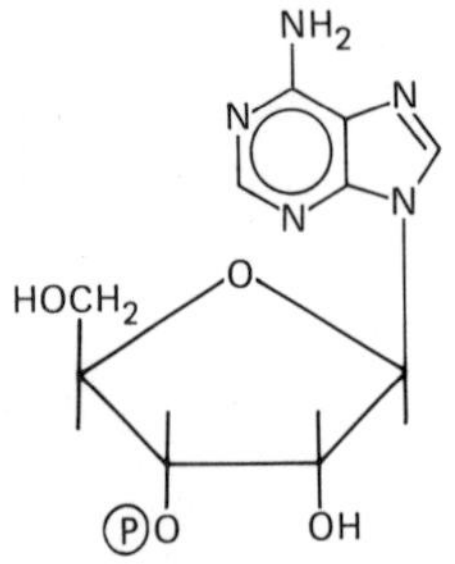	—	Adenosine-3′-phosphate	

and antitumour properties. The role of the nucleotide coenzymes has been widely studied because of their importance as 'vitamins'—necessary dietary constituents for the maintenance of normal growth and health—and their involvment as the active sites of a number of important enzyme-mediated biochemical reactions.

The isolation of nucleosides and nucleotides from nucleic acids and other natural sources will be covered in a later chapter, as will the role of a selection of nucleotide coenzymes in enzyme-mediated reactions. In this chapter some routes to the chemical synthesis of such compounds and their chemical properties will be discussed.

(B) THE CHEMICAL SYNTHESIS OF NUCLEOSIDES

Extensive work has been carried out on the synthesis of nucleosides to complement degradation studies in the elucidation of the structure of naturally occurring compounds and to provide synthetic analogues which are not naturally occurring for biochemical studies and as potential pharmaceutically active compounds.

The most common method of nucleoside synthesis is to react an appropriately activated pyrimidine or purine with a suitable, activated, sugar. It is also possible to construct the purine or pyrimidine ring using an *N*-glycosyl precursor or chemical modification of a natural or preformed nucleoside can give a compound modified in the sugar and/or the heterocyclic base. However, the building of the sugar moeity on to the heterocyclic base has proved to be difficult and to be less useful as a synthetic route, although it has proved to be very useful in C-nucleoside synthesis.

(i) Purine Nucleosides

Fischer[4] carried out the earliest work on nucleoside synthesis and obtained the first synthetic nucleoside by reacting the silver salt of 2,8-dichloroadenine **(9)** with acetobromoglucose **(10)** to give the tetra-acetylnucleoside **(11)**. Subsequent removal of the acetyl groups and dehalogenation gave 9-β-D-glucopyranosyladenine **(12)**.

When the sodium salt of adenine was used it was found to cause dehydrohalogenation of the glycosyl halide rather than nucleoside formation owing to its strong basicity.

By an analogous route, but using tri-*O*-acetyl-D-ribofuranosyl chloride **(13)** Todd and coworkers[5] synthesized the first natural nucleoside, adenosine **(15)**. Guanosine **(16)** was also obtained[5a] by further chemical modification of the intermediate **(14)**.

Although the above reactions led to the first syntheses of nucleosides, the approach was limited because of the low yields obtained and because of interference by substituents, e.g. amino, hydroxy, mercapto, on the purine. However, subsequent modifications have greatly improved the yields and the scope of the reaction. Davoll[6] found that acylation of a purine amino group both protected it and reduced the basicity of the purine. He also found that the use of chloromercuri rather than silver salts gave considerable improvement in the yields obtained.

A further advance in nucleoside synthesis was the fusion reaction of the Japanese workers Shimidate *et al.*[7,8,9] This reaction involves the fusion of the fully acetylated sugar with a substituted purine in the presence of an acid catalyst. The fusion of 6-chloropurine with 1,2,3,5-tetra-*O*-acetyl-β-D-ribofuranose in the presence of *p*-toluenesulphonic acid to give the acetylated nucleoside, which was subsequently modified to give adenosine, illustrates the use of this reaction, which is a considerable advance over the earlier nucleoside syntheses, since it requires neither a glycosyl halide nor a purine silver nor chloromercuri derivative.

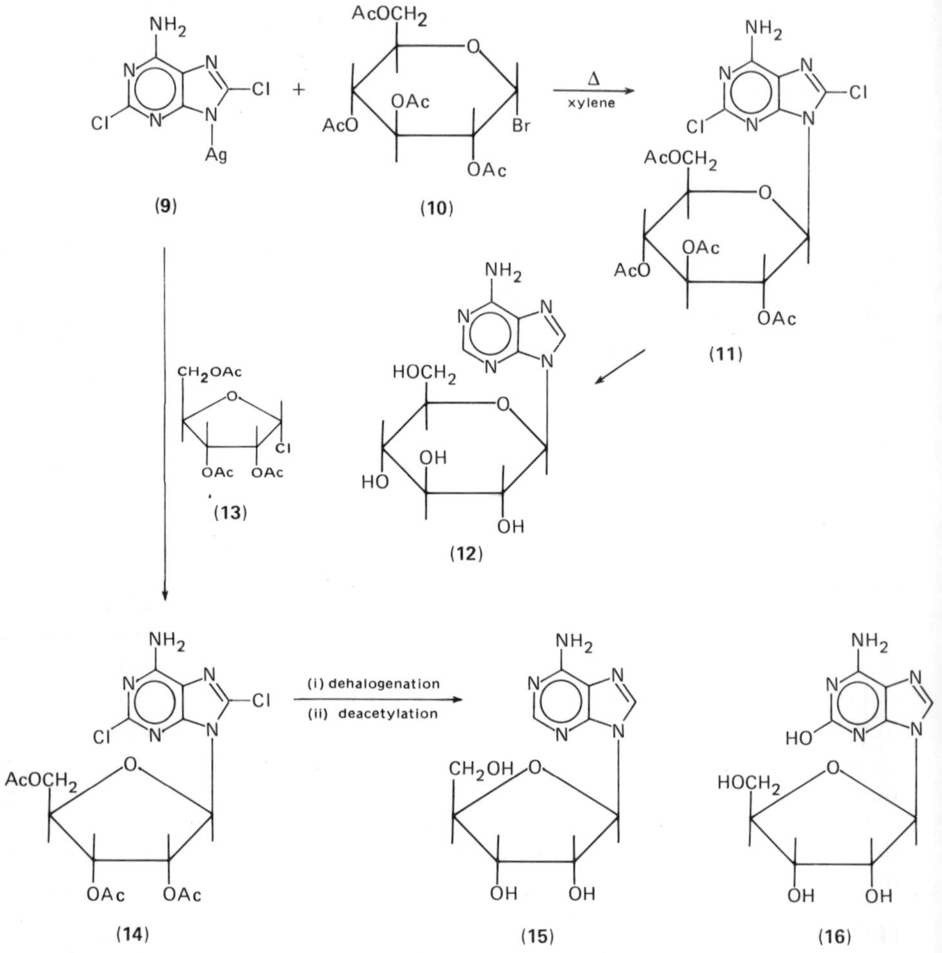

A number of other catalysts can be used for the reaction including zinc chloride, aluminium chloride, polyphosphoric acid, and sulphamic acid, and a number of purines has been successfully used.

In some cases (e.g. 2,6-dichloropurine with tetra-*O*-acetyl-β-D-ribofuranose) no catalyst is necessary and a good yield of the nucleoside is obtained, but theophylline, which gives good yields in acid catalysed reactions, failed to react.[10] In general it has been observed[10,11] that the lower melting bases give higher yields in the fusion reaction, but the presence of substituents, and their position, also seems to be important.

A number of preparative methods for specific purine nucleosides is given in Zorbach and Tipson[12] and in the new edition of this book.[12a]

(ii) Pyrimidine Nucleosides

Although the silver salt of some purines can be reacted with glycosyl halides to form nucleosides, Fischer[13] found that the reaction of the silver salts of 2- or 4-hydroxypyrimidines gave *O*-glycosyl derivatives, products which were less stable and having different properties from the naturally occurring pyrimidine nucleosides. Levene and Sobotka[14] came to the same conclusion and this was later confirmed by Ulbricht.[15,16] However, Ulbricht showed that the labile *O*-glycosyl derivatives could be isomerized to the stable *N*-glycosyl derivatives by heating in the presence of mercuric bromide.

The chloromercuri derivative of uracil[17] also gives an *O*-glycosyl derivative under the catalytic action of the mercuric salt.[15]

By using alkoxypyrimidines, Hilbert and coworkers[18] obtained pyrimidine-*N*-glycosides, this method also providing the first synthesis[19] of a natural pyrimidine ribonucleoside (Fig. 4.1).

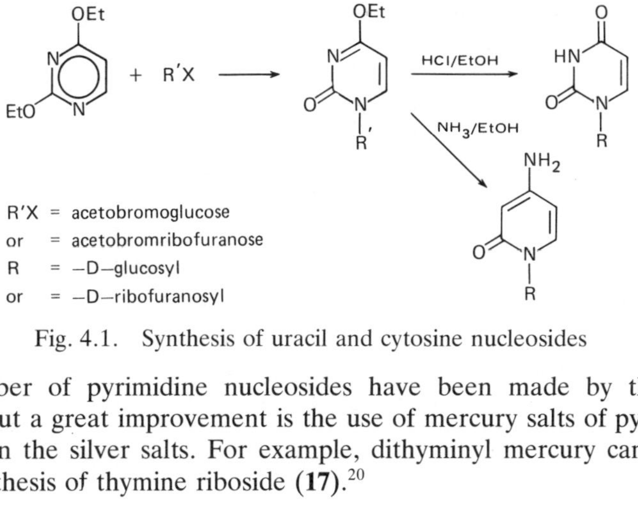

OEt N EtO N + R′X → OEt N O N R′ —HCl/EtOH→ O HN O N R ; —NH₃/EtOH→ NH₂ O N R

R′X = acetobromoglucose
or = acetobromribofuranose
R = —D—glucosyl
or = —D—ribofuranosyl

Fig. 4.1. Synthesis of uracil and cytosine nucleosides

A number of pyrimidine nucleosides have been made by the above method, but a great improvement is the use of mercury salts of pyrimidines rather than the silver salts. For example, dithyminyl mercury can be used in the synthesis of thymine riboside (**17**).[20]

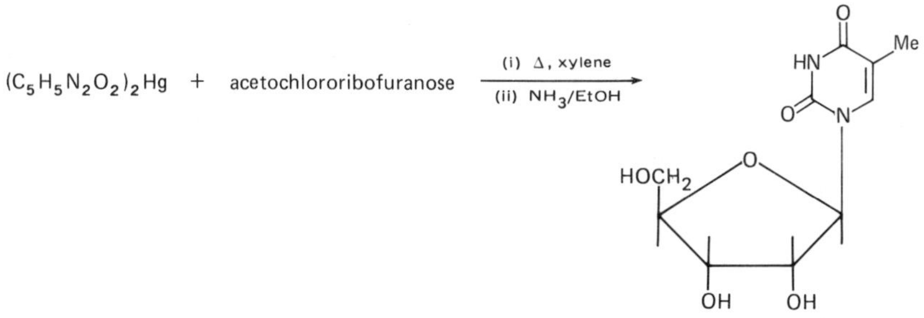

$(C_5H_5N_2O_2)_2Hg$ + acetochlororibofuranose —(i) Δ, xylene; (ii) NH_3/EtOH→ (**17**)

O Me HN O N HOCH₂ O OH OH

(**17**)

The use of mercury and chloromercuri salts of pyrimidines with the appropriate halogenosugar has resulted in the synthesis of a number of

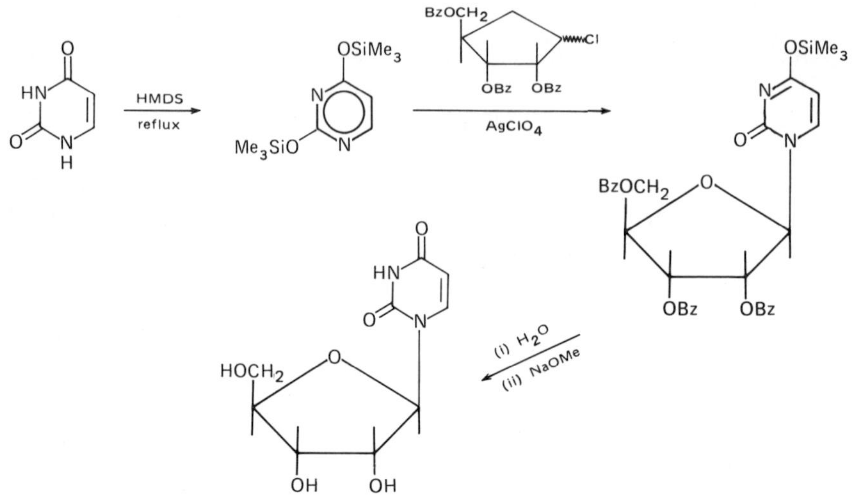

Fig. 4.2. Synthesis of uridine from the bis trimethylsilyloxyprimidine

pyrimidine nucleosides including the ribofuranoside of 5-fluorouracil,[21] thymidine, and deoxycytidine.[22] Ulbricht and coworkers have also synthesized uracil nucleosides in high yield using uracil, glycosylhalides, and mercuric cyanide.[23]

A further useful method for the synthesis of pyrimidine nucleosides involves converting the pyrimidines to their trimethylsilyl derivatives which, on subsequent reaction with a glycosyl halide in the presence of a catalyst such as silver perchlorate and ultimate removal of the protecting groups, gives the nucleoside. This type of reaction was first reported by Birkhofer,[24] Nishimura,[25] and Wittenberg[26] and their coworkers and an example is illustrated in Fig. 4.2.

Further examples of synthetic procedures for the synthesis of nucleosides are given in refs 12, 27 and 28.

Early syntheses of the 2′-deoxyribonucleosides were carried out by transformations of ribonucleosides—for example, nucleophilic displacement of a 2′-tosyl ester by iodide ion followed by reduction. However, the method of choice now is to use crystalline aroylated 2-deoxyribosyl halides, and one of the above methods of reaction of sugar derivatives with an appropriate purine or pyrimidine.

(iii) Stereochemistry and Mechanism of Nucleoside Formation

The stereochemistry of the coupling reaction of a glycosyl halide with a nitrogen base seems to depend on the method of condensation and can vary with the choice of solvent, the catalyst, and be markedly affected by the structure of the reactants. But, in general, early syntheses of ribosides

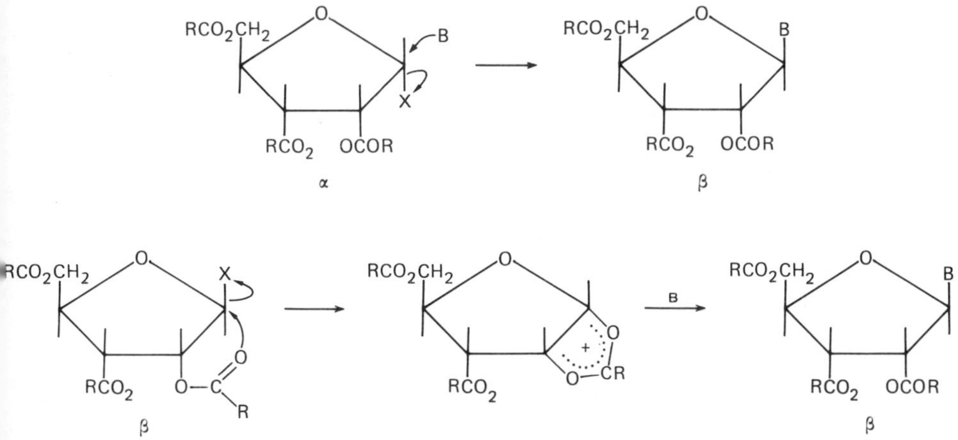

Fig. 4.3. Stereochemistry of nucleoside formation

and glucosides gave β-derivatives, this stereochemistry being independent of the orientation, α- or β-, of the glycosyl halide.

In the case of the α-ribosyl halides, the nucleoside product is formed by a direct S_N2 substitution at C-1. However, it has been proposed (see refs. 27, 29) that in the case of the β-ribosyl halides two S_N2 reactions occur, the first being neighbouring group participation of the 2-acyl- or aryloxy group followed by attack of the base. (Fig. 4.3).

A number of observations support this mechanism involving intermediate orthoester formation.[30]

(C) THE CHEMICAL REACTIONS OF NUCLEOSIDES

Nucleosides undergo virtually all the reactions of their constituent bases and many of the reactions of the constituent sugars.[28] However, there are some reactions which are of particular interest and some which are of importance in nucleoside chemistry, and it is these reactions which will be dealt with here rather than the usual reactions of nitrogen bases and sugars.

(i) Acylation and Alkylation

Nucleosides may be fully acylated on the sugar hydroxy groups, and on any heterocyclic ring amino groups, by an excess of acylating agent, As in the case of simple glycosides, a selectivity for the 5′- primary hydroxyl group is shown on partial acylation,[31] but less work seems to have been done on the selective acylation of nucleosides than such reactions of simple glycosides. However, studies of some ribonucleoside 5′-*O*-acetates have shown that the products of diacetylation are the 3′,5′-diacetylated derivatives rather than the 2′,5′-diacetates which would be expected from the expected reactivities of the respective hydroxyl groups. This is almost certainly due to rapid, facile, acyl migration rather than being due to the specificity of the reaction.[32]

Nucleosides usually react with other reagents, e.g. arylsulphonyl chlorides, trimethylsilyl chloride and aldehydes, to give the usual products of carbohydrate derivatives. Diarylsulphonation does give the 2′,5′-diarylsulphonate as migration is less likely than in the case of acetates.[28] Ribonucleosides react with benzaldehyde to give 2′,3′-*O*-benzylidene derivatives and with acetone to give 2′,3′-*O*-isopropylidene derivatives. Trimethylsilylation also results in reaction in the heterocyclic ring to give silyloxy or silylamino derivatives.

The reaction of nucleosides with triphenylmethylchloride, whilst showing preferential reaction at the 5′- primary hydroxyl groups, is less specific than is the case for simple carbohydrates and can result in the formation of ditrityl derivatives. Uridine for example, forms the 2′,5′-ditrityl product.[33]

Alkylation of nucleosides with diazomethane or with alkyl halides in the presence of alkali or silver oxide results in alkylation of the carbohydrate hydroxyl groups and alkylation on the heterocyclic moiety. For example, the reaction of diazomethane with uridine results in the formation of 3,2′-dimethyluridine (**18**).[34]

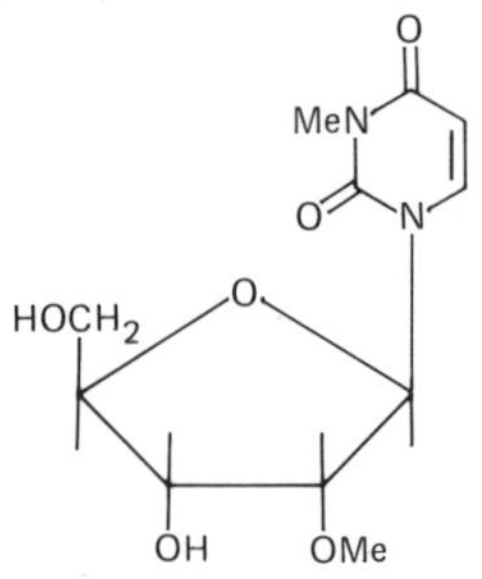

(**18**)

Reaction with diazomethane in aqueous 1,2-dimethoxyethane favours carbohydrate hydroxyl group alkylation, the formation of 2′-*O*-methyl ethers preponderating.[35–38]

(ii) Cyclonucleoside Formation

Methylsulphonyl derivatives of the nucleosides of hydroxypyrimidines form cyclonucleosides on treatment with sodium hydroxide. For example, treatment of 2′,3′,5′-tri-*O*-methylsulphonyluridine (**19**) with one equivalent of sodium hydroxide gives 3′,5′-di-*O*-methylsulphonyl-O^2,2′-cyclouridine (**20**).[39,40]

Boiling this product in aqueous solution gives 3-β-D-lyxofuranosyluracil (**24**) via the intermediates (**21–23**).[39]

The formation of O^2,2′-cyclonucleosides and subsequent ring opening is

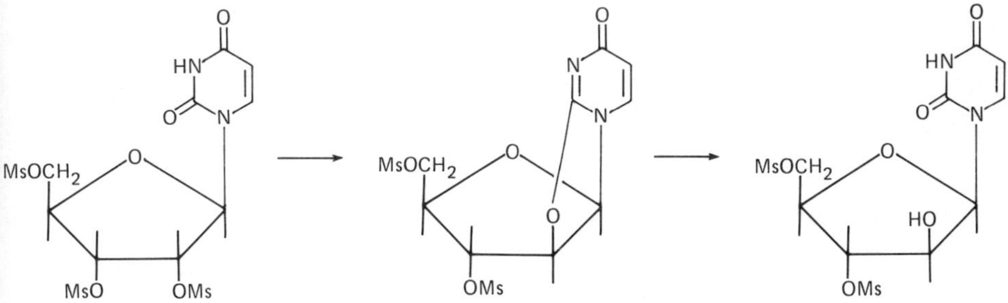

(19) (20) (21)

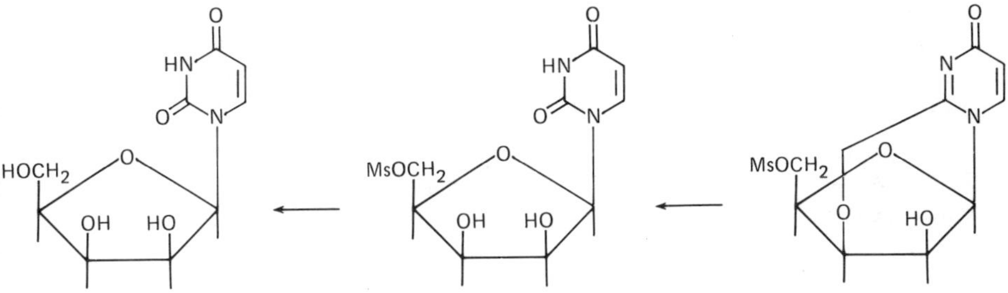

(24) (23) (22)

of great value for the synthesis of glycosyl anomers that would not be readily available by the usual methods of nucleoside synthesis. Further examples are given by Michelson.[28]

In addition to their use for specific inversion of the hydroxyl groups on the sugar moeity, pyrimidine O^2,2′-cyclonucleosides have also been very useful for the synthesis of 2′-deoxynucleosides. For example, the treatment of 5′-*O*-acetyl-O^2,2′-cyclouridine with sodium iodide in acetic acid to give the iodo derivative **(25)**, from which 2′-deoxyuridine is obtained by hydrogenation and deacetylation.[41,42]

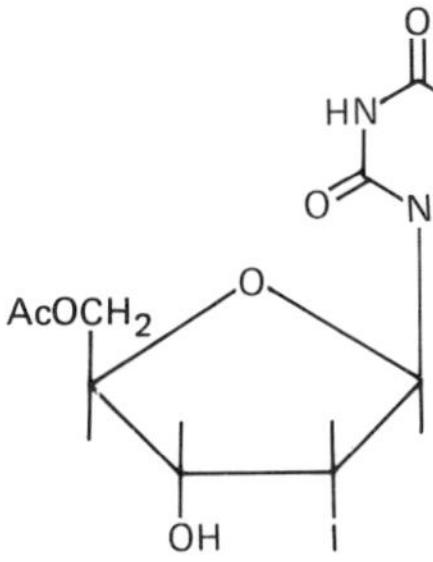

(25) 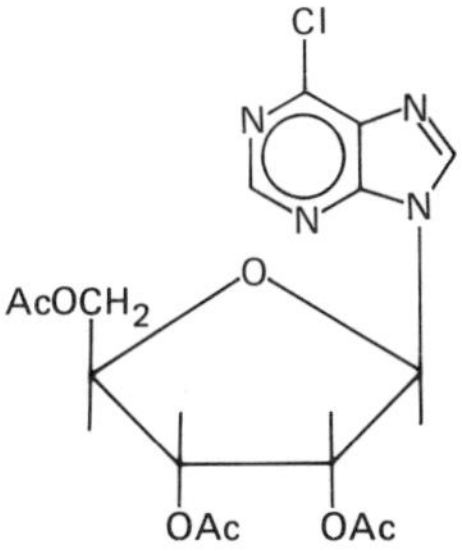 (26)

Reaction of a 5′-arylsulphonyluridine, thymidine, or inosine with sodium iodide results in the formation of the 5′iodo product, but the 5′-arylsulphonyl derivatives of more basic nucleosides form cyclonucleosides on such treatment.

A number of cases have been observed where the heterocyclic portion of the nucleoside affects the reactivity of the sugar hydroxyl groups towards some reagents. For example, whereas most nucleosides form sulphonyl derivatives without difficulty, guanine nucleosides do not.[30]

(iii) Reactions of the Heterocyclic Ring

A number of reactions of the nitrogen heterocyclic ring portion of nucleosides may also be carried out, but in general, it is necessary to protect the hydroxyl groups of the carbohydrate moiety, for example by acetylation.

For example inosine, having been acetylated to give the 2′,3′,5′-triacetate, is readily converted to the corresponding 6-chloropurine-9-β-D-riboside (**26**) by refluxing with phosphoryl chloride in the presence of dimethylaniline.[43]

Guanosine triacetate is brominated by bromine acetic acid at 50°C to give the 8-bromoderivative,[44] and a number of pyrimidine nucleoside acetates may be halogenated to give the corresponding 5-halogeno compounds.

Some pyrimidine nucleosides may be diazo-coupled, nitrosated, or nitrated, the resulting 5-nitrogen containing group being capable of reduction to the corresponding 5-amino product. In these reactions also it is usual to protect the carbohydrate moiety by acylation to avoid destruction of the carbohydrate part of the molecule. The nitration of 2′,3′,5′-tri-*O*-(3,5-dinitrobenzoyl)uridine gives a 5-nitro derivative from which 5-nitrouridine is obtained by deacylation. However, under the same nitration conditions uridine is oxidized to 5-nitrouracil-3-β-D-ribonic acid.[30]

(iv) The Hydrolysis of Nucleosides

In general nucleosides are quite resistant to alkaline hydrolysis but are more susceptible to acid hydrolysis, although the stability of the glycosyl-nitrogen base bond varies widely and depends on the base (and its substitution pattern) and the nature of the carbohydrate.

The deoxynucleosides are generally more labile than the ribonucleosides, and the purine nucleosides more labile than the pyrimidine nucleosides to acid hydrolysis. However, on saturation of the 4,5-bond in pyrimidine nucleosides by hydrogenation, hydration or hydroxybromination, the pyrimidine nucleoside now becomes much more labile towards acid hydrolysis.

Whilst nucleosides are relatively stable to alkaline hydrolysis, some purine nucleosides undergo cleavage of the imidazole ring to give 4,5-diaminopyrimidines and ribosyl derivatives of such compounds. Many pyrimidine nucleosides undergo cleavage of the pyrimidine ring on treatment with hydrazine.

Further mention of degradations of pyrimidines and purines in nucleic acids is made later (Section 7D), and a more detailed account of the chemical reactions of nucleosides is given in ref. 30.

Other examples of the reactions of nucleosides are given by Lister,[45] and in other reviews of nucleoside chemistry and the original literature.

(D) THE SYNTHESIS OF NUCLEOTIDES

In general the most useful methods for the synthesis of nucleotides is by the phosphorylation of the corresponding nucleoside, thus nucleotide synthesis is a case of nucleoside reaction.

The first chemical synthesis of a nucleotide was carried out in 1914 by Fischer, who phosphorylated glucosyltheophylline with phosphoryl chloride in aqueous barium hydroxide.[46] However, the site of phosphorylation was not established. The use of phosphoryl chloride as a phosphorylating agent resulted in the successful synthesis of a number of 5′- nucleotides, for example treatment of 2′,3′-*O*-isopropylideneuridine (**27**) with phosphoryl chloride in pyridine followed by acidic hydrolysis to remove the isopropylidene moiety gave UMP, uridine-5′-phosphate (**28**).[47] However, the yields of nucleotides obtained using the reagent were generally rather poor.

(i) $POCl_3$/ pyridine
(ii) H^+/H_2O

(27) (28)

Better yields of nucleotides are obtained using diaralkyl or diaryl phosphorochloridates, for example the treatment of 2′,3′-*O*-isopropylideneadenosine with dibenzyl phosphorochloridate gave the 5′-dibenzylphosphate from which the benzyl groups were readily removed by catalytic hydrogenolysis. Acid hydrolysis to remove the isopropylidene group then gave a good yield of adenosine-5′-phosphate.[48] Similar reactions have been used to obtain UMP and CMP,[49] and other 5′-phosphates.

Several other phosphorylating agents including polyphosphoric acid, tetra-*p*-nitrophenylpyrophosphate, and *O*-benzyl phosphorous *O*-diphenyl phosphoric anhydride have also been used for nucleotide synthesis but probably the most successful and most widely used in recent years has been the reagent introduced by Tener.[50] In the Tener method a protected nucleoside is reacted with *β*-cyanoethyl phosphate in the presence of di-

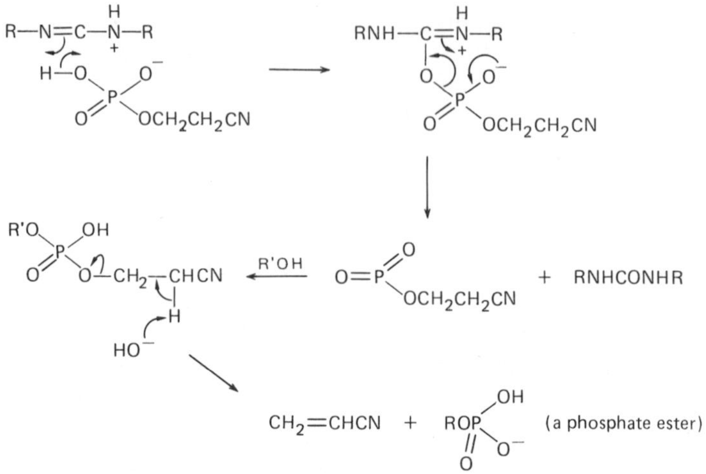

Fig. 4.4 Phosphorylation of an hydroxy compound by the method of Tener

cyclohexylcarbodiimide to give a β-cyanoethyl ester which undergoes base-catalysed elimination of acrylonitrile to give the nucleotide. These steps are shown in Fig. 4.4.

The protection of the 2′-, 3′-hydroxyl groups of ribonucleosides by acetone followed by phosphorylation, gives the corresponding 5′-phosphates. The protection of the 5′-hydroxyl group by tritylation followed by phosphorylation results in mixtures of the 2′- and 3′-phosphates after removal of the protecting groups.[51] The use of a further protecting group at either the 2′- or 3′- position allows specific phosphorylation at either a 3′- or 2′- position. However, after removal of the protecting groups, migration of the phosphate group to a *cis* vicinal hydroxyl group is possible via a cyclic ester. These reactions are reversible and a mixture of 2′- and 3′- phosphates may be obtained (Fig. 4.5).

The nucleoside-2′,3′ cyclic phosphates are readily obtained by treating the nucleoside-2′(3′) phosphates with anhydride reagents, for example, trifluoroacetic anhydride.[52]

In the case of the 2′-deoxyribonucleosides, good yields of the 3′-phosphates can be obtained as there is no extra complication due to the presence of the 2′-hydroxyl group.

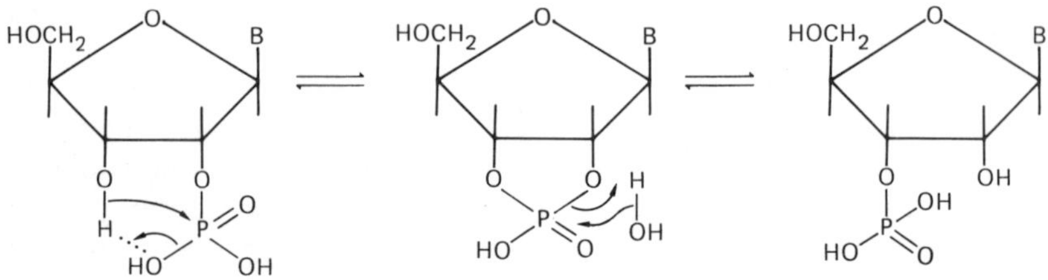

Fig. 4.5. Interconversion of 2′- and 3′-phosphates

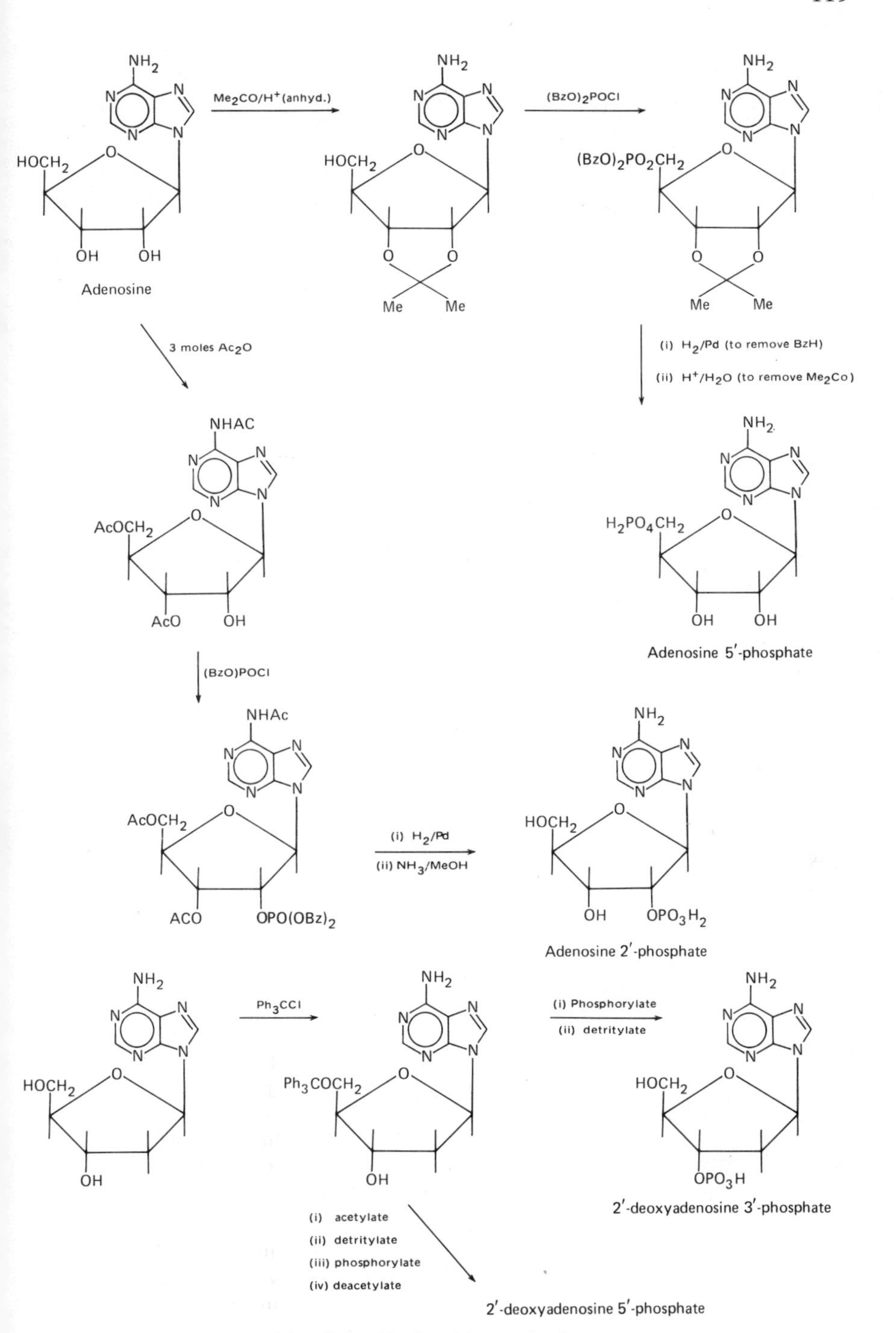

Fig. 4.6. Nucleotide synthesis

The general approaches to nucleotide synthesis are indicated in Fig. 4.6, in relation to the nucleoside derivatives of adenine. Such approaches are obviously also applicable to the synthesis of other nucleotides and a range of examples is given in ref. 30.

(E) CHEMICAL REACTIONS AND PROPERTIES OF NUCLEOTIDES

The nucleotides are nucleoside esters of phosphoric acid and thus have most of the properties and reactions of nucleosides. They undergo many of the heterocyclic ring reactions of nucleosides and the parent bases, for example, uridine-5′-phosphate is readily converted to 5-bromouridine-5′-phosphate by the action of *N*-bromo succinimide.[53] A range of other such reactions is known, some of which are reviewed by Michelson.[28]

The action of some bases on nucleotides results in the formation of cyclonucleotides. For example, thymidine-3′,5′-cyclicphosphate (**29**) is formed at room temperature by treating thymidine-5′-phosphate with dicyclohexylcarbodi-imide in dilute pyridine solution.[54] Good yields of the ribonucleoside-3′,5′-cyclonucleotides have been obtained by using DCCI on the 4-morpholino-*N,N*′-dicyclohexylcarboxamidinium salt of the riboside-5-phosphate or a suitably substituted derivative.[55]

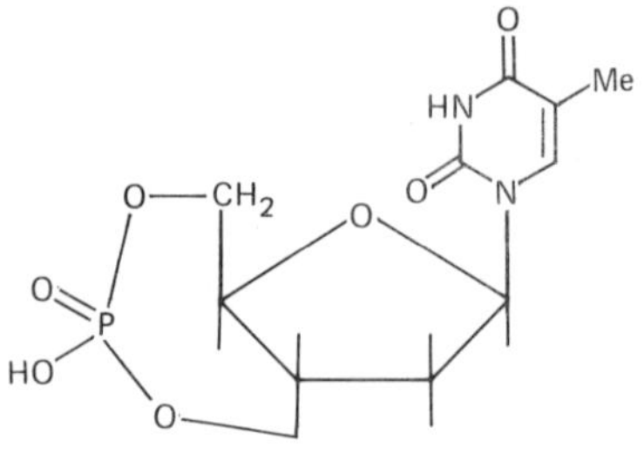

(**29**)

In recent years a considerable effort has been put into the chemical synthesis of oligonucleotides by a number of workers including Khorana, Reese and other workers in the USSR, Japan, as well as in the USA and the European countries. The total synthesis of some genes has been achieved and the total synthesis of a tRNA is in sight.

It is not intended here to go into this aspect of the chemistry of nucleotides. A great deal of work has gone into the development of readily available intermediates containing several protecting groups which can be removed selectively, and the relative merits of the phosphodiester and phosphotriester approaches to oligonucleotide synthesis are still under discussion, although the phosphotriester approach seems to be favoured.

Recent papers (for example C. B. Reese *et al*., *J. Chem. Soc. Perkin I*, 934 (1975), and earlier papers in this series) highlight the problems involved in oligonucleotide synthesis. A recent development in this field has been the use of a solid polymer support for immobilising the growing oligonucleotide chain in a method analogous to the Merrifield method for polypeptide synthesis. A preliminary report of this work by Sheppard and Gait[56] at the MRC Laboratories has appeared and further work was described at a recent meeting.[57] But it is perhaps too early yet to evaluate this method.

In the mononucleotides there is strong interaction between the phosphate group and neighbouring hydroxyl groups, by hydrogen bonding with such groups, stabilizing the anionic forms with the result that the mononucleotides are much stronger acids than the monoalkylphosphates. Adenylic acid has pK_a values of about 0.9 and 6.1, whereas methyl phosphate has values of 1.54 and 6.31 and phophoric acid itself 1.97 and 6.82. Consequently the nucleotides are quite stable in alkali when they are present as the dianion:

$$R{-}O{-}P(=O)(OH){-}OH \underset{}{\overset{-H^+}{\rightleftharpoons}} R{-}O{-}P(=O)(OH){-}O^- \overset{-H^+}{\rightleftharpoons} R{-}O{-}P(=O)(O^-){-}O^-$$

Neutral hydrolysis of the ribonucleotides (catalysed by metal ions) results in the formation of the nucleoside and inorganic phosphate[1,58] since the glycosidic linkage is stable at such a pH.

The monoalkyl phosphates are rapidly hydrolysed in mineral acid, but under these conditions the glycosidic bond of nucleotides is also cleaved to give the base and a sugar phosphate which subsequently degrades to the sugar and inorganic phosphate. However, the pyrimidine nucleotides are much more stable to acid hydrolysis than are the purine nucleotides and the release of the bases is much slower from cytidylic and uridylic acids than from adenylic and guanylic acids. The reduced pyrimidine nucleotides are also more acid labile than the fully aromatic compounds.

Mention of the enzymic cleavage of the nucleotides is left until Chapter 7.

(F) NUCLEOSIDE ANTIBIOTICS

In addition to those nucleosides found as constituents of nucleic acids and nucleotide coenzymes, a number of other nucleosides have been found in nature, several of these showing antibiotic activity. Suhadolnik[59] has given an excellent review of the nucleoside antibiotics and has classed them according to their structural features, according to the carbohydrate moiety and the heterocyclic aglycone that they possess. The nucleoside antibiotics listed by Suhadolnik are shown in Fig. 4.7.

Group (a) represent an important group of nucleoside antibiotics in which there is modification at the 3′-position of the carbohydrate moiety. Cordycepin was the first antibiotic nucleoside to be isolated[60] and puromycin is probably one of the most extensively studied nucleoside antibiotics.

Five of the 3′-deoxyadenine nucleoside antibiotics which have been isolated from micro-organisms or fungi have antibacterial, antitumour and anti-

(a) 3′-Deoxypurine nucleosides

(i) Puromycin

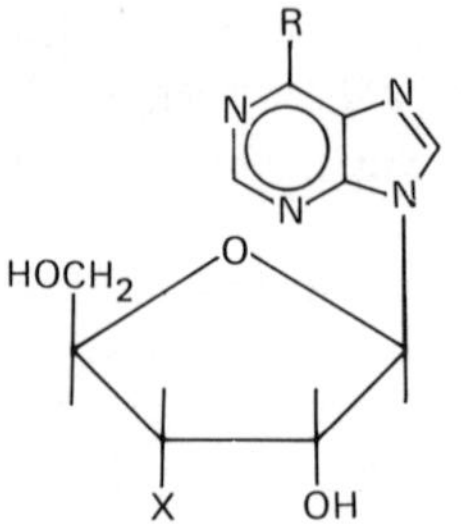

R
N N
N N
$HOCH_2$ O
X OH

R = NMe_2

X = $-NHCOCH(NH_2)CH_2C_6H_4OMe(p)$

(ii) Cordycepin (3′-deoxyadenosine)

R = NH_2

X = H

(iii) 3′-Amino-3′-deoxyadenosine

R = X = NH_2

(iv) 3′-Acetamido-3′-deoxyadenosine

R = NH_2

X = $NHCOCH_3$

(v) Homocitrullyaminoadenosine

R = NH_2

X = $NHCOCH(NH_2)(CH_2)_4NHCONH_2$

(vi) Lysylaminoadenosine

R = NH_2

X = $NHCOCH(NH_2)(CH_2)_4NH_2$

(b) Ketohexose nucleosides

(i) Psicofuranine

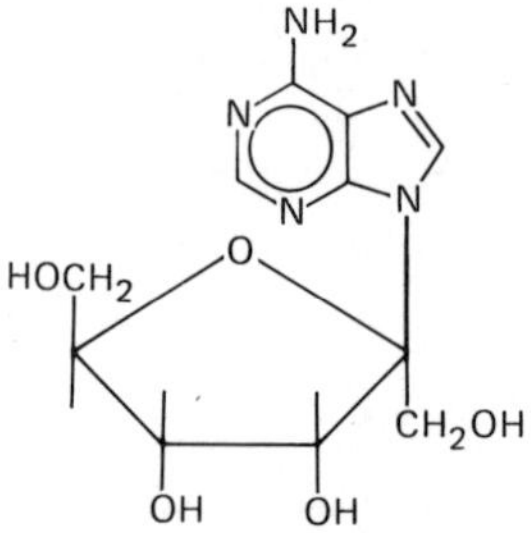

NH_2
N N
N N
$HOCH_2$ O
CH_2OH
OH OH

(ii) Decoyinine

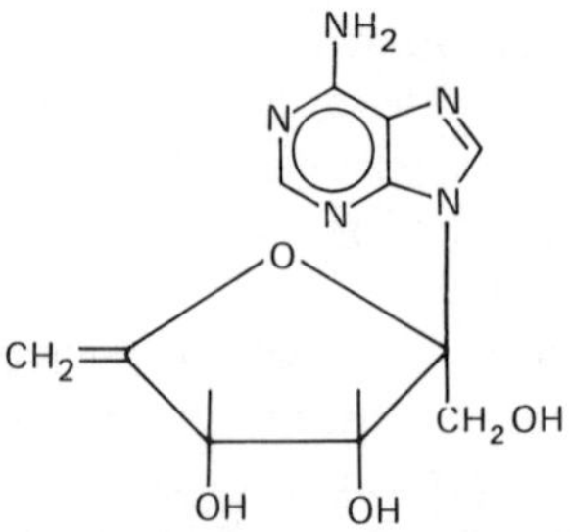

NH_2
N N
N N
O
$CH_2=$
CH_2OH
OH OH

Fig. 4.7. Nucleoside antibiotics

(c) Arabinosyl nucleosides and spongosine

(i) Adenosine arabinoside (ARA-A)

HOCH$_2$ O B HO OH B = NH$_2$ N N N N

(ii) Uracil arabinoside (ARA-U)

B = O HN O N

(iii) Thymine arabinoside (ARA-T)

B = O HN Me O N

(iv) Cytosine arabinoside (ARA-C)

B = NH$_2$ N O N

(v) Spongosine (2-methoxyadenosine)

NH$_2$ N N MeO N N HOCH$_2$ O OH OH

(d) 4-Aminohexose pyrimidine nucleosides

(i) Gougerotin

R N H$_2$NOC O N O OH X OH

R = NH_2

X = $NHCOCH(CH_2OH)NHCOCH_2NHCH_3$

Fig. 4.7. Continued

(ii) Blasticidin S

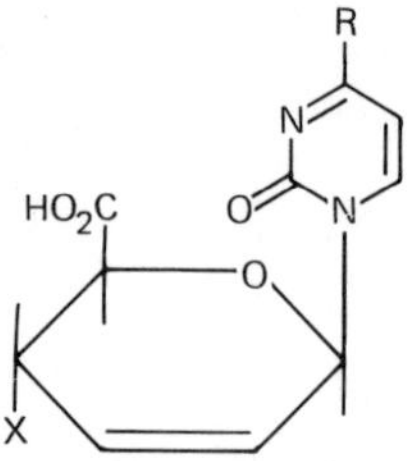

R = NH_2

X = $NHCOCH_2(CHNH_2)CH_2CH_2NMeC:(NH)NH_2$

(iii) Amicetin

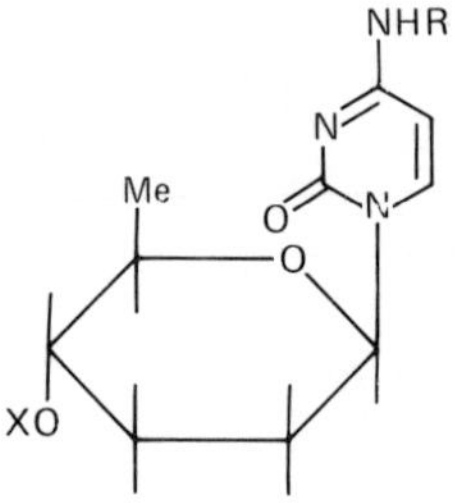

R = CO–C_6H_4–$NHCOC(CH_3)(NH_2)CH_2OH$

X = (Me, Me_2N, OH, OH, H sugar)

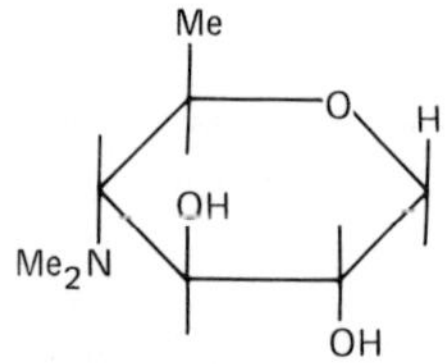

(iv) Plicacetin

R = CO–C_6H_4–NH_2

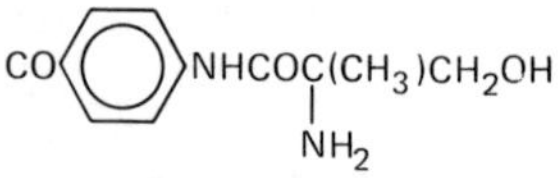

X = (Me, Me_2N, OH, OH, H sugar)

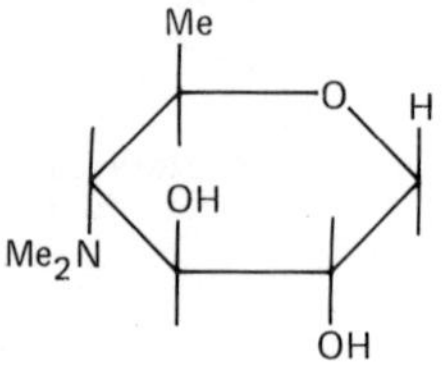

(v) Bamicetin

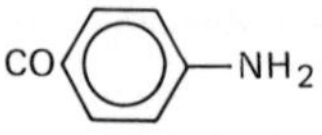

R = CO–C_6H_4–$NHCOC(CH_3)(NH_2)CH_2OH$

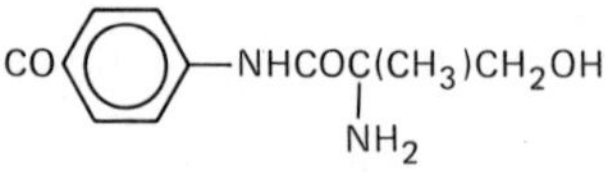

(proposed structure) X = (Me, MeNH, OH, OH, H sugar)

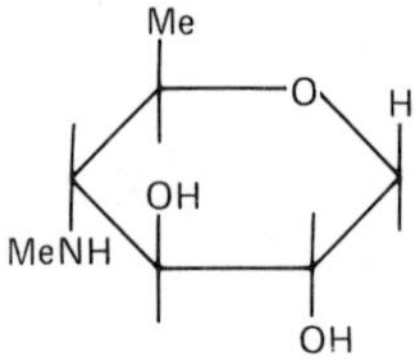

Fig. 4.7 Continued

(e) Peptidyl pyrimidine nucleosides (the polyoxins)

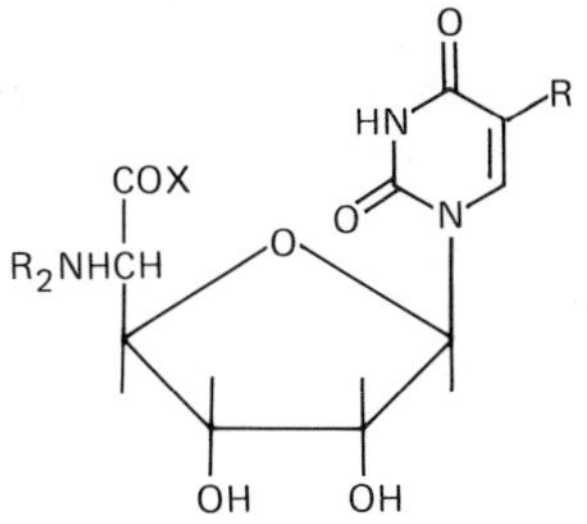

12 compounds, Polyoxins A – L $R = CH_2OH, CO_2H, CH_3, H$

X = 3-ethylidine-L-azetidine-2-carboxylic acid (or OH)

R_2 = 5-O-carbamoyl-2-amino-2-deoxy-L-xylonic acid (or the 3-deoxy derivative)

e.g. Polyoxin C $R = Me$

$X = OH$

$R = H$

(f) Purine nucleosides

(i) Aristeromycin $R = NH_2$

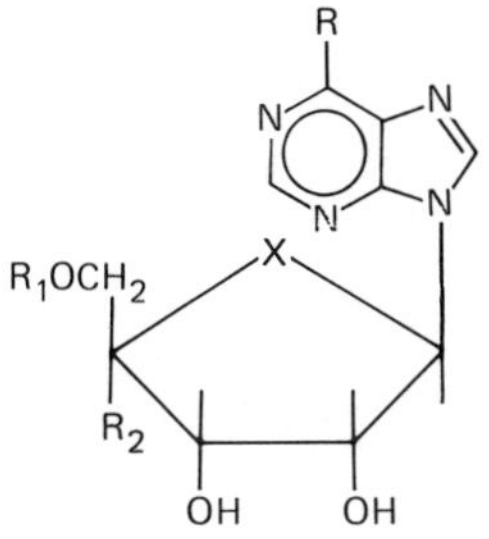

$R_1 = R_2 = H$

$X = CH_2$

(ii) Nucleocidin $R = NH_2$

$R_1 = NH_2SO_2$

$R_2 = F, X = O$

(iii) Nebularine $R = R_1 = R_2 = H, X = O$

(iv) Crotonoside (isoguanosine)

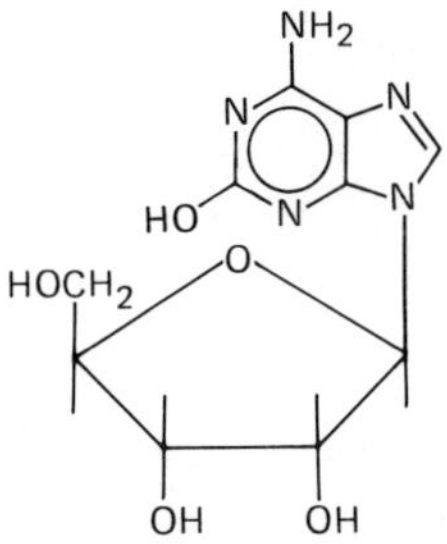

Fig. 4.7. Continued

(v) Septacidin

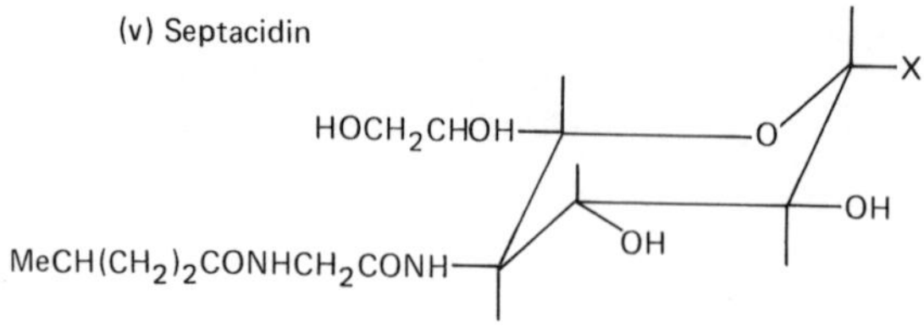
HOCH$_2$CHOH
MeCH(CH$_2$)$_2$CONHCH$_2$CONH
O
X
OH
OH

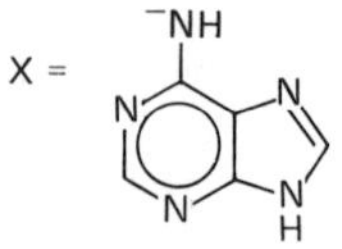
X =
$^-$NH
N
N
N
N
H

(g) Azapyrimidine nucleosides

5-azacytidine

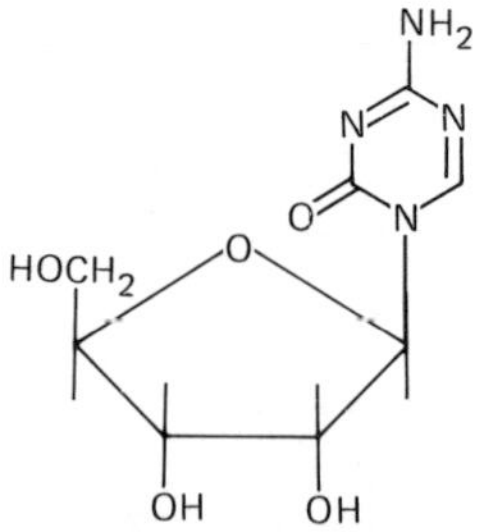
NH$_2$
N
N
O
N
HOCH$_2$
O
OH
OH

(h) Pyrrolopyrimidine nucleosides

(i) Tubercidin

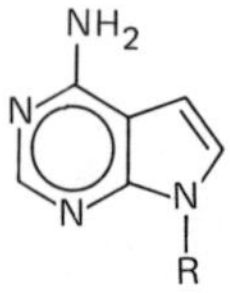
NH$_2$
N
N
N
R

(ii) Toyocamycin

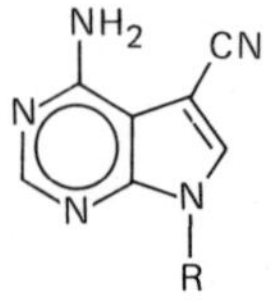
NH$_2$
CN
N
N
N
R

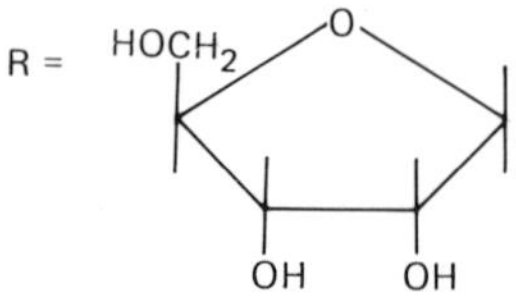
R =
HOCH$_2$
O
OH
OH

(iii) Sangivamycin

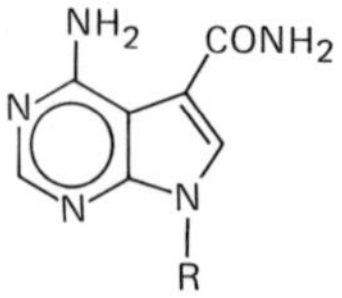
NH$_2$
CONH$_2$
N
N
N
R

(i) Pyrazolopyrimidine nucleosides and coformycin

(i) Formycin

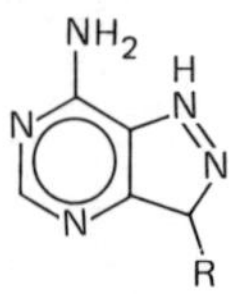
NH$_2$
H
N
N
N
N
R

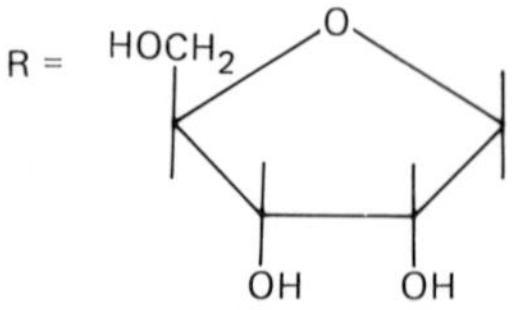
R =
HOCH$_2$
O
OH
OH

Fig. 4.7. Continued

(ii) Formycin B

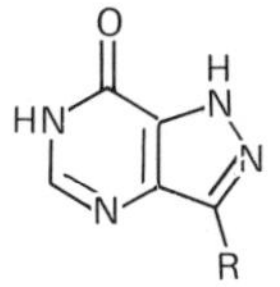

(iii) Oxoformycin B

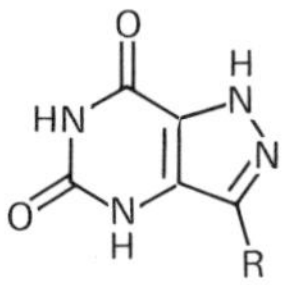

(j) Other nucleosides

(i) Pyrazomycin

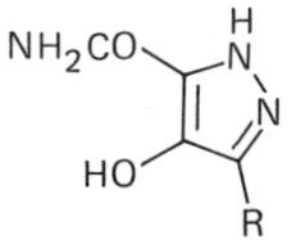

(ii) Showdomycin

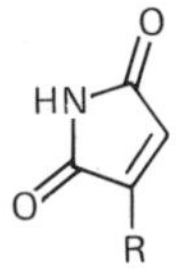

Fig. 4.7. Continued

viral activity, but 3′-acetamido-3′-deoxyadenosine seems to have no such activity.

Puromycin has been widely studied because of its ability to inhibit protein biosynthesis in both mammalian and bacterial cell-free systems. It shows similarity to the aminoacyl end of aminoacyl-tRNA (Fig. 4.8) and by acting as a codon-independent functional analogue of aminoacyl-tRNA catalyses the release of incomplete peptide chains from the peptidyl-tRNA-mRNA-ribosome complex. Puromycin blocks peptide-chain extension by replacing an aminoacyl-tRNA and reacting with the growing polypeptide on the peptidyl-tRNA site on the ribosome, giving a peptidyl-puromycin product. Thus puromycin acts more like an aminoacid antimetabolite rather than an adenosine anologue.

Puromycin is effective *in vivo* against a variety of microorganisms, some viruses, and an adenocarcinoma but it is very toxic in higher animals, the Ld_{50} for mice being 335 mg/kg (intravenous). Therefore the chemotherapeutic use of puromycin is not practical.

Cordycepin inhibits the growth of most strains of *Bacillus subtilis* but it is ineffective against many other species of micro-organism. Guarino and coworkers[61–63] found that the active form of the nucleoside was its 5′-monophosphate ester and that the growth-inhibiting effect of cordycepin 5′-monophosphate could be due to inhibition of phosphoribosyl pyrophosphate amido transferase, i.e. it inhibited *de novo* synthesis of purines.

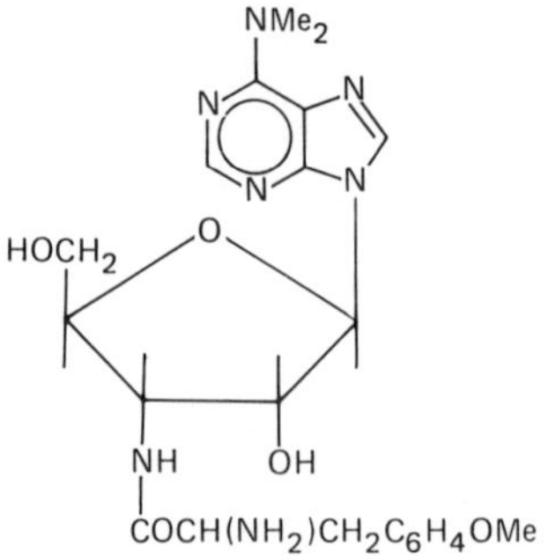

puromycin

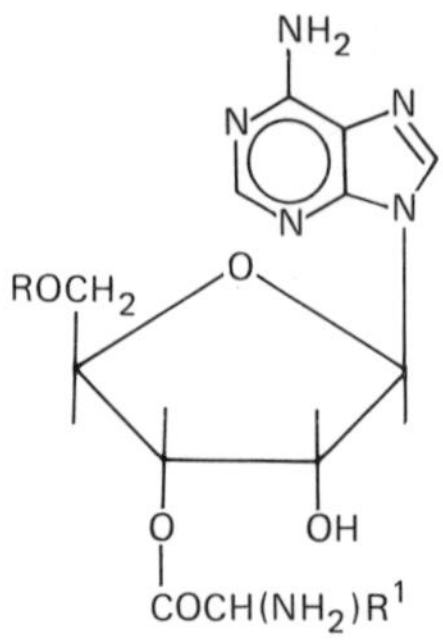

Adenosine end of aminoacyl-tRNA

Fig. 4.8.

Cordycepin (as 5′-monophosphate) has also been found[64] to be cytostatic (but not cytocidal) to human tumour cells grown in culture.

3′-Amino-3-deoxyadenosine is an adenosine antimetabolite which is phosphorylated in Ehrlich ascites tumour cells in culture to 5′-mono-. di-, and triphosphates. It is an inhibitor of both DNA and RNA in such a system both *in vitro* and *in vivo*.[65,66] The nucleoside itself inhibits RNA polymerase but does not seem to inhibit DNA synthesis, it being suggested that this is due to the inhibition of ribonucleotide reductase.[67]

Like puromycin, homocitrullylaminoadenosine inhibits protein synthesis in both bacterial an mammalian systems and acts as an analogue of aminoacyl-tRNA, apparently particularly inhibiting the incorporation of leucine-tRNA.[68] However, 3′-acetamido-3′-deoxyadenosine has neither antibacterial nor antitumour properties, whilst lysylaminoadenosine has, so far, received little attention.

Psicofuranine and decoyinine both have antibacterial activity and effectivity against an adenocarcinoma in rats. Both also seem to exert their action by inhibiting XMP aminase[69] (see also ref. 59, p. 104) but neither seems to have found chemotherapeutic use.

Group (c) includes the compound cytosine arabinoside (ARA-C) which has not been isolated from natural sources and is a synthetic nucleoside, and also the compounds uracil arabinoside (ARA-U) and spongosine, both of which have been isolated from a species of marine sponge but do not have any antibiotic action.

Adenosine arabinoside (ARA-A) has been shown to have cytocidal activity in cell cultures but also has been found to have a broad-spectrum activity against a variety of DNA viruses in culture, but only limited activity against RNA viruses.[70] *In vivo* studies of the activity of ARA-A have confirmed its usefulness against DNA viruses and as it has a low mammalian toxicity (LD_{50} (interperitoneal in mice) 4677 mg/kg)[71] it shows potential as a chemotherapeutically useful antiviral agent.[59]

ARA-A has also been found to be active against two strains of ascites tumours.[72] However, it has been shown to cause chromosomal damage in both human[73] and plant tissue.[74,75]

ARA-C (cytosine arabinoside) has been shown to inhibit certain neoplasms in mammals and also to be active against herpes simplex virus, but ARA-U and ARA-T have much less activity and do not seem to be potentially useful. Both ARA-A and ARA-C have been shown to be phosphorylated *in vitro* to the 5′-triphosphates, which are then incorporated into DNA causing termination.[59,76] However, the precise mechanism of inhibition of DNA synthesis *in vivo* is unknown, although it is suggested that the inhibition of growth of *Escherichia coli* by ARA-ATP might be due to the inhibition of ribonucleotide reductase.[59]

The group (d) nucleoside antibiotics include those in which a 4-aminohexose is linked through N-1 of a pyrimidine or which contain a 4-aminohexose linked through a disaccharide. Gougerotin is the most widely studied member of this group and it shows a wide spectrum of antibacterial activity.[77] Gougerotin is an aminoacyl-tRNA analogue which blocks protein synthesis by acting as an inhibitor of peptide-chain elongation.

Blasticidin S and amicetin also are inhibitors of protein synthesis, acting in a similar way to gougerotin but having rather different specificities, whilst bamicetin and plicacetin probably also have similar biochemical activities, although they have been much less studied than the other aminohexose antibiotics.

Fox *et al.*[78] and Korzybski *et al.*[79] have both reviewed this group of antibiotics.

The polyoxins are a group of peptidyl-pyrimidine nucleoside antibiotics which have an entirely different range of action from the other nucleoside antibiotics. They are very active against fungi but have no biological activities against any other organism. The mode of action of the polyoxins is that of an analogue of UDP-*N*-acetylglucosamine and they block cell wall chitin biosynthesis by inhibiting chitin synthetase.[80,81]

The purine nucleosides (Group f) include the naturally occurring crotonoside (isoguanosine) which has no reported antibiotic activity and septacidin which is unusual in that it has the sugar residue attached through the N-6 amino nitrogen and not through N-9. All of the compounds, except crotonoside, have some biochemical activity.

Aristeromycin has been shown to be cytotoxic to some tumours in culture and to inhibit some micro-organisms, but it does not have chemotherapeutic potential. However, it has been shown to regulate plant growth.[59] The action of aristeromycin is to inhibit *de novo* purine biosynthesis but the precise mechanism is not known and it may act as an inhibitor by interfering with many processes in the cell.

Nucleocidin is an interesting antibiotic as it was the first fluorosugar to be isolated from a natural source.[59] It is an antitrypanosomal antibiotic which also has a fair range of antibacterial activity, the biochemical action being due to its ability to inhibit protein biosynthesis. It acts at the ribosome but the exact mechanism of binding is not known, but it does not seem to block the binding of tRNA to ribosomes nor does it inhibit RNA synthesis.

Nucleocidin is a much more potent inhibitor of protein biosynthesis *in vivo* than puromycin, but seems to have a similar effectivity *in vitro*.[82]

Septacidin has been shown to be inhibitory to some tumour cells in culture and to some fungi but no antibacterial activity has been observed. There is very little information concerning the mode of action but it does not seem to inhibit protein synthesis.[59]

Nebularine also has been found to have antitumour properties as well as some antibacterial action. However, it is very toxic towards mice,[59,83] 20 mg/kg for a three-day period being lethal. There is comparatively little information on nebularine but Suhadolnik[59] and Fox[78] have reviewed this antibiotic.

5-Azacytidine has received much attention as it is a very potent antileukaemic agent and clinical tests have shown potential as a treatment for children with acute leukaemia. 5-Azacytidine is also bacteriostatic and causes chromosomal and genetic mutations in bacteria and plants. The antibiotic seems to interfere with a number of biochemical processes including:[59]

(i) nucleoside kinase activity;
(ii) inhibition of orotidylic acid decarboxylase;
(iii) incorporation into RNA; and
(iv) incorporation into DNA.

5-Aza-2′-deoxycytidine has been shown to have potent antibacterial activity and useful antileukaemic activity in AKR mice[84] although it seems to be more toxic. Suhadolnik[59] gives an excellent summary of 5-azacytidine.

Of the pyrrolopyrimidine nucleosides (group h), Sangivamycin does not have antibacterial properties but all show cytoxicity to cells in culture. Tubercidin has been used in clinical studies for the treatment of patients with advanced neoplastic disease, but a number of problems have arisen concerning its administration, although significant regression of malignancy was observed in several patients.[59]

The pyrrolopyrimidine nucleosides seem to have a number of biochemical actions. All three nucleosides become incorporated at the 3′ end of tRNA and interact with the regulatory sites of several enzymes.

Tubercidin may inhibit protein synthesis at the transcriptional rather than the translational level whilst the other two pyrrolopyrimidine nucleosides may have other activities. Each of the nucleosides is an adenosine analogue and in many of their actions replace adenosine.

The pyrazolopyrimidine nucleosides are unusual since they are C-nucleosides. Coformycin, which has been isolated with formycin,[85] is a related antibiotic which has been assigned the structure (**30**). Also related to the pyrazolopyrimidine nucleosides is pyrazomycin, which is also a C-nucleoside which may be a biochemical precursor to the pyrazolopyrimidine nucleosides.

Formycin, formycin B, and pyrazomycin inhibit a variety of tumour cells, bacteria, fungi and viruses by interfering with a variety of enzymes and biochemical processes.[59] However, oxoformycin B does not seem to inhibit the growth of any organism that has been tested so far. Coformycin broadens the antibiotic spectrum of formycin and is a potent inhibitor of adenosine deaminase. The mode of action of pyrazomycin seems to be related to a competition with uridine metabolism.

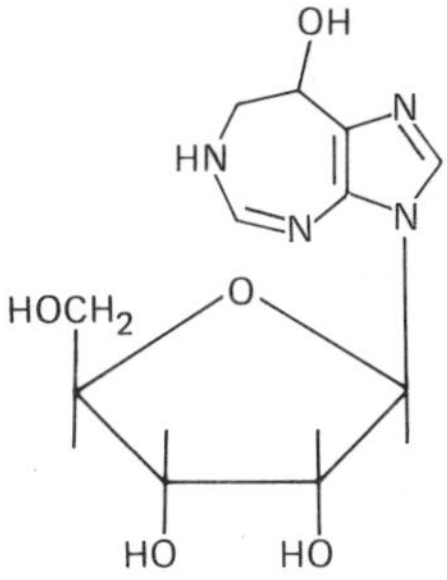

(**30**)

The formycins seem to have a great freedom of rotation about the C-3–C-1′ bond, which may explain some of the unusual biochemical observations on their effects. This group of heterocyclic nucleosides has some particularly interesting properties and will, doubtless, be investigated further in future.

Like the nucleosides antibiotics of groups i and j showdomycin is also a C-nucleoside. Showdomycin is a broad-spectrum antibiotic and also shows activity against some tumour cells. The precise mode of action of showdomycin is unknown, but it does inhibit DNA synthesis and have inhibitory activity against a variety of enzymes. In addition to the review by Suhadolnik[59] showdomycin has been reviewed by Roy-Burman,[86] whilst the C-nucleosides have been recently reviewed by Daves and Cheng.[87]

The heterocyclic nucleoside antibiotics represent a very interesting class of compounds from the chemical, biochemical and chemotherapeutical standpoints, and a considerable extension of studies in this area is likely to occur. This chapter has indicated some of the points concerning the heterocyclic nucleosides of the purine and pyrimidine type, but a fully comprehensive review of the subject is not possible in the space allowed and for further information the reader is directed towards the reviews and books mentioned in the text and to the literature which includes journals on antibiotics as well as a *Journal of Carbohydrates, Nucleosides and Nucleotides.* This journal has recently included a review of nucleoside analogues as antiviral agents.[88]

REFERENCES

1. P. A. Levene and W. A. Jacob, *Chem. Ber.*, **42**, 2475 (1909).
2. P. A. Levene and H. Mandel, *Chem. Ber.*, **41**, 1905 (1908).
3. J. von Liebig, *Annalen*, **62**, 257 (1847).
4. E. Fischer and B. Helferich, *Chem. Ber.*, **47**, 210 (1914).
5. J. Davoll, B. Lythgoe, and A. R. Todd, *J. Chem. Soc.*, **1948**, 967; (a) *ibid*; **1948**, 1685.
6. J. Davoll and B. A. Lowy, *J. Amer. Chem. Soc.*, **73**, 1650 (1951)
7. T. Sato, T. Shimidate and Y. Ishido, *Nippon Kagaku Zasshi*, **81**, 1440 (1960); *ibid*; **81**, 1442 (*Chem. Abs.*, **56**, 11692 (1962)).
8. T. Shimidate, Y. Ishido, and T. Sato, *Nippon Kagaku Zasshi*, **82**, 938 (1961) (Chem. Abs., **57**, 15216 (1962)).
9. T. Shimidate, *Nippon Kagaku Zasshi,* **82**, 1268 (1961); *ibid.,* **82**, 1270 (*Chem. Abs.*, **57**, 16726 (1962)).
10. Y. Ishido, T. Masuba, A. Hosono, K. Fujii, H. Tanaka, K. Iwabuchi, S. Isome, A. Maruyama, Y. Kikuchi, and T. Sato, *Bull. Soc. Chem. Japan*, **38**, 2019 (1965).
11. Y. Ishido, A. Hosono, S. Isome, A. Maruyama, and T. Sato, *Bull. Soc. Chem. Japan*, **37**, 1389 (1964).
12. (a) W. W. Zorbach and R. S. Tipson (eds.), *Synthetic Procedures in Nucleic Acid Chemistry*, Vol. 1, Wiley-Interscience, New York, London, Sydney, Toronto, (1968); (b) L. B. Townsend and R. S. Tipson (eds.), *Nucleic Acid Chemistry*, Wiley-Interscience, New York, London, Sydney, Toronto, (1978).
13. E. Fischer, *Ber.*, **47**, 1377 (1914).
14. P. A. Levene and H. Sobotka, *J. Biol. Chem.*, **65**, 469 (1925).
15. T. L. V. Ulbricht and G. T. Rojers, *J. Chem. Soc.*, **1965**, 6125, 6130.
16. H. G. Garg and T. L. V. Ulbricht, *J. Chem. Soc.* (C), **1967**, 51.
17. T. Ukita, H. Hayatsu and Y. Tomita, *Chem. Pharm. Bull (Japan)*, **11**, 1068 (1963).
18. G. E. Hilbert and T. B. Johnson, *J. Amer. Chem. Soc.*, **52**, 4489 (1930); G. E. Hilbert and E. F. Jansen, *J. Amer. Chem. Soc.*, **58**, 60 (1936).
19. G. A. Howard, B. Lythgoe, and A. R. Todd, *J. Chem. Soc.*, **1947**, 1052.
20. J. J. Fox, N. Yung, J. Davoll, and G. B. Brown, *J. Amer. Chem. Soc.*, **78**, 2117 (1956).
21. R. Duschinsky, E. Pleven, J. Malbica and C. Heidelberger, *Abstr. 132nd Amer. Chem. Soc. Meeting,* **1957**, 19C.
22. J. J. Fox, and N. Yung, I. Wempen, and M. Hoffer, *J. Amer. Chem. Soc.*, **83**, 4066 (1961).
23. G. T. Rojers and T. L. V. Ulbricht, *J. Chem. Soc.* (C), **1969**, 2450, and refs. therein.
24. L. Birkhofer, A. Ritter and H. P. Kuelthan, *Angew. Chem.*, **75**, 209 (1963).
25. T. Nishimura, B. Shimizu and I. Iwai, *Chem. Pharm. Bull*, **11**, 1470 (1963); **12**, 347 (1964).
26. E. Wittenberg, *Zietschrift für Chemie*, **3**, 303 (1964).
27. P. O. P. Tso (ed.), *Basic Principles in Nucleic Acid Chemistry*, Vol. 1., Academic Press, New York and London, (1974).
28. A. M. Michelson, *The Chemistry of Nucleosides and Nucleotides*, Academic Press, London and New York (1963).

29. (a) T. L. V. Ulbricht, *Introduction to Nucleic Acids and Related Natural Products*, Oldbourne Press, London (1965); (b) T. L. V. Ulbricht, *Purines, Pyrimidines and Nucleotides*, Pergamon Press, Oxford, London (1964).
30. B. R. Baker in *Ciba Symposium on Chemistry and Biology of Purines 1956* (1957), p. 122.
31. D. H. Hayes, A. M. Michelson, and A. R. Todd, *J. Chem. Soc.*, **1955**, 808.
32. D. M. Brown, G. D. Fasman, D. I. Magrath, and A. R. Todd, *J. Chem. Soc.* **1954**, 1448.
33. N. C. Yung and J. J. Fox, *J. Amer. Chem. Soc.*, **83**, 3060 (1961).
34. W. Szer and D. Shugar, *Biokhimiya*, **26**, 840 (1961).
35. A. D. Brown and R. K. Robins, *J. Amer. Chem. Soc.*, **87**, 1145 (1965).
36. T. A. Khawaja and R. K. Robins, *J. Amer. Chem. Soc.*, **88**, 3640 (1966).
37. D. M. G. Martin, C. B. Reese and G. F. Stephenson, *Biochemistry*, **7**, 1406 (1968).
38. J. B. Gin and C. A. Dekker, *Biochemistry*, **7**, 1413 (1968).
39. R. Fecher, J. F. Lodington, and J. J. Fox, *J. Amer. Chem. Soc.*, **83**, 1889 (1961).
40. J. F. Codington, R. Fecher, and J. J. Fox, *J. Amer. Chem. Soc.*, **82**, 2794 (1960).
41. D. M. Brown, D. B. Parihar, A. R. Todd, *J. Chem. Soc.*, **1958**, 4242.
42. D. M. Brown, D. B. Parihar, C. B. Reese, and A. R. Todd, *J. Chem. Soc.* **1958**, 3035.
43. J. F. Gerster, J. W. Jones, and R. K. Robins, *J. Org. Chem.*, **28**, 945 (1963).
44. R. E. Holmes and R. K. Robins, *J. Amer. Chem. Soc.*, **86**, 1242 (1964).
45. J. H. Lister, *Fused Pyrimidines, Part II, Purines*, Wiley-Interscience, New York, London, Sydney, Toronto (1971).
46. E. Fischer, *Chem. Ber.*, **47**, 3193 (1914).
47. P. A. Levene and R. S. Tipson, *J. Biol. Chem.*, **106**, 113 (1934).
48. J. Baddiley and A. R. Todd, *J. Chem. Soc.*, **1947**, 648.
49. A. M. Michelson and A. R. Todd, *J. Chem. Soc.*, **1949**, 2476.
50. G. M. Tener, *J. Amer. Chem. Soc.*, **83**, 159 (1961).
51. D. M. Brown and A. R. Todd, *J. Chem. Soc.*, **1952**, 44.
52. D. M. Brown, D. I. Magrath and A. R. Todd, *J. Chem. Soc.*, **1952**, 2708.
53. A. M. Michelson, *J. Chem. Soc.*, **1958**, 1957.
54. G. M. Tener, H. G. Khorana, R. Markham, and E. H. Pol, *J. Amer. Chem. Soc.*, **80**, 6223 (1958).
55. M. Smith, G. I. Drummond, and H. G. Khorana, *J. Amer. Chem. Soc.*, **83**, 698 (1961).
56. M. J. Gait and R. C. Sheppard, *J. Amer. Chem. Soc.*, **98**, 8514 (1976).
57. M. J. Gait in *Chem. Soc. Nucleotide Group 10th Annual Symposium*, Birmingham, 20–21 December, 1977.
58. P. A. Levene and W. A. Jacob, *Chem. Ber.*, **42**, 335, 2469 (1909); **44**, 1027 (1911).
59. R. J. Suhadolnik, *Nucleoside Antibiotics*, Wiley-Interscience, New York, London, Sydney, Toronto (1970).
60. K. G. Cunningham, S. A. Hutchinson, W. Manson, and F. S. Spring, *J. Chem. Soc.*, **1951**, 2299.
61. A. J. Guarino in D. Gottlieb and P. D. Shaw (eds.), *Antibiotics*, Springer-Verlag, New York (1967), p. 468.
62. F. Rottman and A. J. Guarino, *Biochim. Biophys. Acta*, **80**, 632 (1964).
63. F. Rottman and A. J. Guarino, *Biochim. Biophys. Acta*, **89**, 465 (1964).
64. M. A. Rich, P. Meyers, G. Weinbaum, J. G. Cory, and R. J. Suhadolnik, *Biochim. Biophys. Acta*, **95**, 194 (1965).

65. H. T. Shigeura, G. E. Boxer, M. L. Meloni, and S. D. Sampson, *Biochemistry*, **5**, 994 (1966).
66. J. T. Truman and H. Klenow, *Mol. Pharmacol.*, **4**, 77 (1968).
67. R. J. Suhadolnik, S. I. Finkel, and B. M. Chassy, *J. Biol. Chem.*, **243**, 3532 (1968).
68. A. J. Guarino, M. L. Ibershof, and R. Swain, *Biochim. Biophys. Acta.*, **72**, 62 (1963).
69. (a) L. Slechta, *Biochem. Pharmacol.*, **5**, 96 (1960); (b) L. Slechta, *Biochem. Biophys. Res. Commun.*, **3**, 596 (1960).
70. F. M. Schabel, *Chemotherapy*, **13**, 321 (1968).
71. S. M. Kurtz, R. A. Fisken, D. H. Kaump, and J. L. Schardein, *Antimicrobial Agents Chemotherapy*, **1968**, 180.
72. J. J. Brink and G. A. LePage, *Cancer Res.*, **24**, 312 (1964); ibid., **24**, 1042
73. W. W. Nichols, *Cancer Res.*, **24**, 1502 (1964).
74. R. N. Rao and A. T. Natarajan, *Cancer Res.*, **25**, 1764 (1965).
75. B. A. Kihlman and G. Odmark, *Hereditas*, **56**, 71 (1966).
76. R. L. Momparler, *Biochem. Biophys. Res. Commun.*, **34**, 465 (1969).
77. T. Kanazaki, E. Higashide, H. Yamamoto, M. Shibata, K. Nakazawa, H. Iwasaki, T. Takewaka, and A. Miyabe, *J. Antibiotics (Tokyo)*, **15A**, 93 (1962).
78. J. J. Fox, K. A. Wanatabe and A. Bloch, *Progr. Nucleic Acid Res. Mol. Biol.*, **5**, 251 (1966).
79. T. Korzybski, Z. Kowszyk-Gindifer, and W. Kurylowicz, *Antibiotics*, Vol. 1, Pergamon Press, New York (1967).
80. K. Inono and S. Suzuki, 156th Meeting, ACS. Atlantic City, N.J., Sept. 1968, *Abstracts Medi*, 35.
81. A. Endo and T. Misato, *Biochem. Biophys. Res. Commun*, **37**, 718 (1969).
82. (a) J. R. Florini in D. Gottlich and P. D. Shaw (eds.), *Antibiotics*, Springer-Verlag, New York, 1967; (b) J. R. Florini, H. H. Bird, and P. H. Bell, *J. Biol. Chem.*, **241**, 1091 (1966).
83. G. B. Brown and V. S. Weliky, *J. Biol. Chem.*, **204**, 1019 (1953).
84. F. Sorm and J. Vesely, *Neoplasma*, **15**, 339 (1968).
85. (a) T. Tsuruoka, N. Ezahi, S. Amano, C. Uchida, and T. Niida, *Meiji Seika Kenkyu Nempo,* **9**, 17 (1967); *Chem. Abs.,* **68**, 8514 (1968); (b) H. Umezawa, T. Niida, T. Niwa, T. Tsuruoka, E. Nosio, and T. Shonnura, Japan Pat. 70 12, 278; *Chem. Abs.,* **73**, 65025p (1970).
86. P. Roy-Burman, *Recent Results in Cancer Research*, **25**, 80 (1970).
87. G. D. Daves and C. C. Cheng, *Progress in Medicinal Chemistry*, **13**, 351 (1976)
88. E. DeClercq and P. F. Torrance, *J. Carbohydrates, Nucleosides and Nucleotides,* **5**, 187 (1978).

Chapter 5

The Nucleotide Coenzymes and Related Compounds

In the preceding chapter aspects of the chemistry of nucleosides and nucleotides were covered and also details of the nucleoside antibiotics were given. In this chapter reference to the occurrence and involvement of nucleotide derivatives and some related compounds in biochemistry is made. Such compounds include the nucleotide coenzymes and the biological phosphorylating agents. In the case of the nucleotide coenzymes the critical part of the molecule may not be the nitrogen heterocyclic moeity but, nevertheless, the nucleotide part is essential for biological activity. Therefore a brief summary of these topics is given here. The involvement of the nucleoside triphosphates in nucleic acid biosynthesis is discussed in later chapters.

(A) INTRODUCTION : THE B GROUP VITAMINS

It has been known for many years that the constitution of the diet is important to the health of animals. Not only do they require adequate carbohydrate, lipid, protein etc., but also some amino acids were found to be 'essential' for normal growth and that trace elements and accessory dietary factors, which have become known as the *vitamins*, were also essential for normal health and development. The vitamins have been sub-divided into the *fat-soluble* and the *water-soluble* or *B group* vitamins.

The fat-soluble vitamins are steroid or polyisoprenoid compounds, whereas most of the B-group vitamins are heterocyclic compounds. Also, whereas the fat-soluble vitamins tend to have multiple roles in living organisms, some of these roles still being unknown and generally speculative, the B-group vitamins have well-defined coenzyme roles. This chapter will discuss the roles of those heterocyclic coenzymes which include ring systems derived from the pyrimidine, purine or pteridine rings, but folic acid will be considered later (Section 6E).

The compounds which comprise the B-group of vitamins are shown in Fig. 5.1. Some other compounds such as *p*-aminobenzoic acid, inositol,

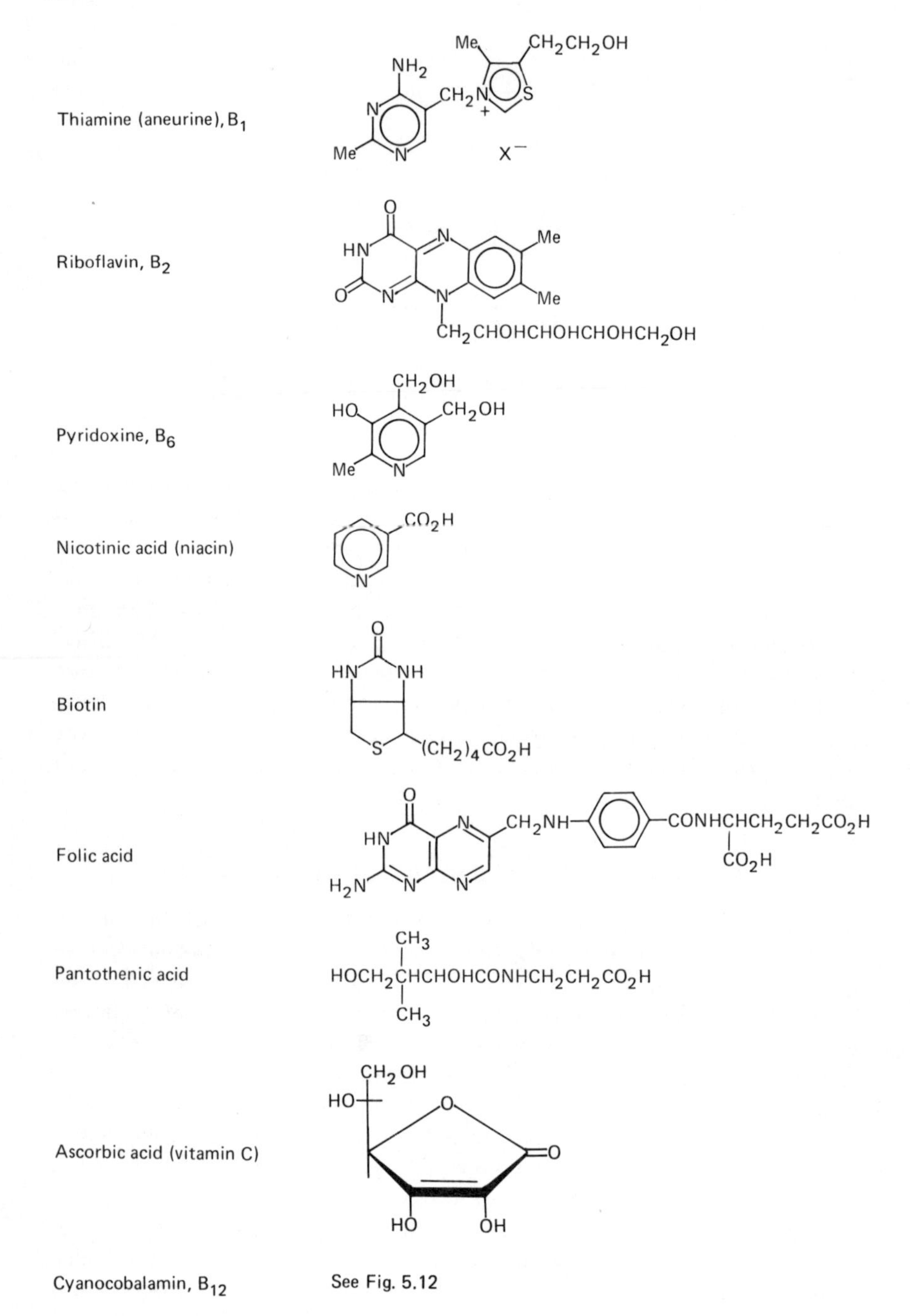

Fig. 5.1. The B group (water-soluble) vitamins

lipoic acid, choline and carnitine are also sometimes included in the list of water-soluble vitamins, but not all nutrionists agree that these are true vitamins. Those compounds listed in Fig. 5.1 are those which are generally accepted as the B-group vitamins.

(i) Thiamine (Aneurine) Vitamin B_1

Thiamine is a pyrimidine derivative which also has a thiazole ring (**1**). A deficiency of thiamine results in the condition known as beri-beri in man whilst animals suffer the condition of polyneuritis.

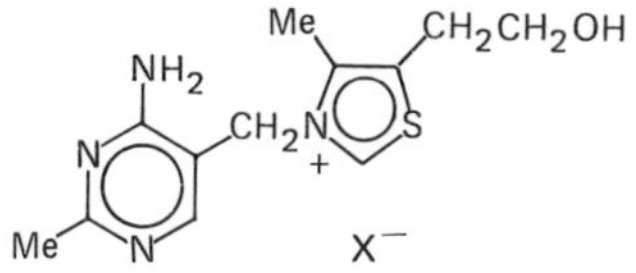

(1)

Lack of thiamine in the diet primarily affects the nervous and circulatory systems, resulting in nervous disorder and paralysis. Beri-beri has been prevalent in Asia where polished rice is the staple diet, the necessary thiamine being in the rejected polishings. However, improved diets and medical aid has greatly reduced the prevalence of dietary deficiency diseases throughout the world.

Thiamine is present in all living organisms but is biosynthesized by plants and some microorganisms. Sources of thiamine are, particularly, green plants, yeast, wholemeal flour and the husks of rice and other grains. Thiamine seems to be synthesized in the leaves of higher plants, the pyrimidine and thiazole fragments being synthesized separately then being joined together. Microorganisms can generally synthesize thiamine, but some require either the pyrimidine or thiazole fragment to be preformed.

All living organisms can pyrophosphorylate thiamine (on the OH group) to give thiamine pyrophosphate (TPP) which predominates in higher animals, although in microorganisms and plants free thiamine predominates. Thiamine can readily pass through cell membranes whereas the pyrophosphorylated form cannot, so TPP is synthesized within the cell which requires it, the liver and kidney being particularly active sites of synthesis, although heart and brain tissue also has high TPP synthetase activity. The principal degradation (i.e. excretory) products of thiamine are the thiazole fragment and the compound thiachrome (**2**).

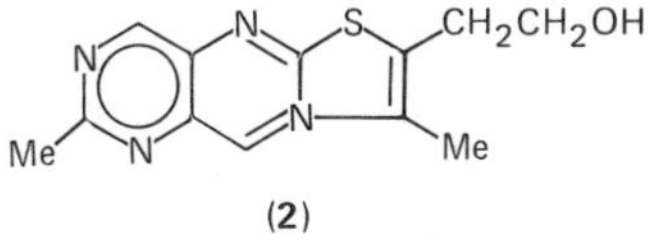

(2)

A feature of the B-group vitamins is that prior to their use as coenzymes in biochemical reactions some transformation is required to produce the

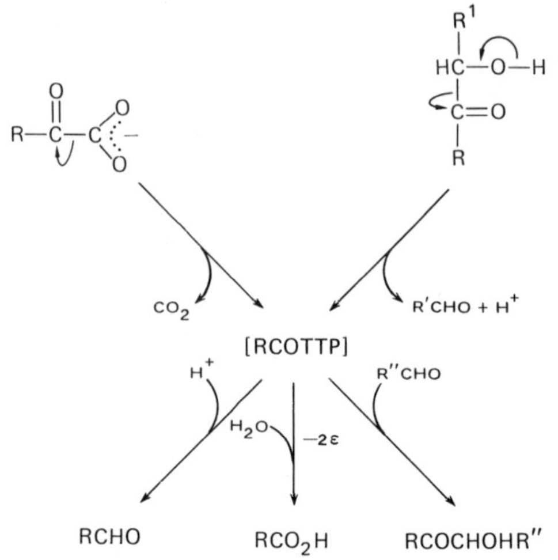

$$R^1$$

Fig. 5.2. Reactions mediated by thiamine pyrophosphate

active coenzyme form. In the case of thiamine it is pyrophosphorylation to give TPP, which is the active form of the coenzyme.

TPP serves as the coenzyme in three types of reaction—oxidation and non-oxidative decarboxylation of α-keto acids and transketolizations of α-hydroxyketones. The basic scheme of such reactions is shown in Fig. 5.2.

Understanding of the mechanism of action of TPP has come from studies of the non-enzymic reactions which TPP can catalyse and the fact that Breslow[1] showed that the hydrogen at the 2-position of the thiazole ring in thiamine very rapidly exchanges with deuterium when in D_2O solution. At pH 7 the reaction is too fast to study by ordinary techniques.[2,3] This observation shows that the zwitter ion (**3**) must be particularly stable.

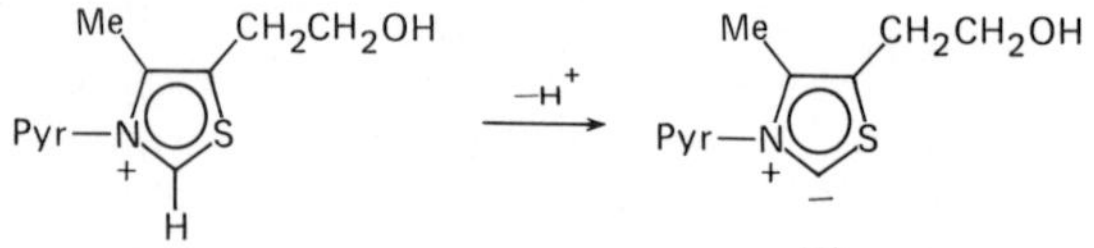

(3)

Suggested mechanisms for the reactions of non-oxidative decarboxylation and transketolization are shown in Figs. 5.3 and 5.4 respectively. In each case the initial reaction is suggested to be attack of the carbanionoid thiazole ring position at the carbonyl carbon atom.

An important example of the role of TPP in oxidative decarboxylation is in the pyruvate dehydrogenase enzyme complex. The role of this enzyme complex is to convert pyruvic acid and coenzyme A to acetyl coenzyme A. This is an important step as it is a control step for several important metabolic pathways and it also represents an interesting example of a multi-enzyme complex. The deficiency symptoms for a lack of thiamine are

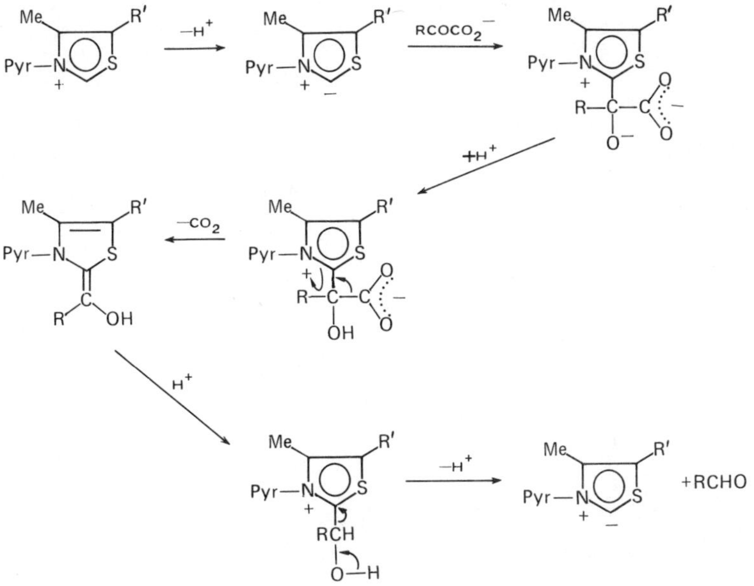

Fig. 5.3.

principally due to an accumulation of excess pyruvic acid in the system since the pyruvate decarboxylase process cannot then be carried out properly.

The overall process of the conversion $AcCO_2H + CoA \rightarrow AcCoA + Co_2$ is shown in Fig. 5.5. which also indicates the enzymes associated with this reaction.

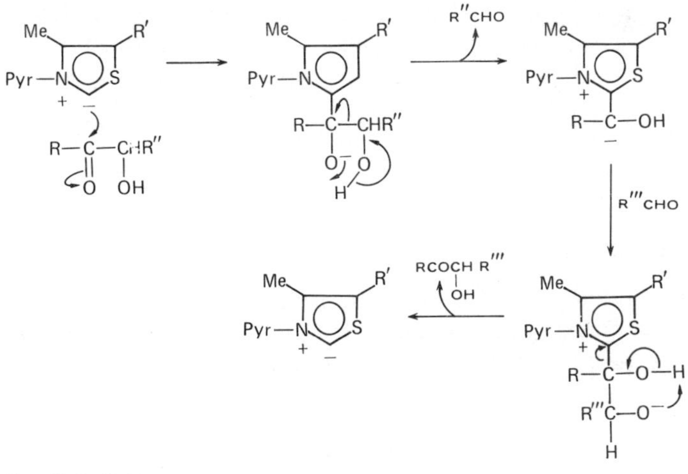

Overall reaction:

$$\text{RCOCHOHR}'' + \text{R}'''\text{CHO} \xrightarrow{\text{TPP}} \text{RCOCHOHR}''' + \text{R}''\text{CHO}$$

Fig. 5.4. Transketolization

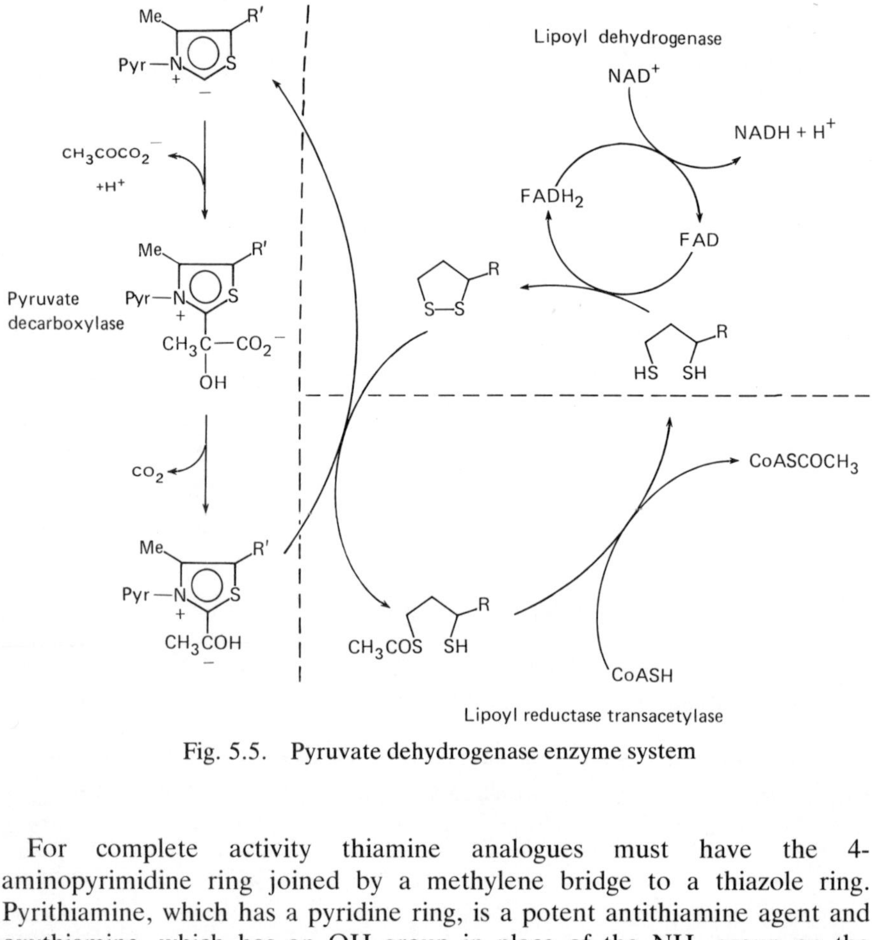

Lipoyl dehydrogenase

Pyruvate decarboxylase

Lipoyl reductase transacetylase

Fig. 5.5. Pyruvate dehydrogenase enzyme system

For complete activity thiamine analogues must have the 4-aminopyrimidine ring joined by a methylene bridge to a thiazole ring. Pyrithiamine, which has a pyridine ring, is a potent antithiamine agent and oxythiamine, which has an OH group in place of the NH_2 group on the pyrimidine is also a thiamine antagonist. The methyl group on the thiazole seems to be important, but that on the pyrimidine ring can be changed, e.g. to ethyl, without loss of activity.

(ii) The Flavin Nucleotides

Riboflavin (Vitamin B_2) is one of a group of compounds known collectively as the flavins, the common feature of which is the substituted isoalloxazine ring system (**4**), i.e. a benzopteridine.

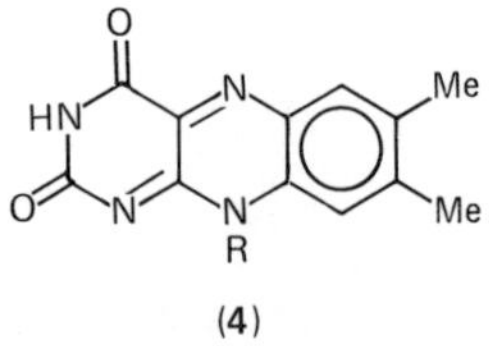

(**4**)

Riboflavin is (**4**), R = CH_2 $(CHOH)_3CH_2OH$, i.e. a combination of the flavin ring system with the carbohydrate ribitol.

Riboflavin was first isolated in 1932 and the structure was elucidated in 1935. Like thiamine, riboflavin is only active in a phosphorylated form (phosphorylated at the ribitol primary OH) and in this form it is commonly known as 'flavin mononucleotide (FMN) although it is not a true nucleotide.

FMN is produced by germinating seeds and by young growing shoots, and most microorganisms also act as a source of this vitamin. The vitamin is also found in milk and egg yolk which become secondary sources, whilst in animals it is mainly found in liver and heart tissue.

In animals flavin adenine dinucleotide (FAD) (**5**) is much more abundant than FMN and is more important as a coenzyme. In the case of the flavin nucleotides it is riboflavin which is the necessary dietary factor and this can be phosphorylated to FMN (intestinal mucosa) or reacted to form FAD (mainly in liver and kidney) by the reaction FMN + ATP $\rightarrow$ FAD + PPi

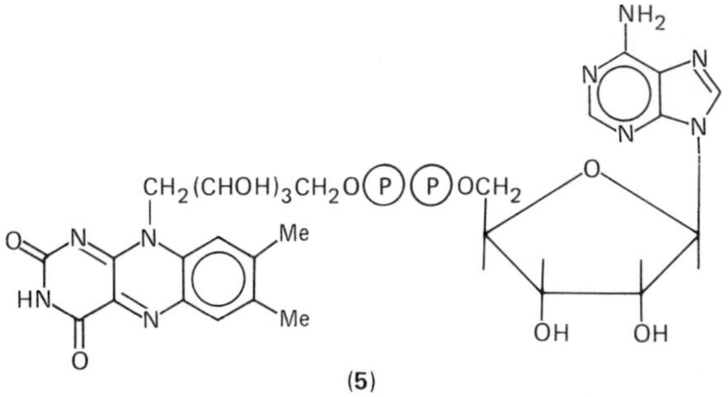

NH_2 N N N N O $CH_2(CHOH)_3CH_2O$(P)(P)OCH_2 O N N Me HN N Me OH OH O

(**5**)

The biosynthesis of riboflavin probably occurs from GTP via a 4,5-diaminouracil, thence to the pteridine (**6**), but the details of the biosynthetic path[4,5] do not seem to have been proved. Similarly comparatively little seems to be known about its catabolism, although in bacteria it is reported that lumichrome (**7**) is one degradation product together with ribitol and an α-pyrone (**8**).

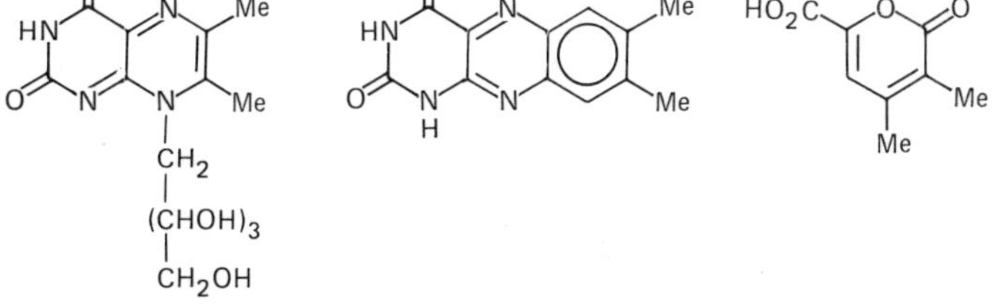

O HN N Me O N N Me CH_2 $(CHOH)_3$ CH_2OH O HN N Me O N N Me H HO_2C O O Me Me

(**6**) (**7**) (**8**)

A deficiency of riboflavin does not lead to any specific disease but it does lead to inflammation of the tongue, lesions at the mucocutaneous junctions of the eyes and lips and several other problems of this kind.

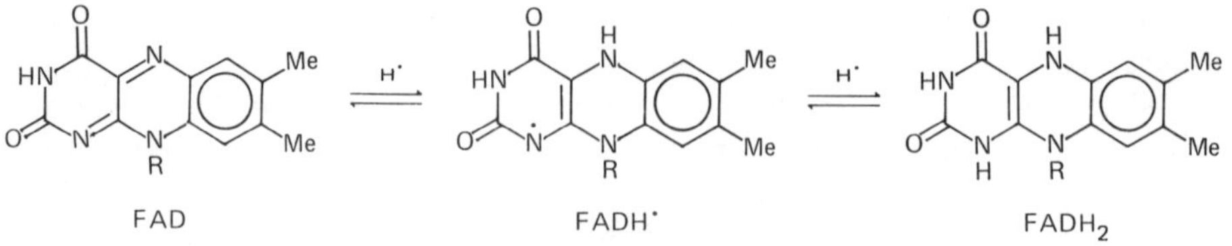

Fig. 5.6. Oxidation-reduction of $FAD/FADH_2$

The flavins are usually firmly bound to the enzymes they activate—i.e. they are prosthetic groups (firmly bound coenzymes)—and the resulting proteins are known as flavoproteins. The number of such flavoproteins having FMN or FAD as prosthetic group is large and they are widely distributed throughout the living world.

The flavins undergo reversible oxidation–reduction reactions, the reduction occurring in two consecutive one-electron transfers involving a flavin semi-quinone intermediate. The reduced flavoproteins ($FADH_2$) may then become substrates for reactions involving other electron acceptors (Fig. 5.6).

A variety of enzymes are activated by FMN or FAD including metabolite-oxidizing flavoproteins and the aerobic dehydrogenases, these including the amino acid oxidases, mono and diamine oxidases, xanthine oxidase and aldehyde oxidases. These enzymic reactions involve electron transfer to oxygen:

$$XH_2 + FAD \rightarrow FADH_2 + X \xrightarrow{O_2} FAD + H_2O_2$$

Hence one observes the production of hydrogen peroxide, the only known biological source of this compound, which is then cleaved by catalase or peroxidase.

The reactions involving electron transport of O_2 are cytoplasmic processes. There are other metabolite oxidizing flavoproteins which involve electron transport to cytochromes: for example, the TCA cycle, electron-transport chain-linked dehydrogenases, succinate dehydrogenase, acyl CoA dehydrogenases. These reactions are mitochondrial processes.

In addition to the above processes, flavoproteins also include the reduced pyridine nucleotide-oxidizing enzymes.

A number of flavin analogues have been prepared, some riboflavin antagonists having antimalarial activity. The activity of riboflavin depends on the molecule having an intact isoalloxazine ring, a pentitol side chain, an ureido (NHCONH) group and at least one methyl (or ethyl group).

(iii) Nicotinic Acid

Nicotinic acid (niacin) or nicotinamide is the essential dietary constituent necessary for the biosynthesis of the oxidation–reduction coenzymes nicotinamideadenine dinucleotide ($NAD^{\oplus}$) (**9**, R = H) and nicotinamide adenine dinucleotide phosphate ($NADP^{\oplus}$) (**9**, R = P)

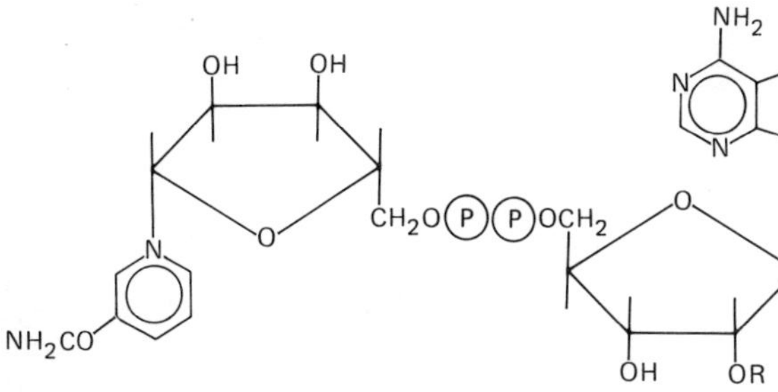

(9)

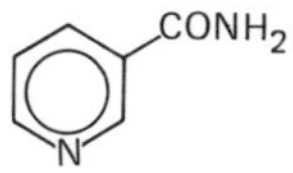

The deficiency disease in man is pellagra, which shows dermatitis, inflammation of the tongue, loss of appetite and other unpleasant symptoms, there being dementia in extreme cases. A similar condition, black-tongue, has been observed in dogs.

Nicotinamide occurs widely either free as NAD$^{\oplus}$ or as NADP$^{\oplus}$, in animal and plant tissues. Yeast and wholewheat are good primary sources, although the animal storage sites such as liver and kidney are good secondary sources of the vitamin. However, it is biosynthesized only by plants and microorganisms. A suggested biosynthetic pathway is shown in Fig. 5.7.

All species have the capability of further metabolising nicotinic acid to give the metabolically active coenzymes. The proposed pathways for the biosynthesis of NAD$^{\oplus}$ are shown in Fig. 5.8.

The two PRPP (phosphoribosylpyrophosphate) reactions are carried out in mitochondria whilst NAD$^{\oplus}$ synthetase is located in cell nuclei. NAD is

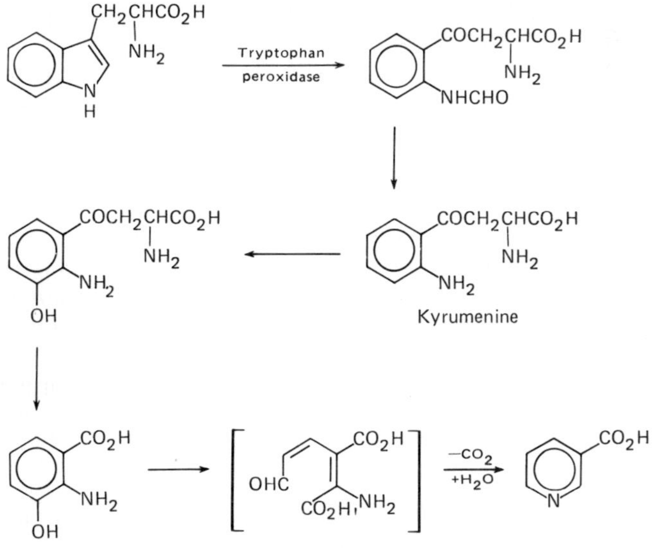

Fig. 5.7. Proposed biosynthesis of nicotinic acid

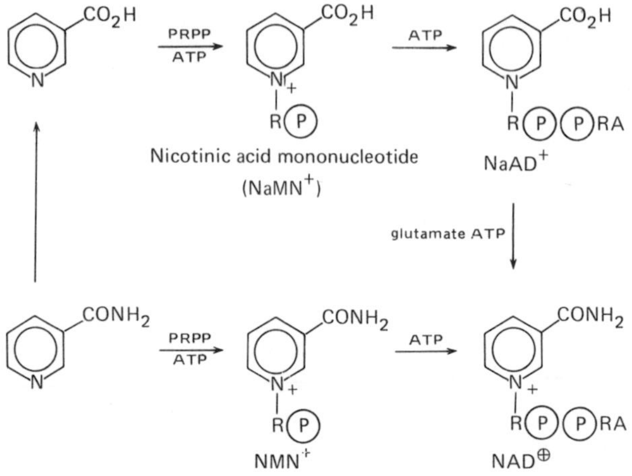

Fig. 5.8. Biogenetic pathways of NAD$^{\oplus}$

possibly implicated in cell growth, since NAD synthetase activity is low in tumours, regenerating liver and rapidly proliferating cells.

NADP$^{\oplus}$ is obtained by an enzymic phosphorylation of NAD$^{\oplus}$. Although the two coenzymes behave identically, they have very specific enzymes for which they each act as the coenzyme, although in some cases there does seem to be a lack of specificity for the particular coenzyme.

It is the nicotinamide portion of the NAD$^{\oplus}$ species which is the hydrogen-accepting site and the adenine part of the molecule appears to play no part, although it is required for activity.

The hydrogen-accepting mechanism of NAD$^{\oplus}$ is shown in Fig. 5.9.

NAD$^{\oplus}$ and NADP$^{\oplus}$ are involved as coenzymes in a large number of oxidation–reduction reactions catalysed by dehydrogenases, most of these enzymes being highly specific for either NAD$^{\oplus}$ or NADP$^{\oplus}$.

The conversion of NAD$^{\oplus}$ (or NADP$^{\oplus}$) to the reduced form is accompanied by a change in the u.v. spectrum, the absorbance at 260 nm decreasing whilst a new band due to the reduced pyridine ring appears having a maximum at 340 nm. This u.v. change is frequently used in assays of NAD$^{\oplus}$-, NADP$^{\oplus}$-dependent enzymes.

The reduction of the pyridine ring in the two coenzymes produces an sp^3 hybridized carbon and the reduction of NAD$^{\oplus}$ with dithionite in deuterium oxide gives two stereoisomers (Fig. 5.10).

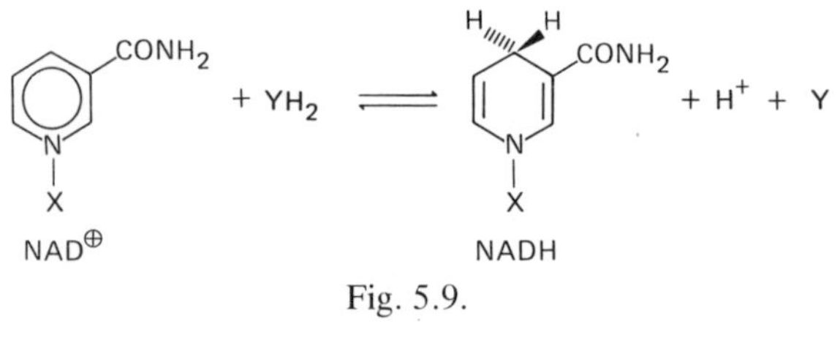

Fig. 5.9.

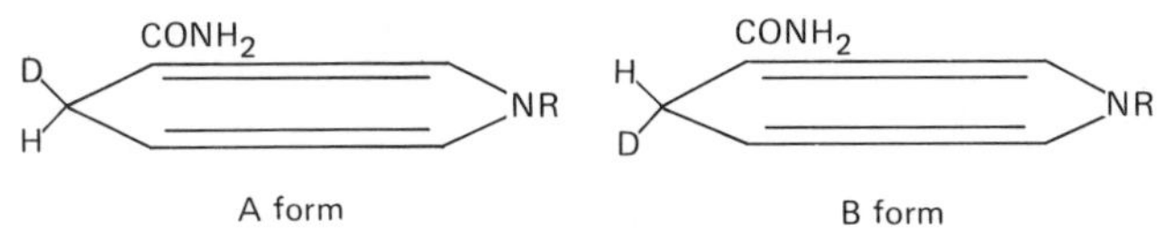

Fig. 5.10. Stereoisomers of NAD2H

Thus the two hydrogen atoms at position 4 in NADH are not equivalent and it is found that in all cases so far bar one (lipoyl dehydrogenase) there is a specificity for the hydrogen transferred. For example, alcohol dehydrogenase and lactate dehydrogenase transfer H to and from the *same side* of the pyridine ring whilst alcohol dehydrogenase transfers H to and from *the opposite side* of the ring from L-glutamate dehydrogenase. There are therefore two types of enzyme having different stereospecificity.

Some of the enzymes activated by $NAD^{\oplus}$ and $NADP^{\oplus}$ and their stereospecificities are shown in Table 5.1.

It has also been found with these enzymes that the transfer of hydrogen between substrate and coenzyme is direct, and although these reactions formally involve a hydride ion transfer, it has not been found whether this reaction proceeds by an actual hydride-ion transfer or by a process involving transfer of a hydrogen atom followed by the transfer of an electron. However, the former mechanism seems more likely.

Table 5.1 Some enzymes having $NAD^{\oplus}$/$NADP^{\oplus}$ as coenzymes and their stereospecificity

Dehydrogenase	Coenzyme	Dehydrogenase	Coenzyme
A specificity		B specificity	
Alcohol	$NAD^{\oplus}$	Glutamate	$NAD^{\oplus}$ (or $NADP^{\oplus}$)
Lactate	$NAD^{\oplus}$	Glucose-6-phosphate	$NADP^{\oplus}$
Glycerate	$NAD^{\oplus}$	6-Phosphogluconate	$NADP^{\oplus}$
Isocitrate	$NADP^{\oplus}$	α-Glycerophosphate	$NAD^{\oplus}$
Dihydrofolate reductase	$NADP^{\oplus}$	UDPG	$NAD^{\oplus}$

(iv) Coenzyme A

Coenzyme A was originally found to be a thermally stable cofactor in a number of biochemical acetylation reactions. It was isolated from yeast by Lynen and his colleagues, and the structure, which was elucidated by Lipmann, Snell and Baddiley, is shown in Fig. 5.11.

The necessary dietary factor which is required by higher animals for the biosynthesis of coenzyme A is pantothenic acid (**10**), which is derived from pantoic acid and β-alanine. β-Alanine seems to be the only known naturally occurring simple β-amino acid.

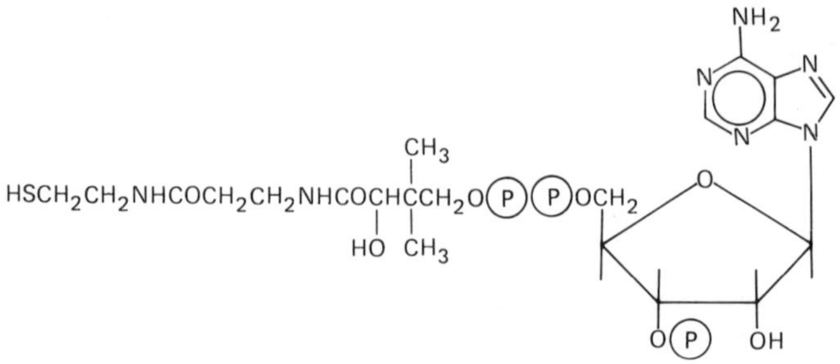

$$HSCH_2CH_2NHCOCH_2CH_2NHCOCH(OH)C(CH_3)_2CH_2O\text{Ⓟ}\text{Ⓟ}OCH_2$$

Fig. 5.11. Coenzyme A

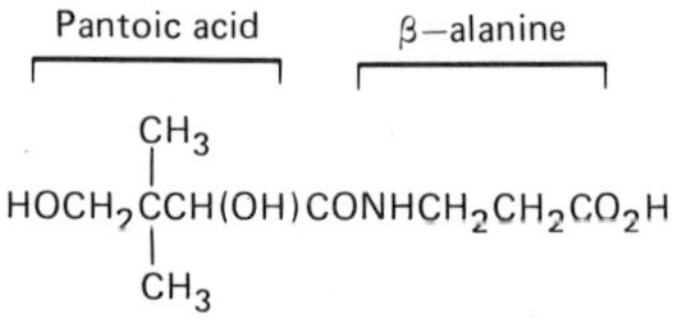

Pantoic acid β–alanine

$$HOCH_2C(CH_3)_2CH(OH)CONHCH_2CH_2CO_2H$$

(10)

Pantothenic acid is widely distributed and is required by all animals which can aquire their needs from the intestinal flora. There is no particular deficiency disease which has been established due to lack of pantothenic acid, although deficiency symptoms are dermatitis, restricted growth, ulceration of the gastro-intestinal tract and loss of appetite.

The stages in the biosynthesis of coenzyme A are shown in Fig. 5.12.

The sulphur-containing aminoacid cysteine is obligatory in this sequence, the reactions of which occur in the cytoplasm, each cell apparently synthesizing its own coenzyme A.

The active site of the coenzyme is the thiol group, and coenzyme A is frequently abbreviated to CoASH.

The types of reaction mediated by CoASH are given below. In each case it is the thiol group which is the site which is involved and essentially the purine part of the molecule plays no part in the reaction.

(1) Formation of acyl derivatives
e.g. succinyl CoA synthetase,
lipoyl CoA transacetylase,
β-ketothiolase, etc.

(2) Reactions of acyl CoA
e.g. citrate synthetase,
acetyl CoA carboxylase etc.

Acetylcoenzyme A is a key intermediate in the TCA cycle and in lipid metabolism.

The major excretory degradation product of CoASH is phosphopantathene.

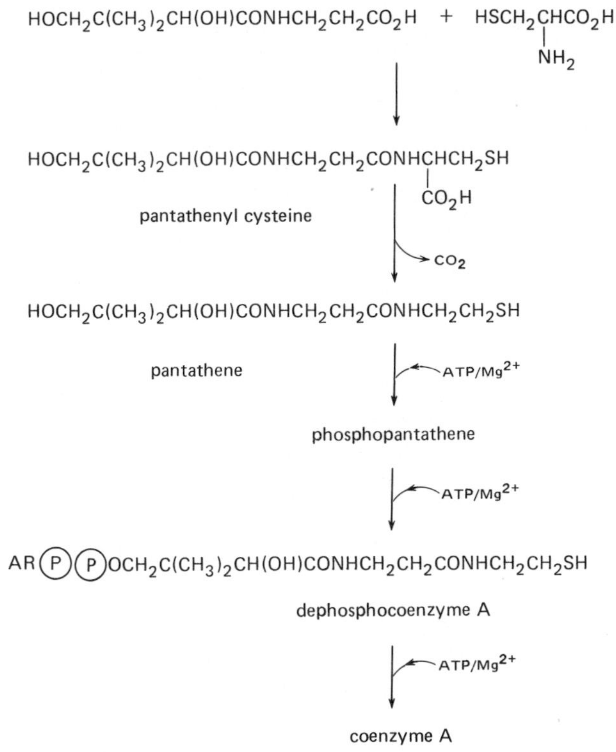

Fig. 5.12. Biosynthesis of coenzyme A

(v) Vitamin B_{12}

It was discovered in 1926 that liver apparently contained a dietary factor which would cure pernicious anaemia, but it was more than twenty years later that the factor was obtained in a pure, crystalline form, [6,7] and the structure was not elucidated for a further nine years.[8] The structure of Vitamin B_{12} is shown in Fig. 5.13.

Vitamin B_{12} contains a highly substituted, reduced corrin ring and a nucleotide which is unusual as it contains an α-glycosidic bond. The corrin ring has cobalt chelated to the four pyrrole nitrogen atoms, to a nitrogen of the 5,6-dimethylbenzimidazole ring and to a cyanide ion, this cyanide ion resulting from the isolation procedure. A variety of other cobalamins have also been prepared, including hydroxocobalamin and nitritocobalamin,[9] and some other B_{12} analogues have been isolated which include purine bases such as adenine and guanine in place of the benzimidazole moeity.

The coenzymatically active forms of Vitamin B_{12} are those having a water molecule in place of the cyanide group (aquocobalamin) and those containing a 5′-deoxyadenosine group in place of the cyanide, these last compounds being first isolated in 1958.[10]

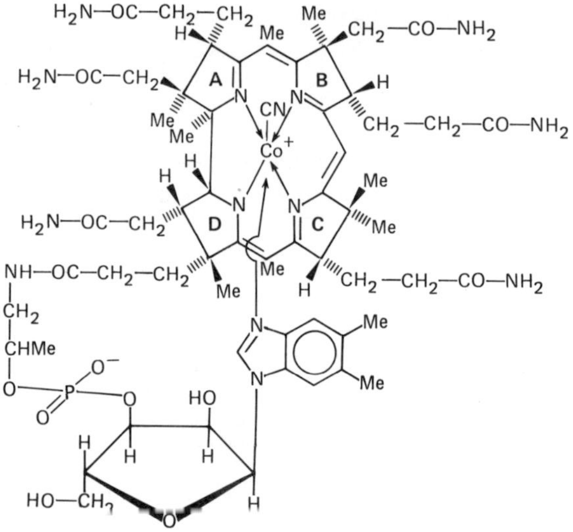

Fig. 5.13. Structure of vitamin B_{12} cyanocobalamin

About six enzymic reactions in which 5′-deoxyadenosyl cobalamin (cobamide) acts as coenzyme are known, each of which involves an intramolecular 1,2-shift of hydrogen with a coupled 1,2- shift of another group. These reactions are listed in Table 5.2. Another reaction in which this coenzyme participates is a ribonucleotide reductase system isolated from certain microorganisms, but this is similar to the other reactions except that the donor and the acceptor of hydrogen in the reaction are different molecules.

The mechanism of the reaction seems to differ according to the reaction catalysed, it being suggested in the case of dioldehydrase that the mode is hydride transfer from substrate to coenzyme, followed by migration of the hydroxyl group, and readdition of the hydride ion, the last process being stereospecific dehydration of the gem diol.[11–13]

Table 5.2 Reactions involving cobamide coenzymes

$$RCH(R')CH_2X \rightarrow RCH_2CH(R')X$$

R	R′	X	Enzyme
H	$CH(NH_2)CO_2H$	CO_2H	Glutamate mutase
H	COCoA	CO_2H	MethylmalonylCoA mutase
CH_3	OH	OH	Dioldehydrase
CH_2OH	OH	OH	Glycerol dehydrase
H	NH_2	OH	Ethanolamine deaminase
H	NH_2	$CH_2CH(NH_2)CH_2CO_2H$	β-lysine isomerase

$$RCH(^{*}OH)\cdot CH_2\cdot OH \longrightarrow RCH(^{*}OH).\overset{+}{C}H_2.OH \longrightarrow R\overset{+}{C}H-CH(^{*}OH)OH \longrightarrow RCH_2CH(^{*}OH)-OH \longrightarrow RCH_2.CHO + H_2O$$

However, some of the reactions involving the cobalamin coenzymes would appear to involve proton abstraction from the substrate with possible formation of an R–CO bond. But there seems to be no firm evidence for the intermediacy of a cobalt–substrate bond in such reactions, although speculative suggestions concerning the cobalt–carbon bond cleavage have been put forward.[14]

The aquocobalamin coenzymes are involved in certain reactions involving methyl group transfer, for example, the coenzyme facilitates the transfer of a methyl group from N-5-methyl THF to homocysteine to give methionine, whilst another important reaction is the formation of acetate by transfer of a methyl group to carbon dioxide.

Comparatively little is really known about the mechanism of action and the roles of Vitamin B_{12}. Only small amounts seem to be required yet without it, a variety of unpleasant symptoms occur—including anaemia and neurological disorders.

The initial symptoms of wasting and anaemia can be shown by ruminants where there is a cobalt deficiency, such cases having been reported from New Zealand. However, pernicious anaemia seems to occur only in humans and would seem to be partly due to the lack of an intrinsic factor concerned with the transfer of the vitamin across the intestinal cell wall and also with the mechanism of storage in the liver. There is also a specific 'releasing factor' for the release of Vitamin B_{12} from the intrinsic factor and the vitamin deficiency can be a result of a lack of the releasing factor.

Biopterin

A compound which should also be mentioned here although it is not considered as one of the B-group vitamins is the pteridine derivative biopterin which, as its dihydro derivative, is an important co-factor in a number of reactions. It participates in aromatic hydroxylation reactions including the hydroxylations of phenylalanine to tyrosine, of tyrosine to 3,4-dihydroxyphenylalanine, and in the biosynthesis of melanins. It is also involved in some other reactions including the biosynthesis of the prostaglandins and of serotonin.

The role of biopterin in the hydroxylation of phenylalanine is shown in Fig. 5.14.

The involvement of biopterin in reaction pathways leading to compounds which act on the central nervous system–adrenalin for example, and 3,4-

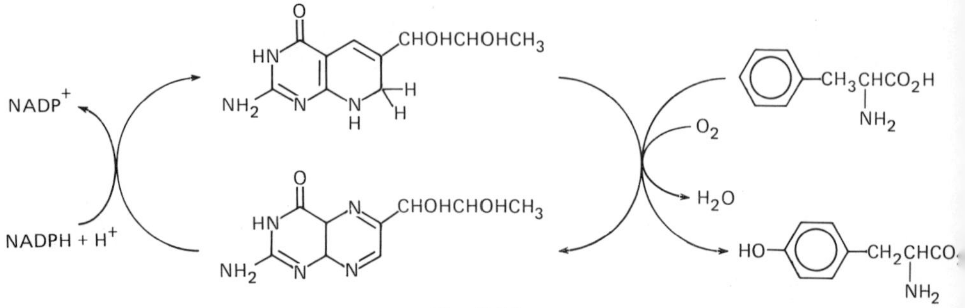

Fig. 5.14.

dihydroxyphenylalanine (DOPA)—has lead to some interesting observations.

Patients who have one of the three types of phenylketonuria show biopterin deficiency, whilst sufferers from Parkinson's disease lack biopterin in the spinal fluid. The possibility of the successful treatment of such patients with biopterin is, however, precluded since biopterin does not pass the blood–brain membrane barrier. Hence the search for pro-drugs which could be metabolized in brain tissue to biopterin but which would pass this barrier.

The chemical synthesis of biopterin and routes to potential biopterin analogues have been developed by E. C. Taylor (Princeton University) details of which are in press.

(B) BIOLOGICAL PHOSPHORYLATING AGENTS

The role of adenosine triphosphate (ATP) in living systems has been extensively investigated and discussed. Its hydrolysis to ADP and phosphate has been found to be coupled with a variety of energy-requiring processes and to be coupled with energy release. For example, it is associated with the biosynthesis of proteins, nucleic acids, complex carbohydrates and lipids and hydrolysis of ATP to ADP accompanies muscle contraction, bioluminescence, electric discharge in specialized organs, and transport of metabolites against a concentration gradient. The synthesis of ATP from ADP + Pi is associated with *energy acquisition*—for example, in photosynthesis and in the electron-transport process accompanying terminal oxidation of carbohydrates, lipids and amino acids – *photophosphorylation* and *oxidative phosphorylation*.

Lipmann[15] put forward the *high-energy bond* concept of ATP in 1941 to explain its role as an *energy store* in biochemical reactions. This idea has stimulated much discussion including a lively series of articles in the early 1970s.[16–21]

It is because of its phosphorylating ability that ATP exerts its *energy-providing* action—by the phosphorylation of substrates providing chemical activation to enable otherwise unfavourable reactions to proceed more readily, and by the phosphorylation of enzymes or other protein species, thereby causing confirmational changes which can result in the effect of transferring ions and metabolites across membranes.

An extended discussion of bioenergetics is not intended here and the reader is directed to the standard texts on bioenergetics and biochemistry for further discussion. However, there are essentially three methods of biosynthesis of ATP namely:

(1) oxidative phosphorylation, i.e. that associated with the re-oxidation of reduced nicotine and flavin adenine nucleotides, this sequence of reactions and the coupled phosphorylation reactions occurring in mitochondria;
(2) photosynthetic phosphorylation, i.e. that associated with the biosynthesis of carbohydrates in photosynthetic organisms in the presence of light and the photon-trapping pigments, this process with the coupled formation of ATP occurring in chloroplasts;
(3) substrate level phosphorylation i.e. phosphate transfer from another biologically active phosphorylating agent to ADP. Only three such cases seem to occur in higher organisms and are catalysed by the enzymes phosphoglycerate kinase, pyruvate kinase, and succinyl thiokinase. In bacteria there is a fourth case of substrate level phosphorylation in which the production of carbamyl phosphate, catalysed by carbamate kinase, can be used to give ATP from ADP by reversal of the process of carbamate synthesis. But the equilibrium for this reaction generally lies in favour of carbamate synthesis and the process only seems to be readily reversible in bacteria.

    The reactions for these four substrate-level phosphorylations are shown in Fig. 5.15.

Such reactions are discussed at greater length, and in a slightly different way, by Racker[22] and are quoted by Henderson and Paterson.[23]

Although ATP is the prime source of utilizable energy for biosynthetic reactions, the other pyrimidine and purine nucleoside triphosphates are also *high-energy* compounds and are also capable of acting as phosphorylating agents. For example, GTP (or ITP) is required for the conversion of oxaloacetate to phosphoenol pyruvate (PEP carboxykinase) in gluconeogenesis:

$$\begin{matrix} COCO_2H \\ | \\ CH_2CO_2H \end{matrix} + GTP \longrightarrow \begin{matrix} CH_2{=}C{-}CO_2H \\ | \\ O\text{Ⓟ} \end{matrix} + GDP + CO_2$$

GTP is also required for protein biosynthesis when it is involved in the translocation step of the ribosome (see Section 8C) and it is a factor

$$\begin{array}{l} CO_2Ⓟ \\ | \\ CHOH \\ | \\ CO_2HⓅ \end{array} \underset{ATP}{\overset{ADP}{\rightleftharpoons}} \begin{array}{l} CO_2H \\ | \\ CHOH \\ | \\ CH_2OHⓅ \end{array}$$

Phosphoglycerate kinase

$$\begin{array}{l} CO_2H \\ | \\ C{-}OⓅ \\ \| \\ CH_2O \end{array} \xrightarrow{ADP \quad ATP} \begin{array}{l} CO_2H \\ | \\ CO \\ | \\ CH_3 \end{array}$$

Pyruvate kinase

$$\begin{array}{l} CH_2CO_2H \\ | \\ CH_2COSCoA \end{array} \xrightarrow[GDP \quad GTP]{H_2O \quad CoASH} \begin{array}{l} CH_2CO_2H \\ | \\ CH_2CO_2H \end{array}$$

Succinyl thiokinase

$$NH_2COⓅ + H_2O \underset{ATP}{\overset{ADP}{\rightleftharpoons}} NH_3 + CO_2$$

Carbamate kinase

Fig. 5.15. Substrate level phosphorylations

required in the biosynthesis of adenylosuccinate (see Section 6B):

IMP + aspartate + GTP → adenylosuccinate + GDP + P_i

The ATP formed by photophosphorylation, oxidative phosphorylation, or substrate-level phosphorylation may be used to phosphorylate any of the other nucleotides in the free nucleotide pool of tissues and phosphate groups may be transferred by readily reversible transphorylation reactions:

$$NMP \rightleftharpoons NDP \rightleftharpoons NTP$$

These transfers involve only the exchange of ***high-energy*** phosphoryl groups and the phosphorylating ability of the nucleoside triphosphates is maintained. A number of transphosphorylating enzymes have been characterized and many of them have a wide distribution among different tissues and in different species. These enzymes may be nucleoside monophosphate kinases:

$$N^1TP + N^2MP \rightleftharpoons N^1DP + N^2DP$$

or nucleoside diphosphate kinases:

$$N^1TP + N^2DP \rightleftharpoons N^1DP + N^2TP.$$

The transfer of a pyrophosphoryl group is very unusual and does not occur between nucleotides, but ATP can act as a pyrophosphoryl transferring agent in some cases, e.g. thiamine + ATP → thiamine PP + AMP (see Section 5 A(*i*)).

(C) ACTIVATION BY DINUCLEOTIDE FORMATION AND ADENYL TRANSFER

In addition to activating potential substrates by phosphorylating them, nucleoside triphosphates can also activate compounds by transferring to them the nucleoside diphosphate moiety. Several such reactions are known and this type of activation is very common for the activation of monosaccharide units in polysaccharide biosynthesis. For example, the biosynthesis of sucrose involves the reactions shown below:

$$\text{glucose–1–P} + \text{UTP} \longrightarrow \text{UDPG (uridine diphosphoglucose)} + PP_i$$

$$\text{UDPG} + \text{fructose–6–P} \longrightarrow \text{sucrose–P} + \text{UDP}$$

The work of L. F. Leloir and his coworkers showed that a number of interconversions and biosyntheses in the area of carbohydrate biochemistry involved such glycosyl esters of aldoses and nucleoside di- (occasionally mono-) phosphates. The derivatives are formed by specific phosphorylases—nucleoside triphosphate: sugar-phosphate nucleotidyl transferases—and then are capable of undergoing further reactions.

Uracil is the predominant base in the nucleoside diphosphate sugars of higher plants, but other bases, e.g. cytosine, guanine, adenine, etc., are occasionally found, their occurrence being more common in lower plants and micro organisms.

Hassid[24] has reviewed the involvement of nucleoside diphosphate sugars in biosynthesis.

The formation of UDPG is shown in Fig. 5.16.

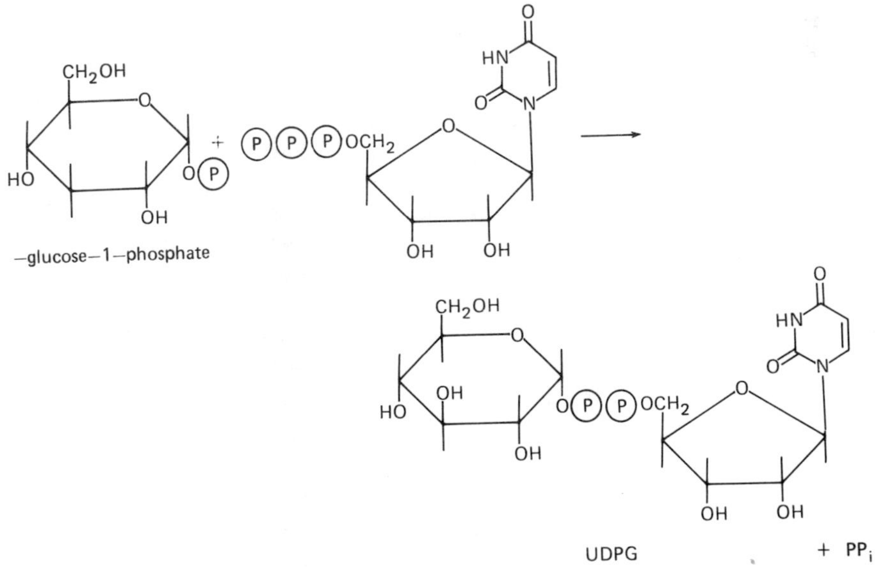

–glucose–1–phosphate

UDPG + PP_i

Fig. 5.16.

Table 5.3 Biosynthetic uses of nucleoside diphosphate sugars

(i) Epimerization
e.g. UDP-glucose ⇌ UDP-galactose
(ii) Glycoside synthesis
e.g. UDP-glucose + phenyl α-D-glucoside → phenyl β-gentiobioside + UDP
(iii) Phospholipid synthesis
CDP-ethanolamine + 1,2-diglyceride → phosphatidyl ethanolamine + CMP
(iv) Formation of sialic acids
e.g. UDP-*N*-acetylglucosamine ↗ glycoproteins ↘ mureins
(v) Polysaccharide synthesis
e.g. UDP-glucose → starch, glycogen

The types of reaction which these compounds undergo are listed in Table 5.3. further examples of such reactions are given by Mahler and Cordes[25] (and other biochemical texts) and in ref. 21.

In addition to monomer activation by nucleoside diphosphate formation enzyme-bound acyl adenylates are common intermediates in some acyl transfer reactions—for example, the formation of acetyl coenzyme A.

But such intermediates are also involved in the activation of fatty acids for glyceride synthesis and the activation of amino acids for protein synthesis:

$$RCO_2H + ATP \longrightarrow RCO{-}AMP\ (\text{enzyme bound}) + PP_i$$

$$RC(H)(NH_2)CO_2H + ATP \longrightarrow RC(H)(NH_2)CO{-}AMP\ (\text{enzyme-bound}) + PP_i$$

Further consideration of protein synthesis is given in Chapter 8.

Two further points of interest relating to adenosine nucleotides which may be conveniently mentioned here are the use of S-adenosylmethionine and cAMP.

The synthesis of S-adenosylmethionine is unusual, since it is a case in which there is cleavage of the C–O bond at the 5′-position of ATP with the release of trimetaphosphate which is cleaved to give phosphate and pyrophosphate:

$$ATP + \text{methionine} \longrightarrow \overset{+}{H_3N}CH(CO_2^-)CH_2CH_2\overset{+}{S}(CH_3)CH_2\text{-ribose(2',3'-OH)-adenine} + P_i + PP_i$$

S-Adenosylmethionine is a *high-energy* compound, but in its reactions acts as a methyl-group donor:

S-adenosylmethionine + substrate → S-adenosylhomocysteine + methylsubstrate.

Several such reactions are known, including the methylation of polynucleotides.

A chapter on nucleotide coenzymes and related compounds would not be complete without a mention of 3′,5′-cyclic AMP (cAMP) and the guanine analogue cGMP.

The cyclic nucleotides are not phosphorylating or activating species like those discussed above. However, cAMP is an activator in a different way, since it is known to influence a variety of cellular processes. It has been proposed to be an 'intracellular second messenger' which mediates the actions of a number of hormones.[26,27] Thus cAMP has the effect of activating a number of enzymes. The hormones whose actions may be mediated by cAMP include glucagon, vasopressin, serotonin, insulin, and the prostaglandins amongst others.[26,27] and the processes influenced by cAMP include lipolysis, gluconeogenesis, insulin release, release of protein from polysomes.

Concentrations of cAMP in the cell depend on two enzymes, adenyl cyclase and cAMP phosphodiesterase:

$$\text{ATP} \xrightarrow{\text{adenylcyclase}} \text{cAMP} + \text{PP}_i \xrightarrow{\text{phosphodiesterase}} \text{AMP}$$

Adenyl cyclase is membrane bound and seems to be the target for hormone action, although the precise reason for the hormonal sensitivity is unknown. A review of the control of the metabolism and the hormonal role of adenosine has recently been published.[28]

The roles of cGMP are at present not understood. A detailed account of this topic may be found in the series edited by Greenwood and Robison.[29]

This chapter has given a brief introduction to the roles of nucleotide derivatives of heterocyclic compounds in biochemical processes, but it is hoped that the reader has gained an insight to the importance of these compounds in living systems, and will consult detailed texts on the aspects covered in this chapter for more specialized knowledge.

REFERENCES

1. R. E. Breslow, *J. Amer. Chem. Soc.,* **79**, 1762 (1957).
2. R. E. Breslow, *J. Amer. Chem. Soc.*, **80**, 3719 (1958).
3. J. Ullrich and A. Mannschreck, *Biochim. Biophys. Acta.*, **115**, 46 (1966).
4. G. W. E. Plant, *J. Biol. Chem.,* **238**, 2225 (1963).
5. A. Bacher, B. Mailänder, R. Baur, V. Eggers, H. Harders and H. Schnepple in W. Pfleiderer (ed.), *Chemistry and Biology of Pteridines*, de Gruyter, Berlin and New York (1975)
6. E. L. Smith and L. F. J. Parker, *Biochem. J.*, **43**, 8 (1948).

7. E. L. Riches, N. G. Brink, F. R. Konisuszy, T. R. Wood, and K. Folkers, *Science,* **107**, 396 (1948).
8. D. C. Hodgkin, J. Kamper, J. Lindsey, M. MacKay, J. Pickworth, J. H. Robertson, C. B. Shoemaker, J. G. White, R. J. Prosen, and K. N. Trueblood, *Proc. Roy. Soc. (London)*, **A242**, 228 (1957).
9. E. L. Smith, S. Ball and D. M. Ireland, *Biochem. J.* **52**, 395 (1952).
10. H. A. Barker, H. Weissbach, and R. D. Smyth, *Proc. Natl. Acad. Sci*., **44**, 1093 (1958).
11. O. W. Wagner, H. A. Lee, P. A. Frey, and R. H. Abeles, *J. Biol. Chem*., **241**, 1751 (1966).
12. B. Zagalak, P. A. Frey, G. L. Karabatsos, and R. G. Abeles, *J. Biol. Chem*., **241**, 3028 (1966).
13. P. A. Frey, M. K. Essenberg, and R. H. Abeles, *J. Biol. Chem*., **242**, 5369 (1967).
14. J. D. Brodie, *Proc. Natl. Acad. Sci*., **62**, 461 (1969).
15. F. Lipmann, *Adv. Enzymol*., **1**, 99 (1941).
16. B. E. C. Banks, *Chemistry in Britain*, **5**, 514 (1969).
17. L. Pauling, *Chemistry in Britain*, **6**, 468 (1970).
18. D. Wilkie, *Chemistry in Britain*, **6**, 472 (1970).
19. A. F. Huxley, *Chemistry in Britain*, (1970).
20. R. A. Ross and C. A. Vernon, *Chemistry in Britain*, **6**, 539 (1970).
21. B. E. C. Banks and C. A. Vernon, *Chemistry in Britain*, **6**, 541, (1970).
22. E. Racker, *Mechanisms in Bioenergetics*, Academic Press, New York (1965).
23. J. F. Henderson and A. R. P. Paterson, *Nucleotide Metabolism*, Academic Press, New York and London (1973).
24. (a) W. Z. Hassid, *Science*, *165*, 137 (1965); (b) W. Z. Hassid, *Metab. Pathways*, **1**, 307, (1969).
25. H. R. Mahler and E. H. Cordes, *Biological Chemistry* (2nd edn.), Harper and Row, New York, Evanston, San Franscisco and London (1966).
26. G. A. Robison, R. W. Butcher and E. W. Sutherland, *Ann. Rev. Biochem*., **27**, 149 (1968).
27. G. A. Robison, R. W. Butcher and E. W. Sutherland, *Cyclic AMP*, Academic Press, New York (1971).
28. J. R. S. Arch and E. A. Newsholme in P. N. Campbell and W. N. Aldridge (eds.), *Essays in Biochemistry*, **14**, 82 (1978).
29. P. Greenwood and G. A. Robison (eds.), *Advances in Cyclic Nucleotide Research*, Raven Press, New York, Vols. 1–8 (1972–).

Chapter 6

The Biosynthesis and Metabolism of the Pyrimidine and Purine Nucleotides

(A) INTRODUCTION

Although some living species require the nucleotide coenzyme vitamins (or appropriate precursors) as essential dietary factors because they do not have the appropriate metabolic capability, the ability to synthesize the purine and pyrimidine ring systems *de novo* is an almost universal biochemical facility. This metabolic capability is necessary to provide the biologically active phosphates ATP, CTP, GTP, etc., as well as to provide these nucleoside triphosphates which are the precursors in nucleic acid biosynthesis.

The most common chemical synthesis of a purine nucleotide would start with a pyrimidine on which the imidazole portion of the molecule would be built, the resulting purine then being reacted with an appropriate carbohydrate derivative to give the nucleoside, which would be subsequently phosphorylated to give the nucleotide. However, the general biochemical pathway to such compounds starts with a phosphorylated carbohydrate, which is aminated to give an aminosugar on which the purine is built, the imidazole part of the molecule being formed before the pyrimidine ring. The purine nucleotide which is formed in this way and which provides the key for the synthesis of the other purine nucleotides is inosine monophophate (IMP).

The key intermediate in the formation of the pyrimidine nucleotides is orotidine monophosphate (OMP) from which the other pyrimidine nucleotides are derived. In pyrimidine nucleotide biosynthesis the pyrimidine ring is preformed before the glycosidic bond is formed. An account of nucleotide metabolism has been given by Henderson and Paterson.[1]

The key carbohydrate starting material in nucleotide synthesis is 5-phosphosibosyl-1-pyrophosphate (PRPP (1)) formed from ribose-5-

phosphate (which is itself formed from glucose by the pentose phosphate pathway) by an ATP-dependent pyrophosphorylation. This is an interesting reaction as it is one of the comparatively few cases in which ATP acts as a pyrophosphorylating agent.

ⓅOCH_2 (ribose, OH, OH, OH) + ATP ⇌ ⓅOCH_2 (ribose, OH, OH, OⓅⓅ) + AMP

(1)

(B) THE BIOSYNTHESIS OF THE PURINE NUCLEOTIDES

Henderson[2] has recently provided a review of the regulation of purine biosynthesis.

Isotopic studies carried out by Buchanan, Greenberg, and others have shown that the origin of the atoms in the heterocyclic ring of biosynthesized inosine monophosphate came from the sources indicated in (**2**).

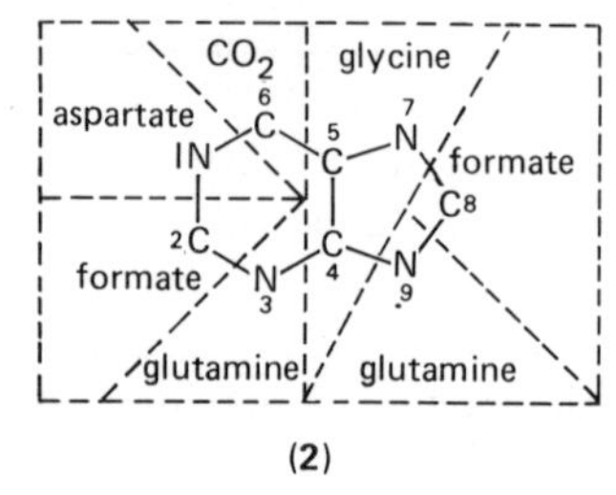

(2)

The reaction scheme leading to inosine monophosphate from 5-phosphoribosyl-1-pyrophosphate is given in Fig. 6.1.

The initial step in the biosynthesis of purine nucleotides is the glutamine-mediated amination of PRPP to give 5-phosphoribosylamine (PRA) the enzyme being phosphoribosyl pyrophosphate amido transferase. This amination involves inversion at C-l of the ribose unit to give the β-aminosugar, so the glycosidic link of the appropriate configuration is introduced at an early stage in the biosynthetic pathway.

The second step in the process is the ATP-dependent conjugation of glycine with the amino group of the sugar to give glycinamide ribonucleotide (GAR), which is then formylated in a step which involves one carbon-atom transfer utilizing a folic acid coenzyme (see Section 6E) the resulting product being *N*-formylglycinamide ribonucleotide (*N*-formyl GAR). A second ATP dependent glutamine mediated amination now takes place to give *N*-formylglycinamidine ribonucleotide (*N*-formyl GAM) which has the structural basis of an imidazole ring. Ring closure now follows via an ATP-dependent dehydration to give 5-aminoimidazole ribonucleotide (AIR) which is the first heterocyclic nucleotide to be formed in the biochemical sequence leading to the purine nucleotides.

The pyrimidine ring is now built on the imidazole nucleotide by a carboxylation with carbon dioxide in equilibrium with bicarbonate—a system which, in pigeon liver, apparently does not require a coenzyme such as biotin or the presence of ATP to give 5-aminoimidazole-4-carboxylic acid ribonucleotide (carboxy AIR), followed by a two-step amination to give the corresponding amide. First an ATP-dependent reaction with aspartate occurs to give 5-aminoimidazole-4-succinocarboxamide ribonucleotide (succino-AICAR), and then this compound subsequently undergoes elimination of fumarate to yield 5-aminoimidazole-4-carboxamide ribonucleotide (AICAR). The purine ring system is completed by another one carbon transfer involving a folate coenzyme to give *N*-formyl AICAR, followed by a dehydration-cyclization to give inosine monophosphate, the starting point for the biosynthesis of the other purine nucleotides.

The routes by which IMP is converted to the other purine nucleotides are indicated in Fig. 6.2.

The amination of IMP to give AMP is a two-step process with the intermediacy of a succino derivative and this is very similar to the process by which AICAR is formed from carboxy-AIR in IMP biosynthesis. It has been suggested that the enzyme catalysing the elimination of fumarate from AMPS (adenylosuccinase) also probably catalyses the second fumarate elimination.

The conversion of IMP to GMP also involves a two-step process but is rather different from the IMP to AMP conversion. IMP is first oxidized to xanthosine monophosphate (XMP) in a reaction which uses $NAD^{\oplus}$ as coenzyme and XMP is subsequently aminated by either ammonia or the amide NH_2 of glutamine in an ATP-dependent step to give GMP. The choice of ammonia or glutamine as amine group source seems to be species dependent.

It should be noted that the conversion of IMP to AMP is GTP-dependent whereas the IMP to GMP is ATP-dependent, so that a regulatory role seems evident: an excess of GTP will direct conversion of IMP to AMP whilst an excess of ATP will direct the synthesis of GMP.

In addition to the above *de novo* synthesis of purine nucleotides, routes involving preformed purines as precursors are also known. The most important reactions of this type are those catalysed by the nucleotide pyrphosphorylase enzymes first investigated by Kornberg[3] and reviewed by Murray.[4] Adenine phosphoribosyl transferase catalyses the reaction between adenine and PRPP to give AMP and hypoxanthine–guanine phosphoribosyltransferase converts hypoxanthine and guanine to IMP and GMP respectively in the presence of PRPP.

The interaction of purines with ribose 1-phosphate to give nucleotides has also been observed,[5] such reactions being catalysed by the nucleoside phosphorylases, and the phosphorylation of nucleosides by nucleoside transferase systems has been described[6] in a variety of species. This phosphorylation involves phosphate transfer from some organic phosphates other than the active phosphates to nucleosides to form 5′-nucleotides.

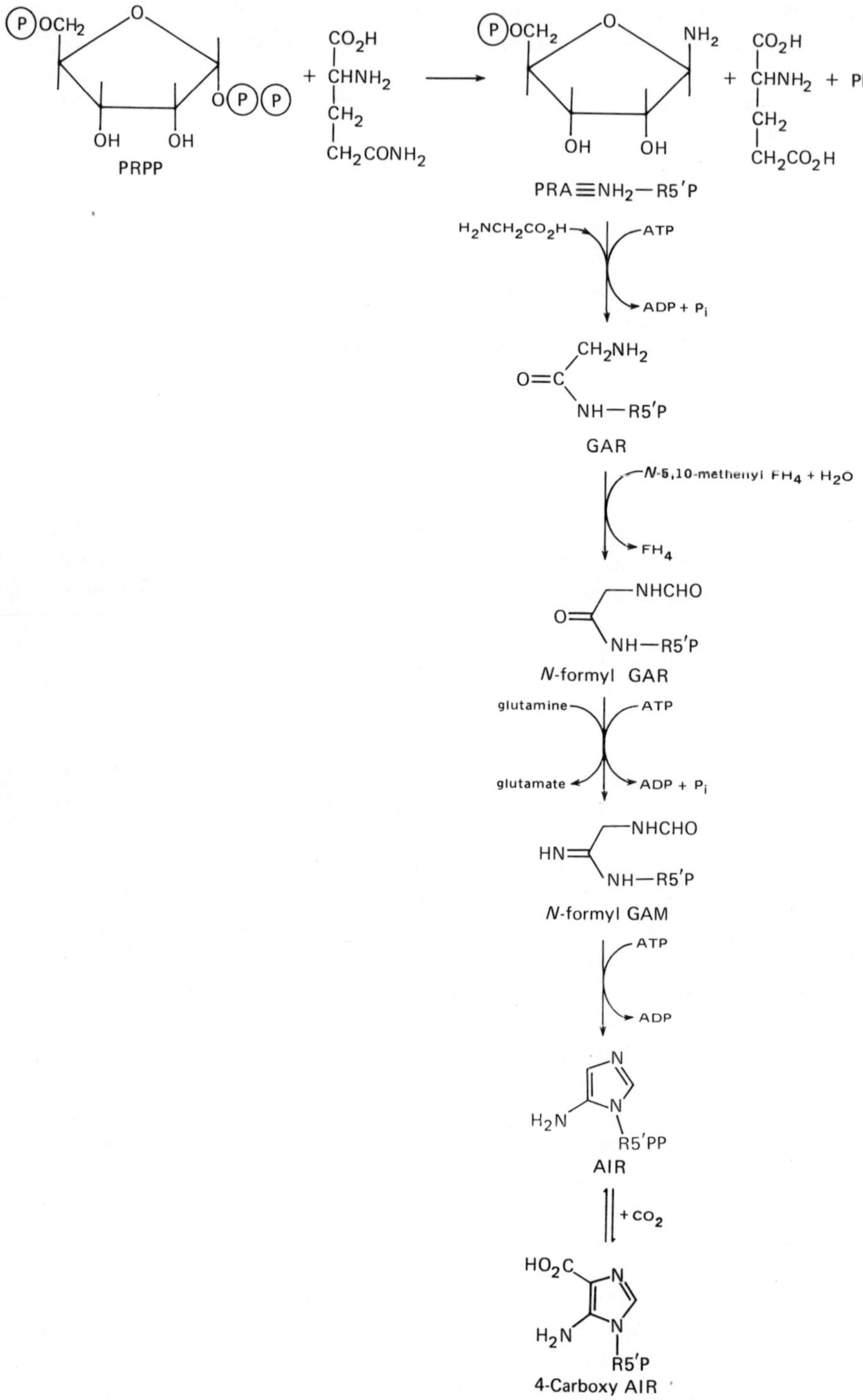

PRPP + CO_2H–$CHNH_2$–CH_2–CH_2CONH_2 ⟶ $PRA \equiv NH_2-R5'P$ + CO_2H–$CHNH_2$–CH_2–CH_2CO_2H + PP_i

$H_2NCH_2CO_2H$, ATP ⟶ ADP + P_i

$O{=}C(CH_2NH_2)NH-R5'P$

GAR

N-5,10-methenyl FH_4 + H_2O ⟶ FH_4

$O{=}C(CH_2NHCHO)NH-R5'P$

N-formyl GAR

glutamine, ATP ⟶ glutamate, ADP + P_i

$HN{=}C(CH_2NHCHO)NH-R5'P$

N-formyl GAM

ATP ⟶ ADP

H_2N, N, R5′PP

AIR

$+CO_2$

HO_2C, H_2N, N, R5′P

4-Carboxy AIR

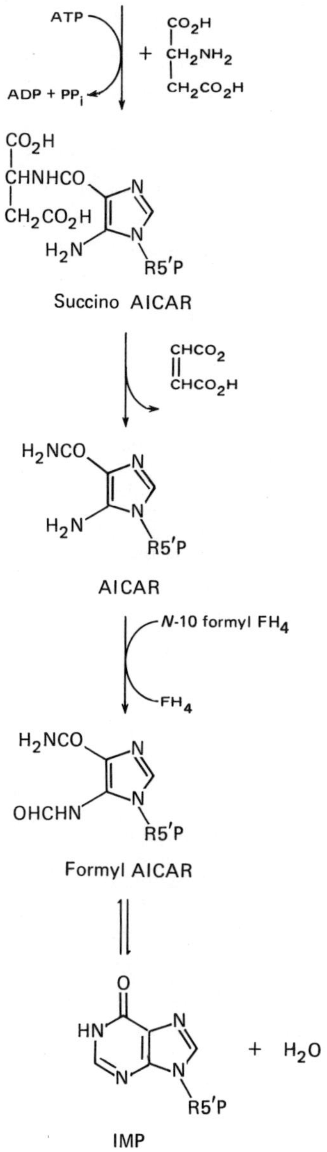
ATP
ADP + PP$_i$
CO_2H
CH_2NH_2
CH_2CO_2H
CO_2H
CHNHCO
CH_2CO_2H
H_2N
N
R5′P
Succino AICAR
$CHCO_2$
$CHCO_2H$
H_2NCO
AICAR
N-10 formyl FH_4
FH_4
OHCHN
Formyl AICAR
O
HN
IMP
+ H_2O

Fig. 6.1. Biosynthesis of IMP

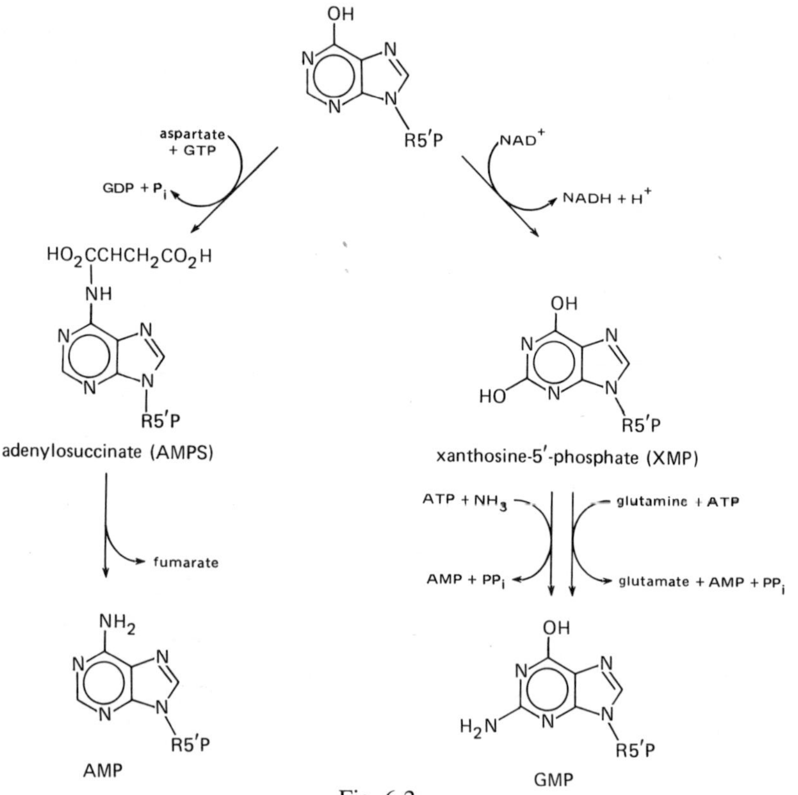

Fig. 6.2.

The interaction of some drugs with the above-mentioned enzyme systems is discussed in Section 9C(iii).

The purine nucleoside di- and triphosphates are formed from the monophosphates by two kinase reactions. The nucleotide kinases which catalyse these reactions are not specific for the nucleotide but do require adenine nucleotides and thus do not result in net synthesis of ATP which is provided by oxidative phosphorylation, photosynthetic phosphorylation, etc.

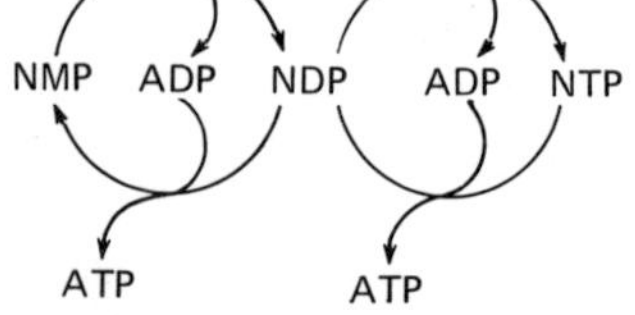

Nucleotide phosphorylation

The biosynthesis of the deoxyribonucleotides seems to occur at the nucleotide level by reduction of the corresponding nucleotide and no deoxy analogue of PRPP has been observed.

In bacterial systems two distinguishable processes have been observed which are species dependent.[7,8] One enzyme system (that from *Escherichia coli*) uses nucleoside diphosphates, whilst that isolated from *Lactobacillus leichmannii* uses the nucleoside triphosphates and requires a cobamide coenzyme. The mechanism which operates in animal systems has not been definitely established but it seems not to require a cobamide coenzyme.[9]

(C) THE BIOSYNTHESIS OF THE PYRIMIDINE NUCLEOTIDES

In the biosynthesis of pyrimidine nucleotides the pyrimidine ring is preformed before the formation of the glycosidic link which occurs when the pyrimidine is reacted with PRPP. The first pyrimidine nucleotide to be formed is orotidine monophosphate (OMP) which is subsequently decarboxylated to give UMP. The steps leading to the production of UMP are given in Fig. 6.3.

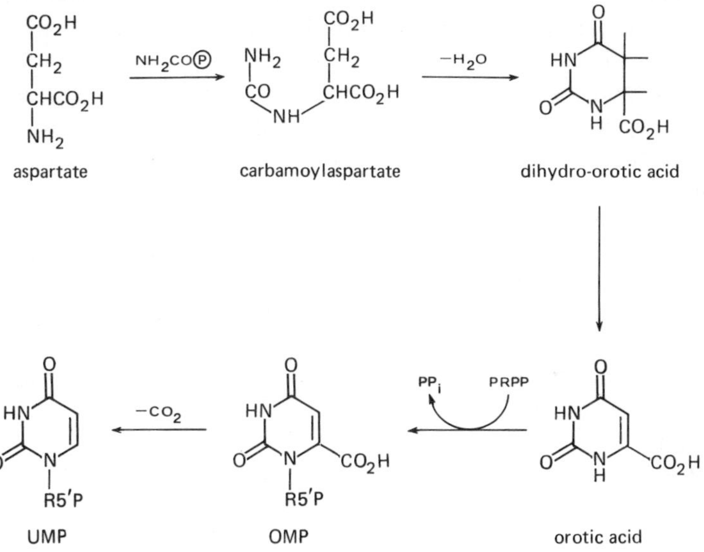

aspartate $\xrightarrow{NH_2CO\text{Ⓟ}}$ carbamoylaspartate $\xrightarrow{-H_2O}$ dihydro-orotic acid $\rightarrow$ orotic acid $\xrightarrow{\text{PRPP} \rightarrow \text{PP}_i}$ OMP $\xrightarrow{-CO_2}$ UMP

Fig. 6.3.

One of the starting materials for pyrimidine biosynthesis is carbamoylphosphate (NH_2COⓅ) which is also involved in urea biosynthesis in a step in which ornithine (**3**) is converted to citruline (**4**).

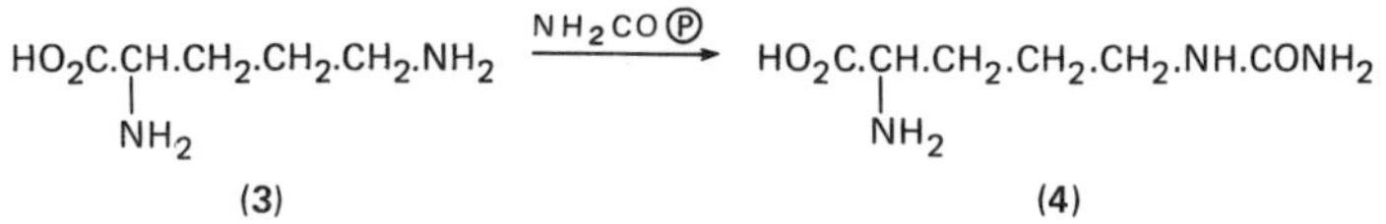

$$\underset{(3)}{HO_2C.CH(NH_2).CH_2.CH_2.CH_2.NH_2} \xrightarrow{NH_2CO\text{Ⓟ}} \underset{(4)}{HO_2C.CH(NH_2).CH_2.CH_2.CH_2.NH.CONH_2}$$

However two different systems are used for carbamoylphosphate synthesis in different cellular locations, that required for pyrimidine biosynthesis being cytoplasmic and using glutamine as the source of the amino group.

$$\text{glutamine} + CO_2 + 2ATP \rightleftharpoons NH_2CO_2\text{Ⓟ} + 2ADP + P_i + \text{glutamate}$$

Carbamoylphosphate and asparate combine under the catalytic influence of aspartate carbamoyl transferase to form carbamoyl aspartate. This enzyme has an interesting structure and is an important regulatory enzyme for controlling pyrimidine biosynthesis.

Carbamylaspartate is converted to the reduced pyrimidine dihydro-orotic acid (dihydro-orotase) which is subsequently oxidized to orotic acid by dihydro-orotic dehydrogenase, an $NAD^{\oplus}$-linked enzyme. After formation of the glycosidic link in a pyrophosphorylase reaction an irreversible decarboxylation occurs to give uridine monophosphate.

The scheme for the biosynthesis of the other pyrimidine nucleotides is given in Fig. 6.4.

The different nucleoside phosphates are readily phosphorylated or dephosphorylated by a group of kinases, but alterations to the pyrimidine ring and the reduction of the ribose moiety to the deoxyribose moiety occur at specific levels of phosphorylation and are subject to careful metabolic control.

The conversion of uracil to cytosine takes place at the nucleoside triphosphate level in an ATP-dependent amination using ammonia (CTP synthetase), this being the only known pathway for the cytidine nucleotides.[10]

The major route to the deoxyribonucleotides is via the ribonucleotide reductase pathway, but an interesting step which is a supplementary process having a control function in the synthesis of dUMP is the deamination of dCMP to dUMP. This reaction is catalysed by the enzyme dCMP deaminase which is activated by dCTP and inhibited by TTP. The direct hydrolysis of dUDP to dUMP seems to be of little importance, the major

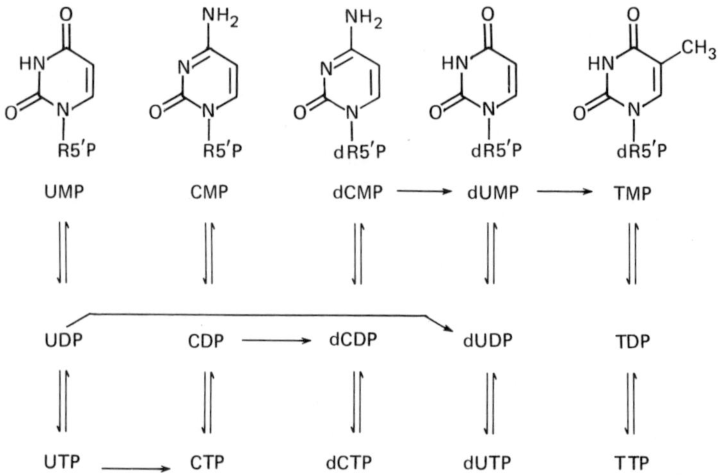

Fig. 6.4.

route to this compound being:

$$\text{UDP} \rightarrow \text{dUDP} \rightarrow \text{dUTP} \rightarrow \text{dUMP}$$

Escherichia coli seems to lack dCMP deaminase and the above route may be the only route to dUMP in this species.[10]

The reduction of the nucleoside diphosphates and the deamination of dCMP require nucleoside triphosphates specific for the particular enzyme—e.g. the reduction of CDP to dCDP requires ATP—and it seems that these triphosphates act as allosteric effectors rather than as phosphorylating agents. Thus the synthesis of nucleotides is controlled by the nucleotide pool already available.

In the deoxyribonucleotide series a pyrimidine which does not normally appear as a ribonucleotide is observed. Thymidine monophosphate (TMP) is biosynthesized from dUMP by a folate-mediated methylation in which *N*-5,10-methylene FH_4 acts both as a one carbon unit carrier and as a reducing agent (Fig. 6.5) (see Section 6E).

In some species of bacteriophage the nucleotide 5-hydroxymethyl dCMP occurs, this being formed from dCMP by an hydroxymethylation involving *N*-5,10-methylene FH_4, but in this case the folic acid coenzyme is not reduced to FH_2.[11]

The biosynthesis of thymidine[12] and its incorporation into DNA,[13] for which it is a specific precursor, has been extensively studied and further reference is made in Section 6E.

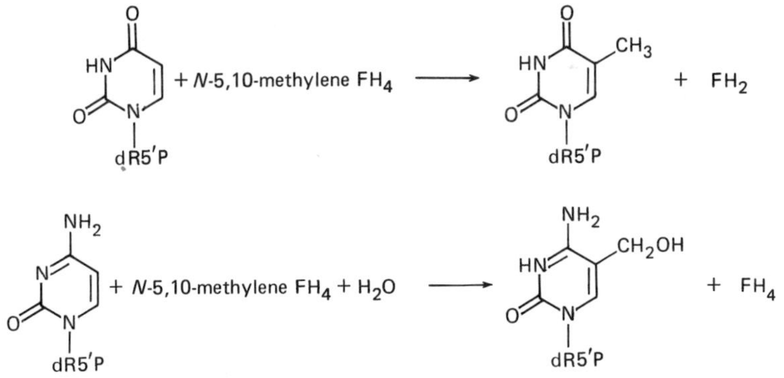

dUMP (dR5′P) + *N*-5,10-methylene FH_4 ⟶ TMP (dR5′P, CH_3) + FH_2

dCMP (dR5′P) + *N*-5,10-methylene FH_4 + H_2O ⟶ 5-hydroxymethyl dCMP (dR5′P, CH_2OH) + FH_4

Fig. 6.5.

(D) THE CATABOLISM OF NUCLEOTIDES

The catabolism of the nucleosides and nucleotides usually involves dephosphorylation of the nucleotide to the nucleoside, then phosphorolytic cleavage of the nucleoside to the pyrimidine or purine base and ribose-1-phosphate (or deoxyribose-1-phosphate).

(i) Purines

The degradation of the purine nucleotides and nucleosides is indicated in Fig. 6.6.

The major route in purine nucleotide and nucleoside catabolism involves deamination, phosphate, and sugar loss to give xanthine, which is oxidized by the enzyme xanthine oxidase to give uric acid. Further metabolism of uric acid is species dependent, but in man the principal purine metabolite is uric acid.

Xanthine oxidase also catalyses the oxidation of hypoxanthine to xanthine and is capable of oxidizing a wide variety of both purines and pteridines to oxo-derivatives. The enzyme has been extensively studied since it has a wide distribution in living tissue, and it has the potential of metabolizing a number of purine and purine-related drugs. Inhibitors of

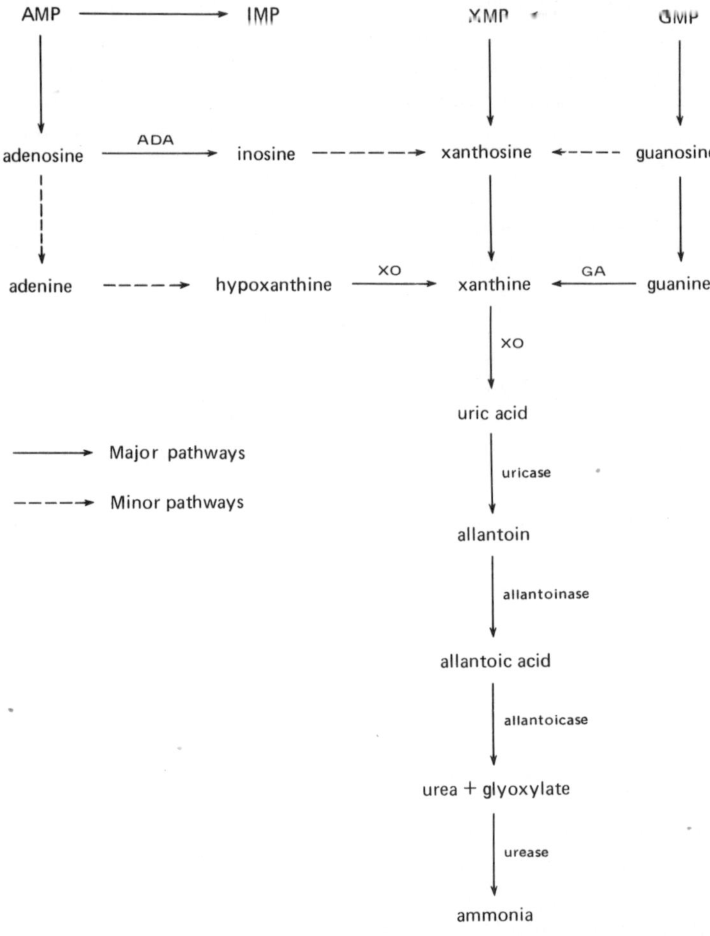

Fig. 6.6. Catabolism of purines

xanthine oxidase have also found use in the treatment of purine metabolism related disorders (see Section 6G). Bergmann and coworkers and other workers have worked extensively on xanthine oxidase, a selection of recent references being given at the end of the chapter.[14–17]

The deamination of AMP to IMP and adenosine to inosine is catalysed by adenylate deaminase and adenosine deaminase respectively, both of these enzymes having a very much greater specificity than that shown by xanthine oxidase. Adenosine deaminase has also been widely studied (e.g. refs. 18–21), but adenylate deaminase has received much less attention (e.g. see refs. 22 and 23 and references therein). The level of serum adenosine deaminase seems to be related to the condition known as severe

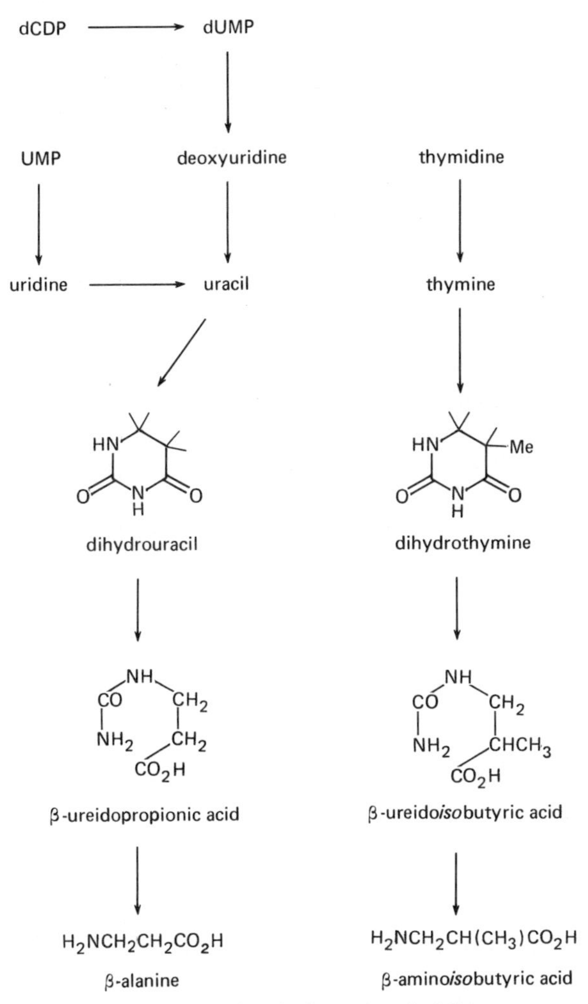

Fig. 6.7. Catabolism of pyrimidines

combined immunodeficiency[24,25] and may be concerned in chronic lymphocytic leukaemia[26] (see Section 6G(*iii*)).

The deamination of guanine to xanthine is catalysed by guanine deaminase (guanase), another enzyme which has been quite extensively studied since it has a wide distribution in animal tissue and also has the potential of metabolizing some purine drugs. Some recent references on the specificity of guanine deaminase are given at the end of the chapter.[27–29]

(ii) Pyrimidines

Like that of the purine nucleotides, the catabolism of the pyrimidine nucleotides also involves dephosphorylation, deamination and cleavage of the glycosidic bond. The general scheme for the catabolism of the pyrimidines is given in Fig. 6.7.

Deoxycytidine is deaminated to deoxyuridine at the nucleoside diphosphate level and in higher animals the only two paths of pyrimidine catabolism are the two parallel degradations of uracil and thymine. However, some bacteria can oxidize uracil to barbituric acid, then cleave this to malonic acid and urea, but this is not a generally available route. The oxidation of thymine is similar.

In contrast to the purines which are catabolized by oxidative processes, the catabolism of pyrimidines occurs via reductive processes to give the β-amino acids, which are subsequently oxidatively deaminated and further metabolized. Virtually all of the degradation products of pyrimidine nucleotides are metabolized this way and only trace amounts of pyrimidines are excreted. The major pyrimidine excreted is orotic acid and orotidine and in cases of orotic aciduria the excretion of orotate (but not orotidine) may be increased 20 times.[30] It is reported that following irradiation by X-rays the urinary excretion of deoxycytidine and thymidine is markedly increased.[31]

(E) THE ROLE OF FOLIC ACID IN ONE-CARBON UNIT TRANSFER

Folic acid is a pteridine derivative which has the structure (**5**) and which is usually numbered as shown.

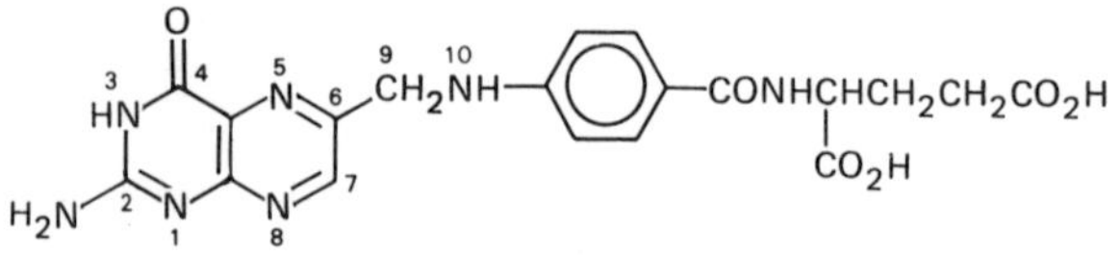

(**5**)

It is biosynthesized and is found in green plants and microorganisms, but it was first discovered when some lactic acid bacteria (*Streptococcus lactis*) were found to require an essential growth factor isolated from green leaves,[32] and other sources, which was ultimately shown to be folic acid.[33]

Folic acid is required by all higher animals but some species obtain sufficient quantities from their intestinal flora. Deficiency of this vitamin produces macrocytic anaemia. However, it is rarely seen due to the provision of folic acid by the intestinal flora.

The biosynthesis of folic acid probably occurs[34] via the route shown in Fig. 6.8 which produces 7,8-dihydrofolic acid (for which the abbreviation FH_2 is used) but it is the 5,6,7,8-tetrahydrofolic acid (FH_4) which is the biochemically active form of the vitamin. Folic acid can be reduced to FH_2 and further to FH_4 by the enzyme dihydrofolate reductase, the reducing species being NADPH.

The important role of FH_4 is that it is concerned with the transfer of one-carbon units—in particular the C-2 and C-8 carbons of purines in purine biosynthesis and in the 5-methyl group in thymidine or the 5-hydroxymethyl group in 5-hydroxymethyluridine.

There are several ways in which FH_4 is activated and the subsequent folate coenzymes are capable of transferring one-carbon fragments at the oxidation levels of methanol, formaldehyde and formate. All of these reactions involve the initial attachment of the one-carbon fragment to either one or both of the *N* atoms at the 5- and 10-sites of the folic acid moiety. These processes and the interconversions of the folate coenzymes are shown in Fig. 6.9.

N-5,10-Methenyl FH_4 provides C-8 of the purine ring system in the purine nucleotides and *N*-10-formyl FH_4 provides C-2 of the purine ring system.

The biosynthesis of thymidine monophosphate (Section 6C) involves

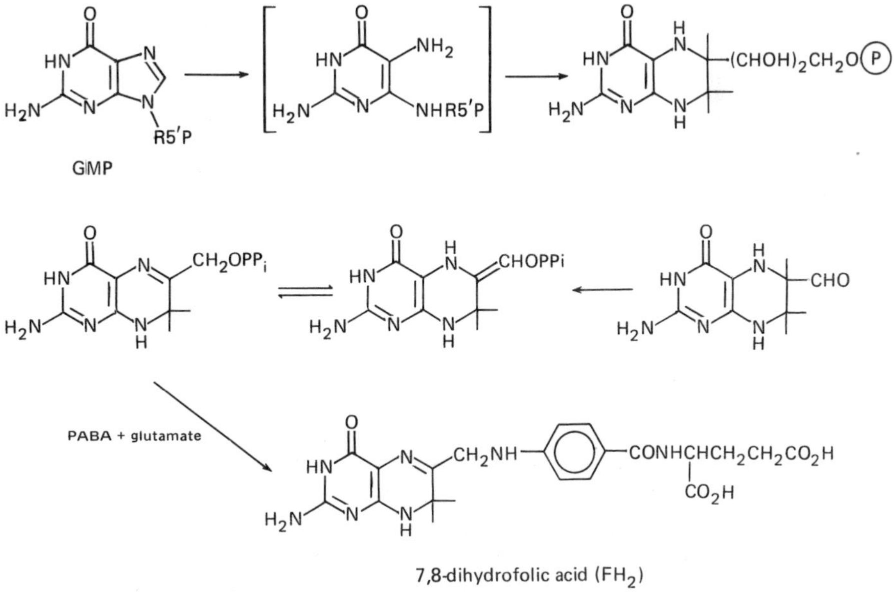

7,8-dihydrofolic acid (FH_2)

Fig. 6.8. The biosynthesis of folic acid

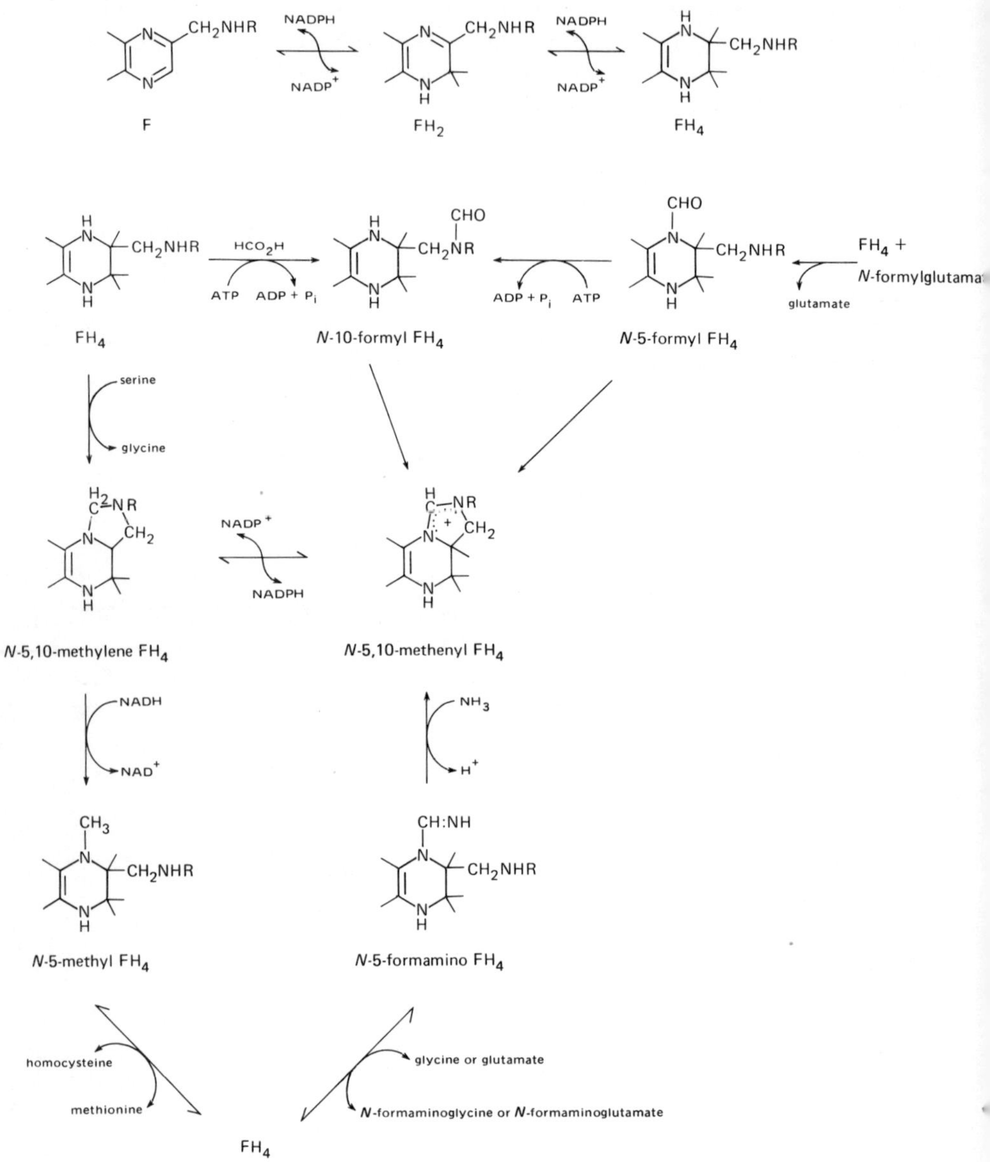

Fig. 6.9. Metabolic interconversions of tetrahydrofolate coenzymes

N-5,10-methylene FH_4 which is converted to FH_2 during the process, the FH_2 being reduced back to FH_4 by DHFR.

The deficiency disease, in man, associated with folic acid is pernicious anaemia, and folic acid has also been shown to be an anti-anaemia factor for chickens and monkeys. This activity may be associated with the role of folic acid as a hydroxymethyl acceptor in the reaction in which serine is converted to glycine. Glycine is necessary for porphyrin, and hence haem, biosynthesis. The presence of folic acid in green leaves may indicate that it has a similar role in mediating the synthesis of chlorophyll, which is also a porphyrin-based molecule.

Folic acid would also seem to play a crucial role in the biosynthesis of methionine from homocysteine[35,36] and may also be concerned with serine biosynthesis from glycine by reversal of the serine transhydroxymethylase reaction which produces *N*-5,10-methylene FH_4.

(F) THE SULPHONAMIDES

The development of the sulphonamides is one of the classic stories in the development of chemotherapy since it was scientifically studied by Ehrlich at the beginning of this century. Although Ehrlich and his coworkers had made substantial advances in the treatment of protozoal infections, they had little success in the treatment of bacterial infections and it was not until 1935 that Domagk reported that the dye *prontosil* (**6**) was effective in combating bacterial infections in animals.

Tréfonël then showed that prontosil was metabolized in vivo giving sulphanilamide (**7**) which was actually the effective agent, whilst Woods (1940) demonstrated that the inhibition of bacterial growth by sulphanilamide was overcome by the presence of excess *p*-aminobenzoic acid. Later it was discovered that some bacteria had a requirement for *p*-aminobenzoic acid as a growth factor, and that *p*-aminobenzoic acid was a component part of folic acid (1945). With further work the role of sulphanilamide and related compounds in the inhibition of the synthesis of bacterial folic acid was realized.

The sulphonamides, having the structure (**8**), thus have one of the prime requirements necessary for chemotherapeutic action. By interfering with a biochemical pathway particular to the invading organism and not the host, they have a specificity of action upon the infecting organism.

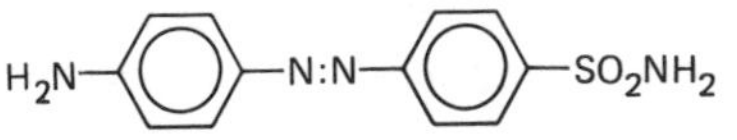

H_2N–C_6H_4–N:N–C_6H_4–SO_2NH_2 (6)

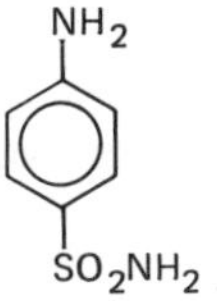

NH_2–C_6H_4–SO_2NH_2 (7)

NH_2–C_6H_4–SO_2NHR (8)

The biosynthesis of folic acid is shown in Fig. 6.6. The final stage in which the dihydropteridine pyrophosphate derivative is condensed with *p*-aminobenzoic acid and glutamate to give FH_4 occurs in two steps. The *p*-aminobenzoic acid is first condensed (condensing enzyme A) to give dihydropteroic acid (**9**) which further reacts with glutamate to give FH_2.

Sulphanilamide and other sulphonamides act as competitive inhibitors of

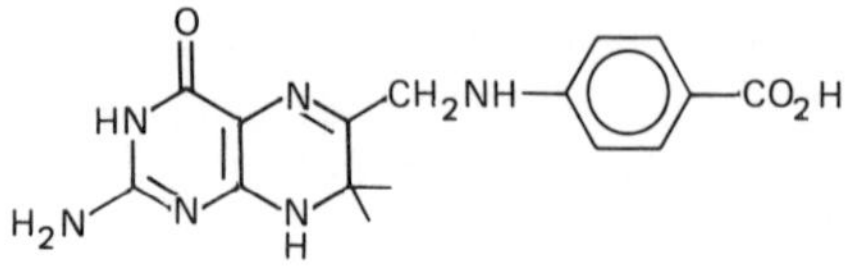

(9)

condensing enzyme A, thus inhibiting folate synthesis and bacterial nucleic acid synthesis. Sulphanilic acid also acts as an inhibitor of the enzyme, but has no antibacterial action because the charged ionic form cannot pass through the bacterial cell wall.

It was at one time thought that the mode of action of the sulphonamides was solely that of a competitive inhibitor of the condensing enzyme A system, since the inhibition was antagonized by the addition of *p*-aminobenzoic acid or *p*-aminobenzoyl glutamic acid[37] and the inhibition of sulphonamides was bacteriostatic rather than bactericidal.[38,39] However, further work[40] showed that ^{35}S-labelled sulphonamides were also incorporated into folate-type compounds in a similar manner to *p*-aminobenzoic acid. The products of such a condensation cannot then, of course, add on the glutamic acid moiety. This type of action, in addition to the competitive inhibition, improves the effectivity of the sulphonamides since the condensation reaction is essentially irreversible.

A further point concerning the mode of action of sulphonamides is that unlike many other antibacterial agents, e.g. the penicillins, their effect does not occur immediately but there is a lag phase of about one to two hours after administration of the drug. This has been shown[41] to be about 3.7 generation times. The same lag time was obtained for different sulphonamides and was interpreted as being the result of the depletion of stored folate.

A considerable literature has now accumulated on the sulphonamides (e.g. refs 42,43) and the reader is directed to such specialist reviews for more detailed information. Many thousand sulphonamides have been synthesized and tested. However, comparatively few are now in general use. Like the first of these to come into general use—sulphapyridine (M and B 693)—most of the most useful compounds are heterocyclic derivatives, some of the more frequently prescribed compounds being listed in Table 6.1.

(G) PATHOLOGICAL ABNORMALITIES OF PURINE METABOLISM

A number of human diseases have been found to be associated with abnormalities of purine metabolism due to deficiencies in the purine-metabolizing enzymes. A brief outline of the biochemical bases of these conditions is given below:

Table 6.1 Some heterocyclic sulphonamides

$H_2N-C_6H_4-SO_2NHR$

R 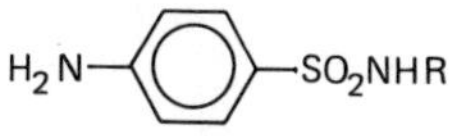	Compound
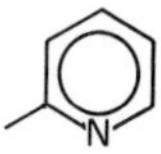 2-pyridyl	Sulphapyridine
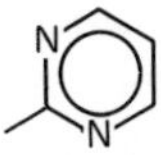 2-pyrimidinyl	Sulphadiazine
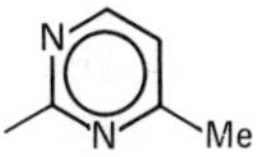 4-methyl-2-pyrimidinyl (N, N, Me)	Sulphamerazine
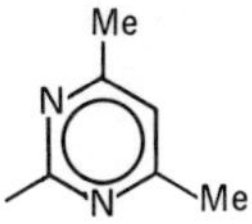 4,6-dimethyl-2-pyrimidinyl (Me, N, N, Me)	Sulphamethazine
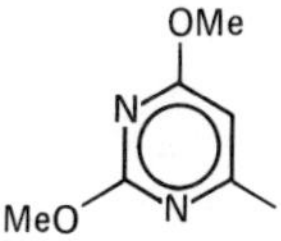 2,6-dimethoxy-4-pyrimidinyl (OMe, N, MeO, N)	Sulphadimethoxine
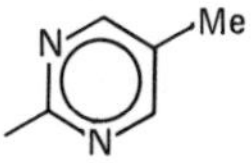 5-methyl-2-pyrimidinyl (N, Me, N)	Sulphamethyldiazine
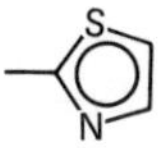 2-thiazolyl (S, N)	Sulphathiazole
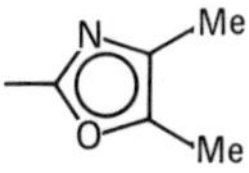 4,5-dimethyl-2-oxazolyl (N, Me, O, Me)	Sulphadimethoxazole
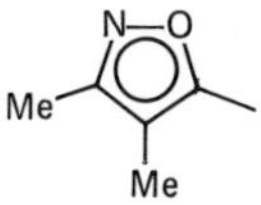 3,4-dimethyl-5-isoxazolyl (N—O, Me, Me)	Sulphisoxazole

(i) Gout

Probably the most common and the best known of the abnormalities of purine metabolism is gout. Gout is really a group of diseases which have in common the precipitation of uric acid (**10**) from the blood or other body fluids. Deposition of uric acid in the joints is common, which may subsequently lead to arthritis, but it also precipitates in the kidneys or other tissues. Reviews of gout[44,45] may be consulted for further details, whilst Henderson also gives some further details. (ref. 2, p. 257).

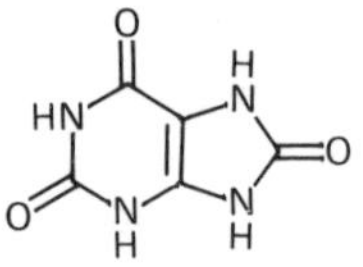

(10) 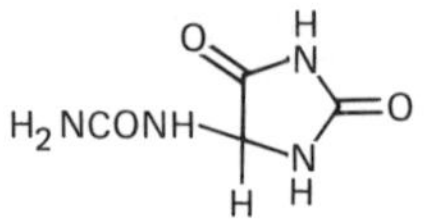 (11) 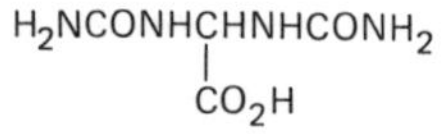 (12)

The catabolism of the nucleic acids, nucleosides and nucleotides is considered in Section 6D. The terminal product of purine metabolism in man is uric acid, which is excreted principally in the urine. However, small quantities of other purines are also excreted (Table 6.2). Most other mammals catabolize uric acid further to give allantoin (**11**) allantoic acid (**12**) etc., but man and the higher primates seem to lack the enzyme uricase. The only other mammal known to excrete uric acid as the end product of purine metabolism is the dalmatian coach hound, although this species has normal liver uricase activity. This phenomenon seems to be due to a deficient reabsorption process in the proximal tubule of the kidney after the plasma uric acid has been completely filtered at the glomerulus.

A wide variety of causes of hyperuricaemia has been observed including accelerated *de novo* purine biosynthesis due to increased PRPP availability in glucose-6-phosphatase deficiency, partial deficiency of hypoxanthine–guanine phosphoribosyl transferase, decreased end-product inhibition of purine biosynthesis and abnormal glutamine metabolism. However, in the majority of gouty patients the cause of the excess uric acid production is still unknown.

Allopurinol (4-hydroxypyrazolo[3,4-*d*]pyrimidine) (**13**) is a potent inhibitor of xanthine oxidase and is used clinically in the treatment of gout

Table 6.2 Purine excretion (urinary) in man[46]

Compound	Daily excretion (mg/day)
Hypoxanthine	9.7
Xanthine	6.1
7-Methylguanine	6.5
8-Hydroxy-7-methylguanine	3.6
Adenine	1.4
6-Succinoaminopurine	1.0
1-methylguanine	0.6
N^2-Methylguanine	0.5
1-Methylhypoxanthine	0.4
Guanine	0.4
Allantoin	20
Uric acid (urine)	400
Uric acid (saliva)	40
Uric acid (gastric juice)	8
Uric acid (bile)	4

and in other conditions of rapid purine catabolism. But in addition to inhibiting purine catabolism, allopurinol also inhibits *de novo* purine biosynthesis, although the mechanism of inhibition of purine biosynthesis remains in doubt (see ref. 2, p. 238).

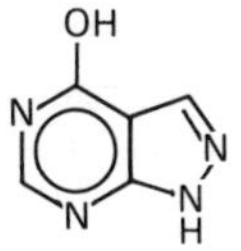

OH

(13)

(ii) Lesch–Nyhan syndrome

The Lesch–Nyhan syndrome was first recognized[47] as a disease in 1964, although a few cases had been described previously, and is characterized by a variety of effects including developmental retardation, aggressive behaviour and self-mutilation. Biochemically the Lesch–Nyhan syndrome is characterized by a very high urinary excretion of uric acid—up to six times the normal amount—and an accelerated rate of *de novo* purine biosynthesis of up to twenty times the usual rate. There is also a greatly increased (tenfold) urinary excretion of aminoimidazole carboxamide ribotide (AICAR).

The Lesch–Nyhan syndrome only occurs in males and is inherited as an X-linked recessive trait, the principal cause of the disease being a deficiency of hypoxanthine–guanine phosphoribosyl transferase activity. Seegmiller *et al.*[48] showed that there was no detectable activity of this enzyme in the erythrocytes of patients with the Lesch–Nyhan syndrome, this being subsequently confirmed by many other workers, and assay of this enzyme activity is now used for diagnostic purposes.

Attempts have been made to treat patients suffering from this disease with 6-methylmercaptopurine riboside (**14**), a drug that inhibits *de novo* purine biosynthesis, and with azathioprine (1′-methyl-4′-nitro-5′-imidazolyl-6-mercaptopurine) (**15**). However, these attempts have been unsuccessful.

Reviews of the Lesch–Nyhan syndrome have been written by Henderson[49] (see also ref. 2, p. 252 and ref. 50).

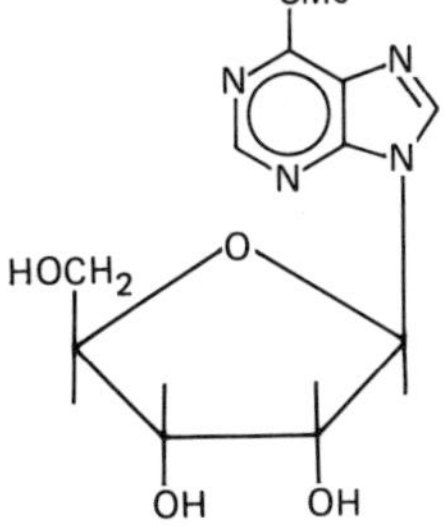

(14) 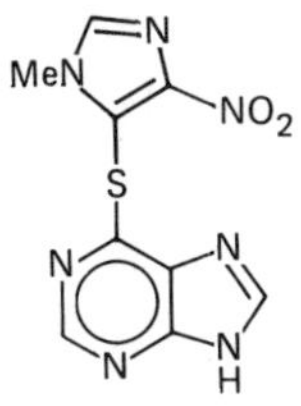 (15)

(iii) Other Diseases

Several other cases of neurological abnormalities have been described which also have been associated with increased *de novo* purine biosynthesis. These include a girl who exhibited the characteristics of the Lesch–Nyhan syndrome although the usual cause of this condition, hypoxanthine–guanine phosphoribosyl transferase deficiency, is an X-linked inherited trait associated only with boys. However this, and some other cases of malfunctions of purine metabolism, currently remain clinical curiosities and information concerning these cases is too sparse to merit further consideration here. As clinical studies proceed no doubt further information and new interesting cases will emerge.

Recent work has shown that a serious inherited disorder characterized by a virtually complete lack of B and T lymphocytes known as *severe combined immunodeficiency* is associated with very low levels of adenosine deaminase activity in the erythrocytes of the patients.[24,25,51] A similar but not so severe deficiency in the activity of this enzyme has been found in the lymphocytes of patients suffering from chronic lymphocytic leukaemia.[26]

The interpretation of the results of the studies into these conditions is still not complete. However, in the case of one variant of SCID the results have been interpreted in terms of the genetically controlled production of an adenosine deaminase inhibitor.[52] Thus there are several interesting conditions resulting from malfunctions of purine metabolism and no doubt in the future other cases will be discovered.

REFERENCES

1. J. F. Henderson and A. R. P. Paterson, *Nucleotide Metabolism, an Introduction*, Academic Press, New York and London (1973).
2. J. F. Henderson, *Regulation of Purine Biosynthesis*, American Chemical Society Monograph No. 170 (1972).
3. A. Kornberg, I. Lieberman, and E. S. Sims, *J. Amer. Chem. Soc.*, **76**, 2027 (1954).
4. A. W. Murray, *Ann. Rev. Biochem.*, **40**, 811 (1971).
5. H. Kalckar, *Symp. Soc. Exp. Biol.*, **1**, 38 (1947).
6. G. Brawerman and E. Chargaff, *Biochim. Biophys. Acta.*, **15**, 549 (1954).
7. P. Reichard, *European J. Biochem.*, **3**, 259 (1968).
8. A. Larsson and P. Reichard, in J. N. Davidson and W. E. Cohn, (eds.) *Progress in Nucleic Acid Research and Molecular Biology*, Vol. 7, Academic Press, New York (1967) p. 303.
9. H. Gershman, S. Marcia, and R. H. Abeles, *Biochim. Biophys. Acta.*, **246**, 169 (1971).
10. J. F. Koerner, *Ann. Rev. Biochem.*, **39**, 291 (1970).
11. J. G. Flaks and S. S. Cohen, *Biochim. Biophys Acta.*, **25**, 667 (1957).
12. M. Friedkin, *Ann. Rev. Biochem.*, **32**, 185 (1963).
13. *Davidson's Biochemistry of the Nucleic Acids*, (8th edn.), revised by R. L. P. Adams, R. H. Burdon, A. M. Campbell, and R. M. S. Smellie, Chapman and Hall, London (1976).

14. F. Bergmann, L. Levene, I. Tamir, and M. Rahat, *Biochim. Biophys. Acta.*, **480**, 21 (1977).
15. F. Bergmann, L. Levene, H. Govrin, and A. Frank, *Biochim. Biophys. Acta.*, **480**, 39 (1977).
16. L. I. Hart, M. A. McGartoll, H. R. Chapman, and R. C. Bray, *Biochem. J.*, **116**, 851 (1970).
17. B. R. Baker and W. F. Wood, *J. Medicin. Chem.*, **11**, 644, 661 (1968).
18. H. P. Baer, G. I. Drummond, and J. Gillis, *Arch. Biochem. Biophys.,* **123**, 172 (1968).
19. L. N. Simon, R. J. Bauer, R. L. Tolman, and R. K. Robins, *Biochemistry*, **9**, 573 (1970).
20. R. P. Agarwal, S. M. Sagar, and R. E. Parks, *Biochem. Pharmacol.*, **24**, 693 (1975).
21. C. L. Zielke and C. H. Suelter in P. D. Boyer (ed.) *The Enzymes*, Vol. 4 (3rd edn.), Academic Press, New York (1971).
22. C. Y. Lian and D. R. Harkness, *Biochim. Biophys. Acta*, **341**, 27 (1974).
23. C. J. Coffee and C. Solano, *J. Biol. Chem.*, **252**, 1606 (1977).
24. H. J. Mennissen, R. J. Pickering, B. Pollara, and I. A. Porter (eds.), *Combined Immunodeficiency Disease and Adenosine Deaminase Deficiency, A Molecular Defect*. Academic Press, New York (1975).
25. R. P. Agarwal, G. W. Crabtree, R. E. Parks, J. A. Nelson, R. Keightley, R. Parkman, F. S. Rosen, R. C. Stern, and S. H. Polmar, *J. Clin. Invest.*, **57**, 1025 (1976).
26. R. Tung, R. Silber, F. Quagliata, M. Conklyn, J. Gottesman, and R. Hirschhorn, *J. Clin. Invest.*, **57**, 756 (1976).
27. B. R. Baker, *J. Medicin. Chem.*, **10**, 59 (1967).
28. B. R. Baker and H. U. Siebenick, *J. Medicin. Chem.*, **14**, 802 (1971).
29. C. Silipo and C. Hansch, *Mol. Pharmacol.*, **10**, 954 (1974).
30. M. Latz, H. J. Fallon, and L. H. Smith, *Nature*, **197**, 194 (1963).
31. C. D. Gun and L. J. Cole, *Clin. Chem.*, **14**, 383 (1968).
32. H. K. Mitchell, E. E. Snell, and R. J. Williams, *J. Amer. Chem. Soc.*, **63**, 2284 (1941).
33. R. B. Angier, J. H. Boothe, B. L. Hutchings, J. H. Mowat, J. Semb, E. L. R. Stokstad, Y. Subba Row, C. W. Waller, D. B. Cosulich, M. J. Fahrenbach, M. E. Hultquist, E. Kuh, E. H. Northey, D. R. Seeger, J. P. Sickels, and J. M. Smith, *Science*, **103**, 667 (1946).
34. K. Iwai and M. Kobashi in W. Pfleiderer (ed.), *Chemistry and Biology of Pteridines*, de Gruyter, Berlin and New York (1975) and refs. therein.
35. C. D. Whitfield and H. Weissbach, *Biochem. Biophys. Res. Commun.*, **33**, 996 (1968).
36. R. T. Taylor and H. Weissbach, *Arch. Biochem. Biophys.*, **129**, 728, 745 (1969).
37. R. Tschesche, *Arzneimittel-Forsch.*, **1**, 335 (1951).
38. R. D. Muir, V. J. Shamleffer, and L. R. Jones, *Proc. Soc. Exptl. Biol. Med.*, **47**, 77 (1941).
39. J. Hirsch, *Schweiz Med. Wochschr.*, **73**, 1470 (1943).
40. L. Bock, G. H. Miller, K. J. Schaper, and J. K. Seydel, *J. Medicin. Chem.*, **17**, 23 (1974).
41. E. R. Garret and O. K. Wright, *J. Pharm. Sci.*, **56**, 1576 (1967).
42. E. H. Northey, *Sulphonamides and Allied Compounds*, Reinhold, New York (1948).
43. J. K. Seydel, *J. Pharm. Sci.*, **57**, 1455 (1968).
44. J. E. Seegmiller in P. K. Bondy and L. E. Rosenberg (eds.), *Duncan's Diseases of Metabolism*, (6th edn.), Saunders, Philadelphia (1969), p. 516.

45. W. N. Kelly and J. B. Wyngaarden in J. B. Stanbury, J. B. Wyngaarden, and D. S. Fredrickson (eds.), *The Metabolic Basis of Inherited Disease* (3rd edn.), McGraw-Hill, New York (1972), p. 889.
46. B. Weissmann, P. A. Bromberg, and A. B. Gutman, *J. Biol. Chem.*, **224**, 407 (1957).
47. M. Lesch and W. L. Nyhan, *Amer. J. Med.*, **36**, 561 (1964).
48. J. E. Seegmiller, F. M. Rosenbloom, and W. N. Kelley, *Science*, **155**, 1682 (1967).
49. J. F. Henderson, *Clin. Biochem.*, **2**, 241 (1969).
50. Proceedings of the Seminars on the Lesch–Nyhan Syndrome, *Fed. Proc. Fed. Amer. Soc. Exp. Biol.*, **27**, 1017 (1968).
51. (a) E. R. Giblett, J. E. Anderson, F. Cohen, Bollera and H. J. Mennissen, *Lancet*, **2**, 1067 (1972); (b) K. Dissing and B. Knudsen, *ibid*, **2**, 1316.
52. P. P. Trotta, E. M. Smithwick and M. E. Balis, *Proc. Natl. Acad. Sci.*, **73**, 104 (1976).

Chapter 7

The Nucleic Acids

(A) INTRODUCTION

The origins of nucleic acid chemistry stem from the work of Miescher who in 1868 isolated the nuclei from pus cells obtained from discarded surgical dressings and found that they contained a phosphorus-containing material which he called nuclein. In later work Miescher used salmon sperm as a source of nuclein and in 1872 he showed that the material isolated from sperm contained an acidic compound *nucleic acid* and a base *protamine*. It was later shown that nucleic acids were normal constituents of all cells.

The thymus gland was shown to be a good source of nucleic acid and much of the early work was carried out using this material which was found to yield, on hydrolysis, adenine, guanine, cytosine, thymine, phosphoric acid and a sugar shown to be 2-deoxy-D-ribose. Thus this nucleic acid was called deoxyribonucleic acid (DNA). However, another source of nucleic acid, yeast, gave a material which, on hydrolysis, yielded adenine, guanine, cytosine, uracil, phosphoric acid and the sugar D-ribose. It differed from DNA in that it contained uracil instead of thymine and D-ribose instead of 2-deoxy-D-ribose and was called ribonucleic acid (RNA).

These early workers thought that DNA was characteristic of animal tissue and RNA was characteristic of plant tissue and it was not until the 1920s that the occurrence of both types of nucleic acid was observed in both animal and plant tissue. However, it was not until the 1940s that unequivocal evidence for the occurrence of both types of nucleic acid as general constituents of all types of living cell was obtained from the work of Caspersson,[1] Brachet,[2] and Davidson.[3,4]

In recent years the study of nucleic acids has become a branch of science in its own right—molecular biology—and a considerable volume of work has been carried out leading to the model for the structure of DNA (Watson and Crick, 1953[5]), the structure of yeast phenylalanine transfer RNA (Robertus *et al.*[6] Kim *et al.*[7]) and many other major breakthroughs in understanding the structures and roles of the nucleic acids, including recent work on the structure of chromatin and the structural elucidation of viroids. Much of current literature is of great interest, but for reviews of the state of knowledge the reader is directed to the excellent works of Davidson,[8] Chargaff and Davidson,[9] and to the other reviews and current literature on the topic.

(B) THE OCCURRENCE OF NUCLEIC ACIDS

Early workers in the field of nucleic acid chemistry had assumed that the nucleic acids were essentially constituents of the nucleus of eucaryotic cells, and it was not until the 1930s that it was shown[10] that the cytoplasm contained substantial amounts of RNA. In fact the nucleus only contains about 10% of the total cellular RNA,[11,3] whilst it contains by far the largest portion of the DNA although some DNA has been observed in the cytoplasm of other sub-cellular particles. For example, DNA has been found in the white of hen's eggs,[12] in the chloroplasts of spinach[13,4] and other photosynthetic species[15,17] and in mitochondria[18]. In the nucleus the DNA is found in the nucleolus, a roughly spherical particle which also contains a large portion of the nuclear RNA, in the chromatin which comprise the chromosomes which contain the genes. In cell nuclei the majority of the DNA is associated with the basic proteins—the histones and protamines. The precise function of these associated proteins is unclear, but there is evidence that the histones prevent the transcription of segments of the DNA chain and thus exert a 'masking' effect and influence protein biosynthesis. The association between genes and histones is suggested[19] to be different in different tissues, thus leading to the production of proteins characteristic of each organ or tissue. This would explain how, in an individual multicellular organism having the same genetic material in the nucleus of each cell, *differentiation* of metabolism between different cells is possible.

In bacteria and other procaryotes there is generally no nucleus, but the DNA appears to be present in aggregates (nucleoids), about one to three in number, usually attached to the cell membrane. There are also some DNA viruses. The amount of DNA present in the nuclei of animal cells is about 2 mg per gram of tissue, which represents of the order of 6×10^{-12} g per nucleus. This would also represent of the order of 50×10^8 nucleotide pairs in the double-helical DNA genome and a length of the order of 1 m. The DNA of *E. coli* and other bacteria is of the order of 10^{-3} of this amount, i.e. representing about 1 mm of DNA ($\sim 4 \times 10^{-3}$ pg per nucleus).

The majority of RNA in any type of cell is found in the cytoplasm and about 80% of it is found in the sub-cellular particles called the ribosomes, which are found either free in the cytosol or attached to the membrane system known as the endoplasmic reticulum.

The ribosomal RNA (rRNA) is metabolically stable and is of high molecular weight, and from the two ribosomal subunits are obtained separate rRNAs which are different from each other and can hybridize with different sites on the genome. From *E. coli* the larger ribosomal subunit (50 S) provides rRNA, having a molecular weight of 1.1×10^6 approximately.[19] The ribosomal subunits of mammalian cells (80 S) give rRNAs of molecular weight 1.8×10^6 (from the larger subunit) and 0.7×10^6 (from the smaller subunit) respectively.[20]

Also found associated with the ribosomes are two RNAs of much lower molecular weight, an RNA of about 120 nucleotides found associated with the larger ribosomal subunit of both bacterial and animal cells, and an RNA of about 130 nucleotides which has been found in the larger subunits of mammalian ribosomes.[8]

About 15% of the cellular RNA occurs as small units of molecular weight 2.3 to 2.8×10^4 having about 80 nucleotide units. These are the transfer RNAs (tRNA), each being specific for a particular amino acid, some amino acids apparently having more than one tRNA.

The rest of the RNA ($\sim$ 5%) occurs as the metabolically labile messenger RNA (mRNA) whose role is the carrying of the 'genetic message' from the DNA in the nucleus to the ribosomes, the site of protein synthesis, in the cytoplasm. Values for the molecular weight of mRNA show wide variation,[21] but a value of 0.5×10^6 seems to represent the lower limit.

In addition to the chromosomal DNA and the species of RNA described above, some other species of nucleic acids have been found in a number of cells. Some of these types are of unknown function or are precursors to the materials of established function. These relatively unknown types of nucleic acid will not be considered here and for further information concerning these the reader is directed to the standard texts on nucleic acids and to the original literature.

(C) THE STRUCTURES OF THE NUCLEIC ACIDS

(i) DNA

DNA has a polydeoxyribonucleotide structure, the phosphate groups providing 3′–5′ links between the individual nucleotides. However, it has been difficult to obtain accurate values for the molecular weights of DNA since the long polymeric molecules are very susceptible to mechanical shear during isolation and to the action of the nuclease enzymes. However, as work continues better nucleic acid preparations are being made and higher values for molecular weights are being obtained. The values for DNA from eukaryotic cells are not yet known. This material is almost certainly heterogeneous and so satisfactory separation into the component species would first have to be effected. The chromosome from *E. coli* has been found to have a molecular weight of 2.2×10^9, to represent 3×10^6 nucleotide pairs and have a length of about 1 mm, and to be a cyclic duplex molecule.[22]

Early work on the chemical degradation of DNA had shown that the degradation products were the pyrimidines cytosine and thymine, the purines adenine and guanine, phosphate and carbohydrate material shown by Levene and Mori[23] to be a deoxypentose. Later workers established[24] that the sugar isolated from the DNA of a bacterium was 2-deoxyribose by comparison with a synthetic sample, and the sugar of some fish roe DNA has also been established as being 2-deoxyribose.[25]

In addition to the above pyrimidine bases many DNAs also contain small amounts of other bases, for example, wheat germ contains quite a high molar proportion of 5-methylcytosine whilst bacteriophages frequently have alternative pyrimidine bases including uracil and 5-hydroxymethyl cytosine.

A variety of other investigatory work showed that in DNA the point of attachment of the sugar to the heterocyclic base was at N-1 for the pyrimidines and N-9 for the purines, that the stereochemistry of the glycosidic link was β, and that it was at positions 3′ and 5′ that the phosphate groups were attached, thus providing DNA with a sugar–phosphate backbone with attached heterocyclic bases. The structures of nucleosides and nucleotides have been considered in Chapter 4.

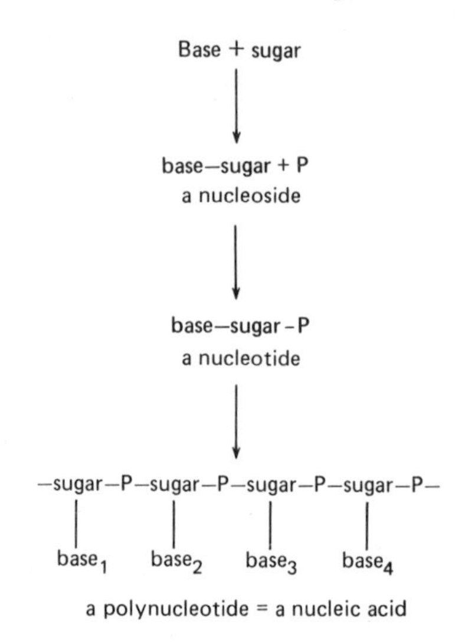

Different methods of representing this primary structure of DNA are shown in Fig. 7.1.

Methods for determining the molar proportions of the nitrogen bases in DNA by acid hydrolysis and enzymatic hydrolysis and subsequent chromatography have been reviewed by Bendich and other workers.[26]

Some of the results obtained from the DNA isolated from a variety of sources are shown in Table 7.1.

Chargaff in 1950 had pointed out that the results of base content analyses of DNA showed regularities,[27–29] that the ratio of the sum of the

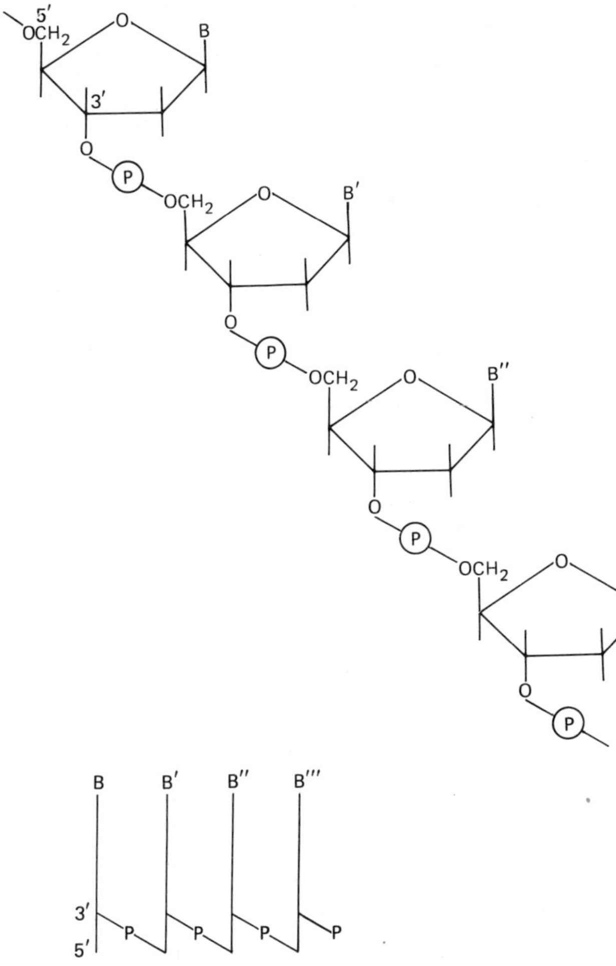

BpB′pB″pB‴p

B–B′–B″–B‴p

Fig. 7.1. Methods of representing the primary structure of DNA and of oligonucleotides (illustrating 3′-phosphate ending)

pyrimidines was unity, as was the ratio of adenine to thymine and guanine to cytosine and the ratio of the 4(6) amino-bases (A and C) to 4(6) hydroxy bases (G and T).

Doty[30,31] considered the physical properties of a number of DNAs and pointed out that the physical properties of DNA are dependent on the G + C content of the DNA. There is a linear relationship between the

Table 7.1 Base ratio and molar proportion for a number of different DNAs. (Data from various sources)

Source	A	G	C	T	$\frac{A+T}{G+C}$	$\frac{A}{T}$	$\frac{G}{C}$	$\frac{Pu}{Py}$	$\frac{A+C}{G+T}$
E. coli K12	26.0	24.9	25.2	23.9	1.00	1.09	0.99	1.08	1.05
Salmonella paratyphi A	24.8	24.9	25.0	25.3	1.00	0.98	1.00	0.99	0.99
Staph. aureus	30.8	21.0	19.0	29.2	1.50	1.05	1.11	1.07	0.99
Sacc. cerevisiae	31.7	18.3	17.4	32.6	1.80	0.97	1.05	1.00	0.97
Wheat germ	27.3	22.7	22.8*	27.1	1.19	1.01	1.00	1.00	1.00
Bovine thymus	28.2	21.5	22.5*	27.8	1.27	1.01	0.96	0.99	1.03
Bovine spleen	27.9	22.7	22.1*	27.3	1.23	1.02	1.03	1.02	1.00
Bovine sperm	28.7	22.2	22.0*	27.2	1.26	1.05	1.01	1.03	1.03
Human thymus	30.9	19.9	19.8	29.4	1.52	1.05	1.01	1.03	1.03
Human liver	30.3	19.5	19.9	30.3	1.54	1.00	0.98	0.99	1.01
Herring testes	27.9	19.5	24.3*	28.2	1.28	0.99	0.80	0.92	1.09
Rat-bone marrow	28.6	21.4	21.5*	28.4	1.33	1.00	1.00	1.00	1.00
Paracentrotus lividus	32.8	17.7	18.4*	32.1	1.80	1.02	0.96	1.00	1.03
ϕX 174	24.3	24.5	18.2	32.3	1.32	0.75	1.34	0.97	0.75

*Includes 5-methylcytosine

G + C content and the buoyant density of different DNAs and also a linear relationship between the G + C content and the *melting temperature* (T_m) of the DNA.

The Tm refers to the characteristic of most DNAs that when heated in neutral solution the 260 nm u.v. absorbance increases rapidly between quite narrow temperature limits around a mean value—the *melting temperature* or *transition temperature*. Coincident with this increase in u.v. absorbance there is a decrease in the specific optical rotation at 589 nm.

The data of Chargaff concerning base ratios (*Chargaff's rule*) coupled with the evidence provided by the X-ray diffraction studies of Astbury,[32] Franklin,[33] and Wilkins[34] and with their knowledge of the helix-forming characteristics of polypeptides led Watson and Crick to propose the double-stranded helical structure for DNA.[5]

The model for the secondary structure of DNA proposed by Watson and Crick is still considered to be essentially that for many samples of DNA in aqueous solution. However, in addition to double-stranded linear DNA, double-stranded cyclic DNA, single-stranded linear DNA, and single-stranded cyclic DNA have now been isolated.

The Watson–Crick model for the secondary structure of DNA is indicated in Fig. 7.2. It consists of two strands of linear polydeoxyribonucleotides in the form of a right-handed helix with the two strands wound round the same axis, the bases of the nucleotides being at the centre of helix with their planes at right angles to the axis of the molecule. The two strands are held together by hydrogen bonds and other non-bonding interactions between the bases on the two separate strands, the bases being

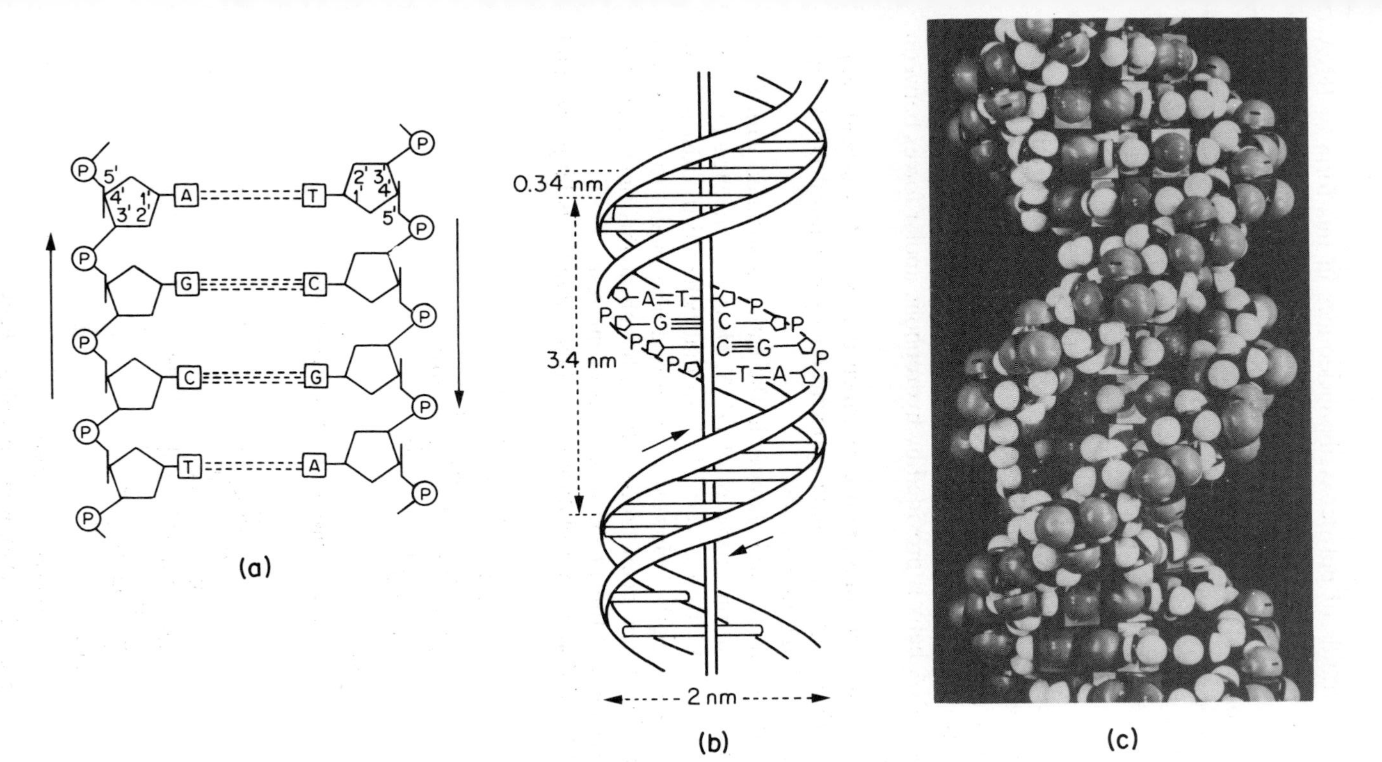

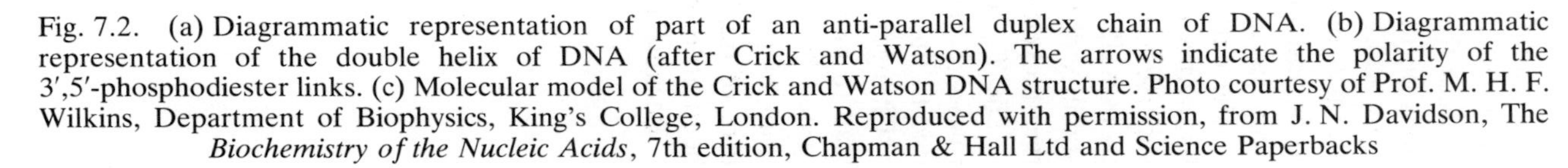

Fig. 7.2. (a) Diagrammatic representation of part of an anti-parallel duplex chain of DNA. (b) Diagrammatic representation of the double helix of DNA (after Crick and Watson). The arrows indicate the polarity of the 3′,5′-phosphodiester links. (c) Molecular model of the Crick and Watson DNA structure. Photo courtesy of Prof. M. H. F. Wilkins, Department of Biophysics, King's College, London. Reproduced with permission, from J. N. Davidson, The *Biochemistry of the Nucleic Acids*, 7th edition, Chapman & Hall Ltd and Science Paperbacks

arranged in distinct pairs of adenine–thymine and guanine–cytosine (Fig. 7.3).

Thus the two strands of polydeoxyribonucleotides in this form of DNA are 'complementary' to one another, the order of bases in one strand automatically determining the order in the other strand as long as the principle of the complementarity of the bases is conserved. There are no restrictions on the sequence of base pairs along the molecule of DNA.

In addition to the two strands of the DNA being complementary, they are also of opposite polarity, i.e. the internucleotide link in one strand is 3′–5′ and in the other is 5′–3′. This means that in the duplex DNA the helix has a 'wide groove' and a 'narrow groove' (Fig. 7.2).

The earlier idea that the two strands of DNA were held together solely by hydrogen bonding between the complementary bases has now been superseded, and although the hydrogen bonds provide some additional stability as well as providing specificity, it is now thought that the major part of the stability of the DNA helix is provided by hydropholic apolar forces between the stacked bases.[35]

The T_m represents the temperature at which the two DNA strands separate—i.e. the DNA 'denatures'—the hypochromicity of natural DNA being due to the base-stacking interactions.

Changes in the humidity and changes in the cation associated with DNA lead to modifications in the structure of DNA. However, the structure suggested by Crick and Watson is very close to that believed to exist in solutions of low ionic strength.[36] However, recently a new model for the structure of duplex DNA has been proposed:[37] although at this time there has been little discussion of this proposition, it is of interest.

Because of the difficulty of isolating intact DNA due to its large size and

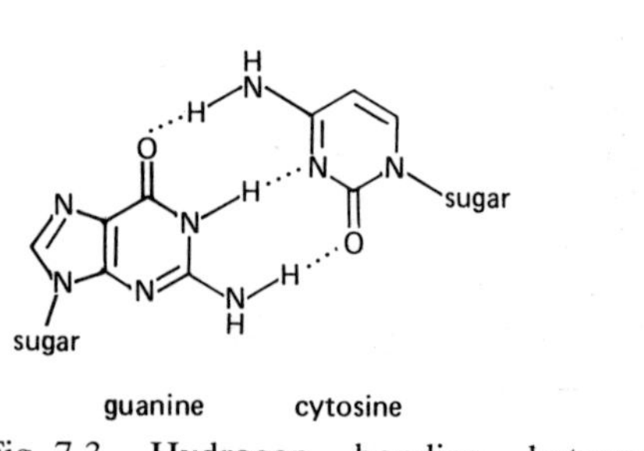

adenine thymine (R = Me), uracil (R = H)

guanine cytosine

Fig. 7.3. Hydrogen bonding between complementary bases

sensitivity to mechanical and enzymic degradation, it has not been possible in most cases to investigate the tertiary structure of DNA, i.e. the way in which the duplex helix is further coiled and structured. However, there is no doubt that in the eukaryotic chromosome the DNA is extensively folded, and supercoiling[38] and kinking[39] have been suggested to account for this. The DNA of a number of viruses, mitochondria, and some other sources has been further investigated and various forms has been observed, for example double-stranded cyclic DNA,[40] but further consideration of these types will not be made here. Davidson[8] includes a chapter on nucleic acids in viruses and plasmids.

(ii) RNA

Like DNA, RNA has a linear, unbranched, structure, in this case being a polyribonucleotide, phosphate groups forming 3′–5′ links between the nucleotides.

The relative molar proportions of bases in different RNAs vary widely from one source to another, and the equivalence shown by the bases in DNA is not generally seen. Another striking difference between RNA and DNA is that in RNA the pyrimidine uracil takes the place of thymine, also, particularly in tRNAs, a variety of modified bases is found usually in low molar ratios. One of these bases which occurs in RNA is the C-nucleoside pseudouridine (1) whilst the other modified bases are usually methylated bases. Some of these modified bases which have been isolated from RNA are shown in Fig. 7.4.

RNA does not normally exist in the duplex helical structure which has been established for DNA, and the nature of the secondary and tertiary structure of mRNA and rRNA is less clear, although some information is available.[14] However, the structure of tRNA is well established (see below). Certainly there are helical regions in RNAs, this being indicated by the high values of T_m found for some RNAs, their hypochromicity, and a measure of equivalence between the A : U and C : G ratios. The rRNAs would seem to be closely associated with the ribosomal proteins and have a compact, well-ordered structure, but mRNA does seem to have the appearance of long strands.

A critical factor in the elucidation of the structure of RNAs has been the ability to determine the primary structure, i.e. the sequence of bases in the chain, of these.

The methods which are used involve the controlled enzymatic and chemical degradation of RNA followed by chromatographic and electrophoretic separation of the products. This has been difficult in the case of DNA due to the size of the molecules and the lack of base-specific deoxyribonucleases with a narrow range of specificity, but this problem is now being overcome. The degradation of nucleic acids is considered in Section 7D. However, since the pioneering work of Holley and his coworkers,[42,43] who sequenced yeast alanine tRNA ($tRNA^{Ala}$) (Fig. 7.5), about forty tRNA's

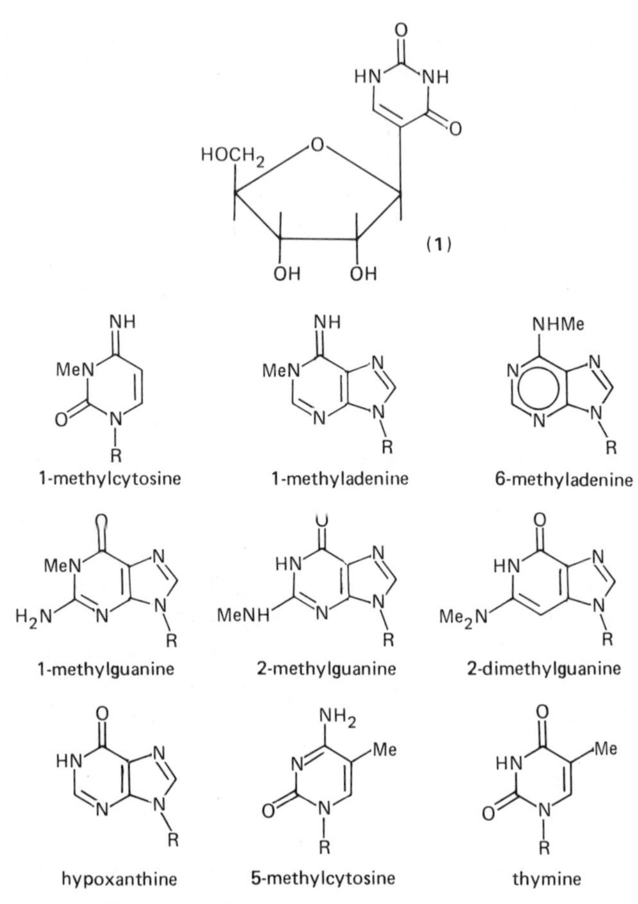

Fig. 7.4. Some modified bases found in RNA (for other examples see R. H. Hall, *The Modified Nucleosides in Nucleic Acids*, Columbia University Press, New York, 1970)

have now been sequenced and also a number of other RNA species. Refinements to the sequencing techniques involving the use of ^{32}P-labelled RNA and ^{32}P-labelled oligonucleotides have been used for example by Sanger[44,45] and co-workers who determined the primary structure of the 120 nucleotide for *E. coli* rRNA, and have since carried out many other sequence determinations.[46]

Each of the tRNAs which have been sequenced have structures which can be fitted to a general secondary structure of loops and short helical regions indicated in Fig. 7.6.

The general features are:

(a) An amino acid arm (A), which is a short helical region of about seven base pairs terminating with the sequence –C–C–A–3′-OH, the amino acid being attached to the terminal adenosine.
(b) A dihydrouracil loop (B) containing about 8 to 11 nucleotides.

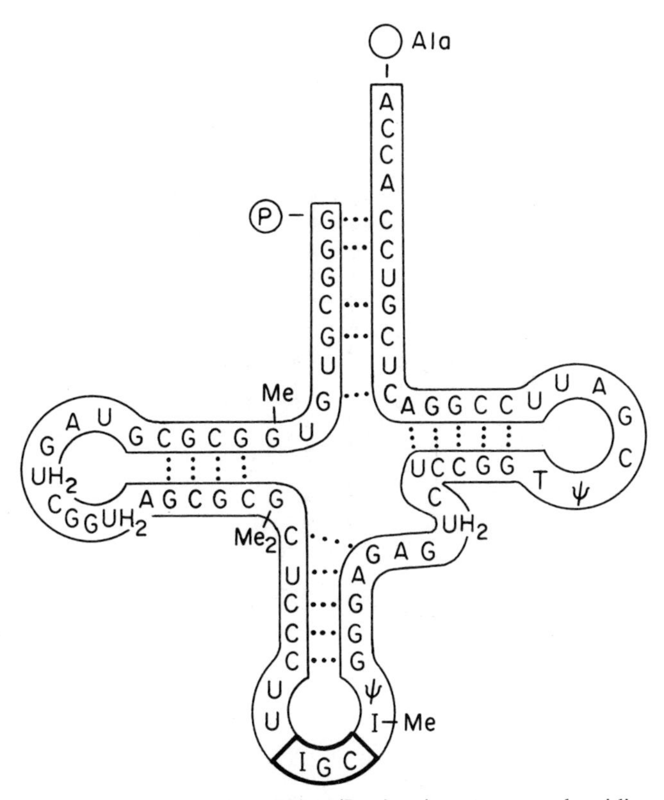

Fig. 7.5. Yeast alanine tRNA (I = inosine, ψ = pseudouridine, T = ribothymidine, UH_2 = dihydrouracil, Me—I = methylinosine, Me—G = methylguanine, Me_2G = dimethylguanine). (After R. W. Holley, J. Apgar, G. A. Everett, J. T. Madison, M. Marquisee, S. H. Merrill, J. R. Penswick, and A. Zamir, *Science*, **147**, 1462 (1965)

(c) An anticodon loop (C) containing about seven nucleotides, a group of three nucleotides at the exposed turn being the anticodon.

(d) An extra arm (D) which varies greatly from species to species but usually contains between 3 and 18 nucleotides.

(c) An anticodon loop (C) containing about seven nucleotides, a group of three nucleotides at the exposed turn being the anticodon.

(d) An extra arm (D) which varies greatly from species to species but usually contains between 3 and 18 nucleotides.

(e) A pseudouridine–ribothymidine loop (E) containing about 7 nucleotides which includes the sequence T–ψ–C.

Considerable data have been gathered which confirm this two-dimensional clover-leaf structure, and recently X-ray crystallographic analysis has provided a full three-dimensional structure which includes the general features indicated above.[6,7] Further recent work[7b,c] supports this tertiary structure which has recently been reviewed.[7d]

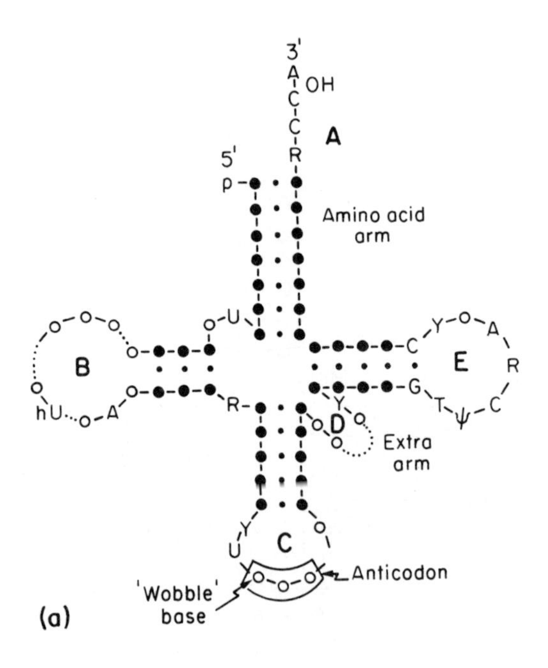

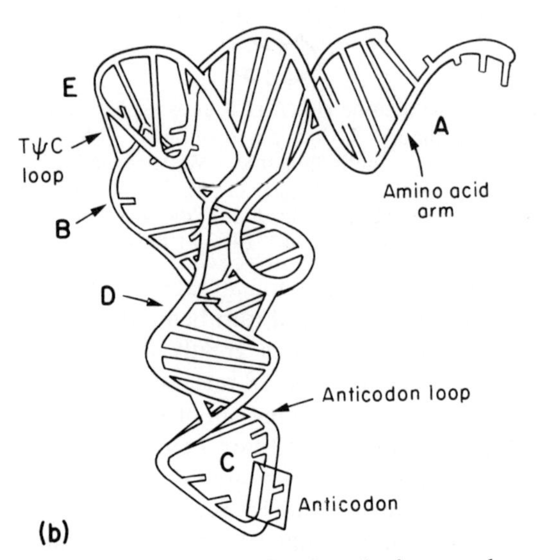

Fig. 7.6. (a) General cloverleaf secondary structure of tRNA. (b) Schematic model of tertiary structure of tRNA. Reproduced, with permission, from Kim, S. H., *et al.*, *Science*, **185**, 435–439. Copyright 1974 by the American Association for the Advancement of Science

(D) CHEMICAL AND ENZYMIC DEGRADATION OF NUCLEIC ACIDS

(i) DNA

Polydeoxyribonucleotides are stable to the action of alkali but when DNA is subjected to dilute acid (pH < 3) at temperatures greater than 60°C or to formic acid at 37°C, the phosphodiester chain remains virtually intact but the sugar-purine bonds are hydrolysed to give a product polydeoxyribnucleotide having no purine residues. This *apurinic acid* now has free anomeric positions in the depurinated sugars and is readily degraded by alkali. This method of DNA degradation has been extensively studied by Chargaff, Burton and others who have been concerned with the sequencing of nucleic acids (for examples see Burton[47]).

The treatment of DNA with hydrazine causes a loss of the pyrimidine residues selectively to give a product *apyrimidinic acid*,[48] whilst treatment of DNA with permanganate gives a product in which all the guanine, cytosine and thymine residues are oxidized to ureido groups. Treatment of this product with alkali gives oligodeoxyribonucleotides containing only adenine.[49]

A variety of enzymes capable of degrading nucleic acids and having differing actions have been described. In the case of the DNA degrading enzymes—the deoxyribonucleases, which may be endonucleases or exonucleases—the following have been described, among many others:

(i) Pancreatic deoxyribonuclease (DNase I) which cleaves DNA into oligonucleotides of about 4 units with a free 3′-hydroxyl and a 5′-phosphate. The enzyme is most active at pH 6.8–8.2 and hydrolyses native DNA more rapidly than denatured DNA.

(ii) Deoxyribonuclease II (DNase II) which degrades DNA into oligonucleotides of about 6 units having a free 5′-hydroxyl and a 4′-phosphate.

(iii) Endonuclease I (from *E. coli*) which gives oligonucleotides of about 7 units having a 5′-phosphate and having a greater specificity for native DNA.

(iv) Endonuclease II (from *E. coli*) which acts on double-stranded alkylated DNA to produce cleavage on a single strand. At least four different deoxyribonucleases have been characterized from Streptococci including

(v) Streptodornase which gives oligonucleotides of various lengths having 5′-phosphate groups.

Other types of endonucleases include ATP-dependent endonucleases, virus-induced, and mammalian virus endonucleases. However, much interest recently has been in the area of restriction enzymes and enzymes involved in recombination processes.

Restriction enzymes are endonucleases isolated from bacterial sources which act upon specific foreign modified DNAs. A review of restriction enzymes has been given by Nathans and Smith.[50] Such restriction enzymes have become the principal agents for the sequencing of DNAs.

Amongst the exonucleases which have been described are:

(a) Exonuclease I (from *E. coli*) which hydrolyses denatured single-stranded DNA from the 3′-hydroxy end releasing deoxyribonucleoside 5′-monophosphates.
(b) Exonuclease II and III (from *E. coli*) which also act on the 3′-hydroxy end of a DNA chain to release 5′-phosphate mononucleotides but are specific for double-stranded DNA.

A number of other exonucleases have been described, details of which may be found in books by Davidson,[8] the review by Lehman,[51] and elsewhere.

(ii) RNA

Polyribonucleotides are degraded by alkali, e.g. 0.3 M KOH at 37°C for 16 h, to give 2′- and 3′-ribonucleotides. The cyclic 2′,3′-phosphates are formed as intermediates in the reaction by participation of the cis 2′-hydroxyl group, and as a result a mixture of 2′- and 3′-phosphates is formed. DNA is resistant to alkaline degradation since it has no 2′-hydroxyl group to participate in the reaction.

The mechanism of degradation is indicated in Fig. 7.7.

The action of strong acid at 100°C (e.g. perchloric acid for 1 h) hydrolyses both RNA and DNA to liberate the bases, phosphate, and sugar degradation products.

RNA is hydrolysed by dilute acid to give the two purines (adenine and guanine) ribose and phosphoric acid but the pyrimidine nucleotides are more resistant to such hydrolysis and require strong acid or heating in a sealed tube. Under these conditions there is a tendency for cytosine to be deaminated to uracil.[52]

However, the elegant RNA sequencing work of Holley, Sanger and other

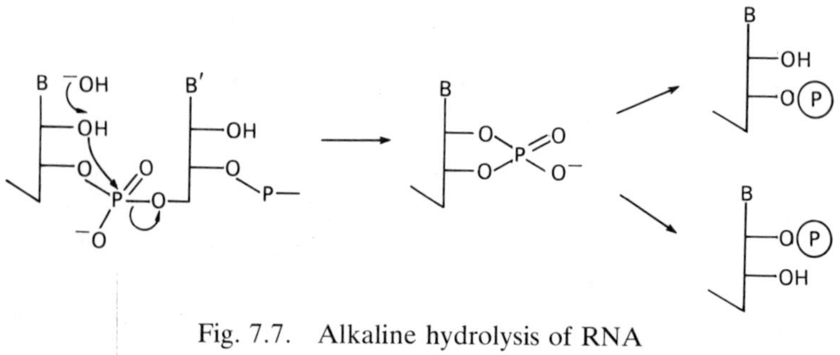

Fig. 7.7. Alkaline hydrolysis of RNA

workers was considerably aided by the use of a variety of ribonucleases obtained from a number of sources, these enzymes having a greater degree of specificity than the deoxyribonucleases.

Some of the more commonly used ribonucleases are:

Pancreatic ribonuclease. An enzyme from the pancreas capable of digesting yeast RNA was isolated[53] as early as 1920 and was crystallized[54] in 1940. This enzyme has been extensively investigated and has been sequenced,[55] the active site and the mechanism of action has been determined,[56] the complete structure has been elucidated using X-ray crystallography[57,58] and the total chemical synthesis has been achieved.[59,60]

Pancreatic ribonuclease is an endonuclease which cleaves the bond between a 3′-phosphate in a pyrimidine nucleotide and the 5′-position of the adjacent nucleotide. The enzyme is highly specific, the essential action being intramolecular nucleophilic attack of the 2′ hydroxyl group on the 3′-phosphate to form the 2′,3′-cyclic phosphate. Further enzymic action results in the formation of the pyrimidine 3′-phosphate as a free nucleotide or as a terminal nucleotide residue in an oligonucleotide chain.

Ribonuclease T_1 is obtained from *Aspergillus oryzae* and is an enzyme which has a high specificity for hydrolysing the bond between a 3′-GMP unit and the 5′-hydroxy group of the adjacent nucleotide.

Ribonuclease T_2 is also isolated from *Aspergillus oryze* and has a preference for cleaving off 3′-AMP units. However, it will digest tRNA almost completely to give 3′-monophosphates.[61]

Enzymes having actions similar to T_1 and T_2 seem to be common in fungi and microorganisms as well as in the tissues of higher organisms, and a variety of such enzymes has been described. For example, rat liver contains several ribonucleases[62–64] and endonuclease activity has been found in HeLa cells[65] and human KB cells.[66]

Some ribonucleases cleave the internucleotide link to give 5′-monophosphates, for example the ribonucleases IV and P of *E. coli*, and ribonuclease H from calf thymus.

Ribonuclease II from *E. coli* is an RNA exonuclease which acts on single-stranded RNA starting from the 3′- end to give 5′-mononucleotides. An oligoribonuclease having a similar action to ribonuclease II has also been found[67] which acts on low molecular weight RNA resistant to RNase II.

Several other exonucleases have been described from a number of sources including the enzyme polynucleotide phosphorylase, which catalyses the reversible reaction in which polyribonucleotides and inorganic phosphate react to give ribonucleotide diphosphates.

(iii) Nucleases Acting on Both DNA and RNA

Micrococcal nuclease is an enzyme isolated from Staphylocci species which acts upon both DNA and RNA to give 3′-monophosphates and oligonucleotides with 3′-phosphate ends. Other non-specific endonucleases have been isolated from *Neurospora crassa*,[68] *Aspergillus oryzae*,[69,70] and mung bean.[71]

The two most important non-specific exonuclease are venom phosphodiesterase which gives 5′-phosphates starting at the 3′-hydroxyl terminus,[72] and spleen phosphodiesterase which gives 3′-monophosphates starting at a 5′-hydroxy terminus.[73]

(E) REPLICATION

Replication is the process whereby, during cell division, new DNA is biosynthesized using DNA of the original cell as the template. As DNA biosynthesis proceeds the duplex double helix is separated and two new DNA strands are formed to be complementary to the template strands. Thus two new duplex helical DNAs are biosynthesized which are identical to the original DNA, so the principle of heredity is maintained.

This idea of the original DNA strands becoming separated, one of each going to each of the two daughter cells, is *semiconservative* replication and some classic work by Meselson and Stahl[74] provided convincing evidence for such a mechanism of replication in the case of *E. coli*. They cultured *E. coli* in a medium having $^{15}NH_4Cl$ as the sole source of nitrogen in order to label the DNA as completely as possible with ^{15}N. The microorganism was harvested and transferred to a medium having $^{14}NH_4Cl$ as the sole source of nitrogen. Samples of the microorganism were removed at intervals for several generations, the samples were lysed, and the resulting suspensions were centrifuged in caesium chloride (140 000 g for 20 h) when the DNA reached sedimentation equilibrium.

The results showed that at the start of the experiment the DNA sedimented as a single band corresponding to ^{15}N-labelled DNA. After one generation time in the ^{14}N medium, ^{14}N–^{15}N hybrid DNA was the only species present. Successive generations gave mixtures of ^{14}N–^{15}N DNA and ^{14}N only DNA, and after further generation times the unlabelled DNA accumulated. Also when the ^{14}N–^{15}N hybrid DNA species was denatured and the two strands were separated it gave a ^{14}N-unlabelled strand and a ^{15}N-labelled strand.

The principle of these results and the principle of semi conservative replication is shown in Fig. 7.8.

Further experiments using *E. coli*, but at the chromosomal level, also support the principle of semi-conservative replication,[75] and also experiments using plant chromosomes[76] show semiconservative replication in that system. So the principle of the role of DNA as the genetic material can be explained.

The precise details of the mechanism of replication are not known and the problems involved in considering strand separation of a long DNA duplex helix are enormous. For complete strand separation the body of the

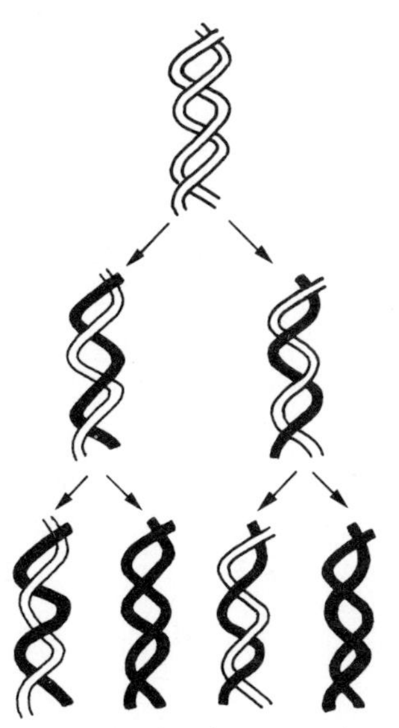

Fig. 7.8. Semiconservative replication

DNA must turn about its axis once for every 10 base pairs. The single chromosome of *E. coli* is known to be circular duplex helical DNA having a length of 1 mm and having 3×10^6 base pairs, and it is more difficult to imagine the rotations required to separate these two strands. Yet the minimum time required for *E. coli* to undergo chromosomal replication seems to be about 40 min, indicating a rate of synthesis of about 1700 base pairs per minute.[75]

Currently much work is being carried out into the precise mechanisms of replication, and it is not intended here to have a prolonged consideration of the various models and systems that have been proposed. For detailed information the reader is directed to the current literature and reviews, but an excellent summary is provided by Davidson.[8] Here a short account of the currently most widely accepted model of replication will be presented.

On each replicating DNA duplex strand, be it circular or linear, there is a specific site for the initiation of replication and there are specific proteins required in order that replication may begin. Replication occurs as the strands separate in a replication 'fork'. The two DNA strands in duplex are antiparallel, but the new DNA is not synthesized in a different direction for each of the initial strands. The evidence shows that the new strands grow in the 5′→3′ direction and that one strand is extended backwards towards the replication fork, being formed in short segments which are subsequently joined together. This view of replication is that proposed by Okazaki who has carried out many experiments which support this view.[77–79]

Thus DNA biosynthesis seems to occur in a discontinuous manner which

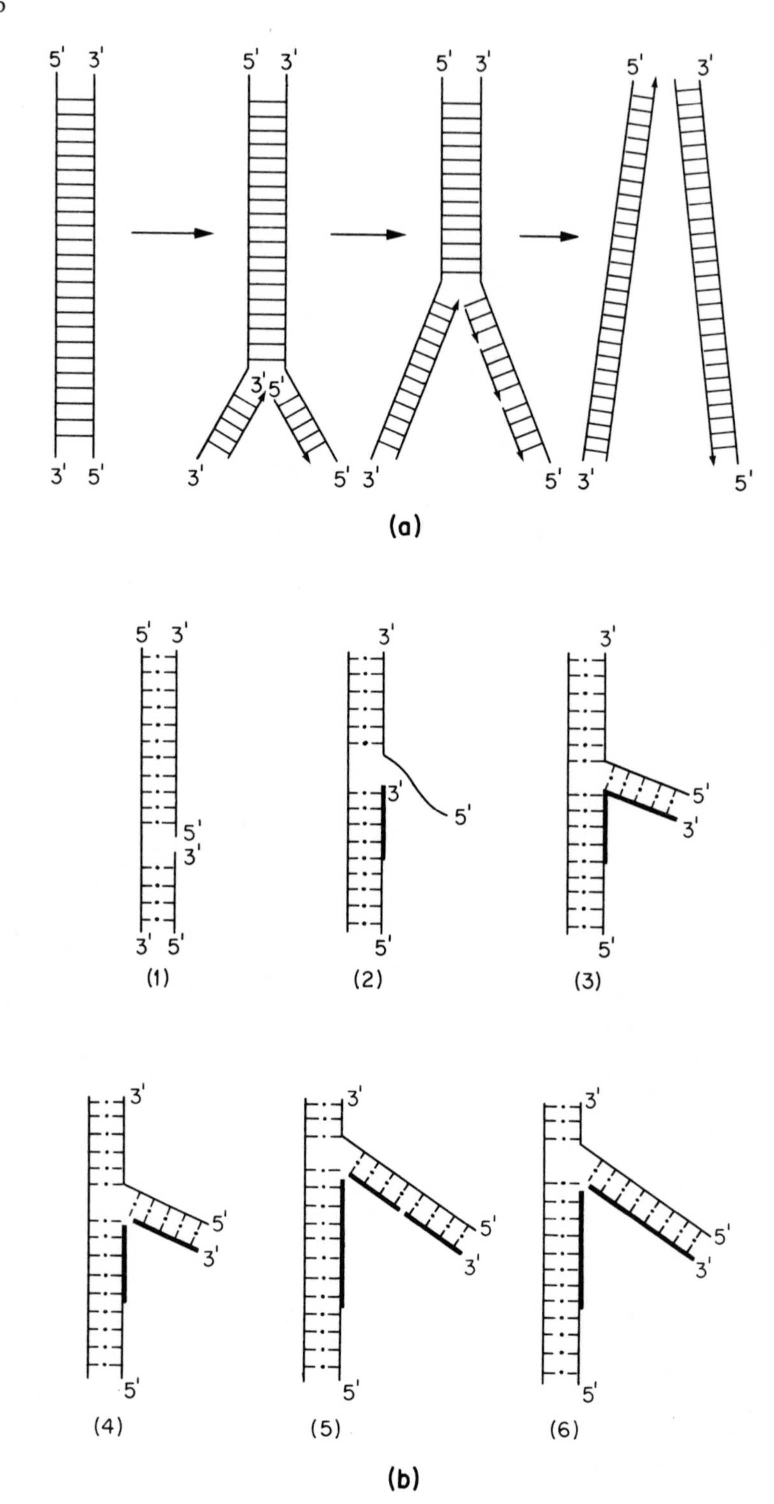
5'
3'
3'
5'
(a)
(1)
(2)
(3)
(4)
(5)
(6)
(b)

is consistent with the experimental data which have been acquired to date, but other models have also been proposed.[8]

The Okazaki fragment discontinuous synthesis model may be summarized thus: in DNA synthesis, synthesis occurs in the 5′→3′ direction in an antiparallel fashion. On the exposed 3′-ended DNA the new DNA may be synthesized in a continuous or discontinuous way, but on the exposed 5′-ended DNA strand the new DNA is synthesized in short sections, beginning at some point on the parental DNA and being extended towards the replication fork until the space is filled. A DNA ligase then joins the Okazaki fragments. This scheme is illustrated diagrammatically in Fig. 7.9.

(F) DNA POLYMERASES

A number of enzymes capable of catalysing the biosynthesis of DNA have been isolated from different sources but the general mechanism of action seems to be common to them all—that is, the formation of a phosphodiester link between a free 3′-hydroxyl at the end of a growing DNA strand and the 5′-phosphate of an incoming deoxyribonucleoside triphosphate, the incoming nucleotide being controlled by the principle of base-pairing complementarity with the template DNA strand.

The first enzyme which was isolated which had the property of catalysing the biosynthesis of DNA-like material was obtained from *E. coli* by Kornberg and co-workers.[80] The principal features of the Kornberg enzyme reaction are that:

(i) a template section of single-stranded DNA is required;
(ii) a 'primer' DNA which can base-pair with the template is required;
(iii) the reaction is dependent on the presence of a divalent metal ion (Mg^{2+} preferably);
(iv) a supply of the deoxynucleoside triphosphates is required.

The enzyme has a binding site for the template strand having one base

Fig. 7.9. Replication of DNA via Okazaki fragments. (a) The replication fork showing Okazaki fragments which are subsequently joined by polynucleotide ligase. (b) Kornberg's suggested scheme (ref. 81) for unidirectional replication of a duplex chain. (1) A nick is introduced into one strand, possibly at a specific site. (2) DNA polymerase binds at the nick and replication proceeds by covalent extension of the 3′-OH end. The 5′ end is meanwhile displaced. (3) Replication switches to the complementary strand to form a fork. (4) The fork is then cleaved by an endonuclease. (5) The process is then repeated by further covalent extension with fork formation. (6) The segments formed on the branch strand are formed by ligase. Reproduced with permission, from J. N. Davidson, *The Biochemistry of the Nucleic Acids*, 7th edition, Chapman & Hall Ltd and Science Paperbacks

at the active site and a few bases on each side of this, a site for the growing primer strand (base-paired to the template), a site for the 3′-hydroxyl end of the primer which has adjacent a site for a deoxynucleoside triphosphate (this site being adjacent also to the template active site), and a site at which the $5' \rightarrow 3'$ cleavage of a 5′-phosphate terminated chain may occur.[81]

The enzyme catalyses the net synthesis of DNA material on the DNA primer which is complementary to the template, the synthesized strand growing in the $5' \rightarrow 3'$ direction in the opposite direction to the polarity of the template.

However, it is not thought that the Kornberg enzyme has the role of the replication enzyme and it is suggested that it is only concerned in the maintenance and repair of DNA. The *E. coli* DNA polymerase I cannot replicate duplex DNA in a semiconservative way and mutants of *E. coli* have been isolated which seem to have only 1% of the normal DNA polymerase I activity yet still show normal growth.

Further DNA polymerases II and III have been separated from cultures of *E. coli* and of these DNA polymerase III seems to be essential for DNA synthesis. It shows a high rate for the polymerization of nucleotides and the best template for the enzyme seems to be double-stranded DNA with small gaps having 3′-OH ends.[8] This enzyme may thus be that which joins the Okazaki fragments.

Enzymes having similar DNA polymerase activities to those above have been demonstrated in other microorganisms, and similar enzymes are present in eukaryotic cells. Thus it seems that a variety of enzymes, including DNA ligases, are required for DNA-dependent DNA biosynthesis, and the picture is by no means clear yet.

Good surveys are provided by Davidson,[8] Ingram,[82] Kornberg[83] and others.

(G) TRANSCRIPTION

In the process of transcription a single strand of DNA acts as the template for the biosynthesis of RNA. In duplex DNA *only one strand* acts as the transcriptional strand; the other seems to act in a structural role for the maintenance of the stability of DNA. The base sequences on the DNA strand dictate the base sequence of the RNA strand; thus through the base-pairing complementarity the genetic information is transcribed to mRNA. Each RNA chain is initiated at a specific site on the DNA template (the promoter region) and there is another site (a terminator) which acts as a chain-terminating signal.

Transcription is mediated by the DNA-dependent RNA polymerases, and such enzymes have been isolated from a variety of sources, but the enzyme isolated from *E. coli* has been the most extensively investigated. However, all of the enzymes which have been studied seem to have a simi-

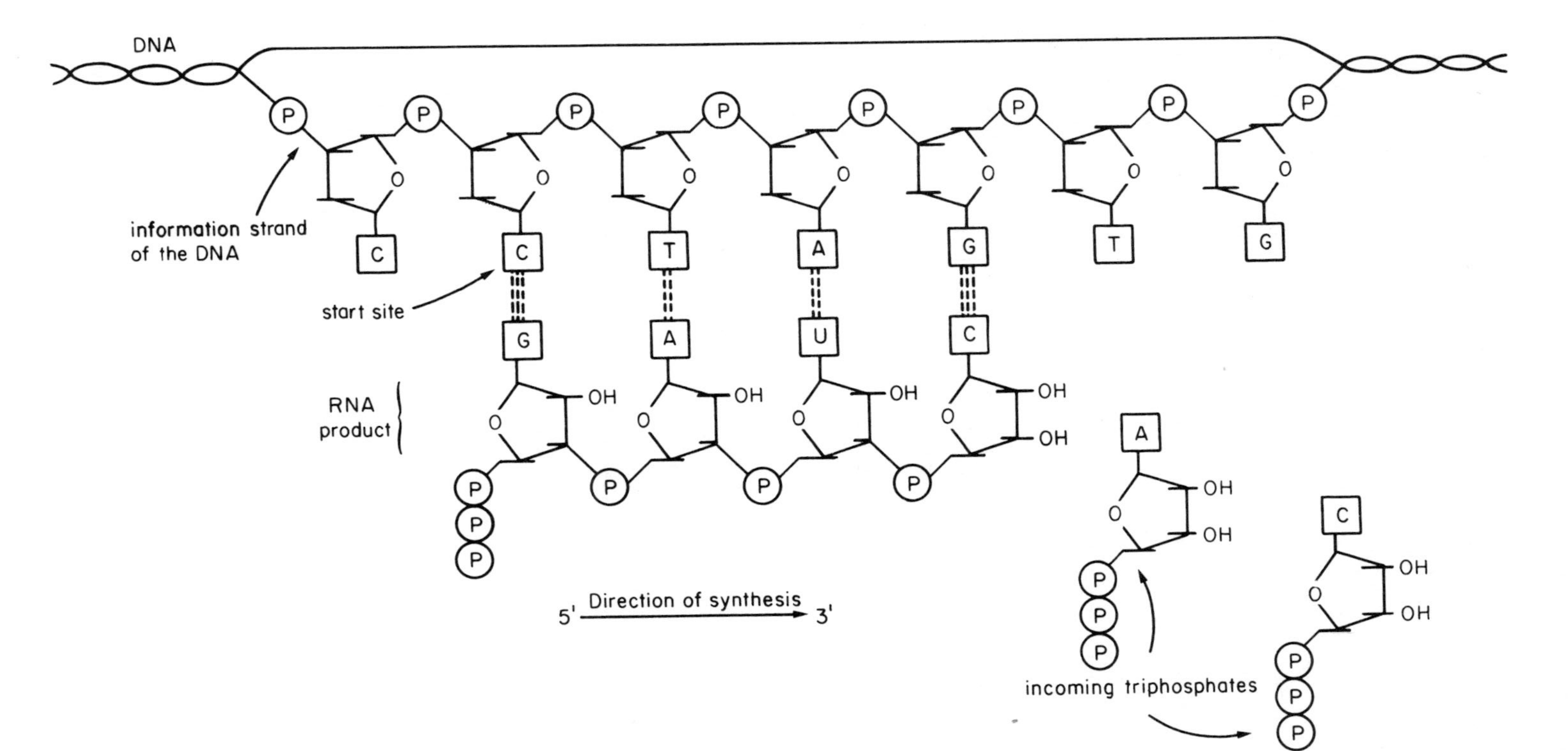

Fig. 7.10. Diagrammatic representation of the biosynthesis of RNA on one strand of DNA acting as template (Reproduced by permission from R.L.P. Adams, R.H. Burdon, A.M. Campbell, and R.M.S. Smellie (eds.), Davidson's The Biochemistry of the Nucleic Acids (8th edn.); Chapman and Hall, London (1976))

lar action: namely, the polymerization of nucleoside triphosphates using single-stranded, or partially uncoiled duplex. DNA to give complementary polyribonucleotide species. The chain elongation proceeds in the 5′→3′ direction and inorganic pyrophosphate is produced (Fig. 7.10).

There is some similarity between RNA biosynthesis and DNA biosynthesis, since the nucleoside triphosphates are the subtrates, the reaction requires the presence of divalent metal ions (Mg^{2+} , Mn^{2+}), the reaction requires a template which determines the next nucleotide to be inserted into the growing chain, and the direction of growth is 5′→3′.

The events which occur during transcription are:[8,84,85] (i) recognition of an initiation site in the promoter region of DNA by the transcription enzyme followed by a change in the conformation of the DNA, thought to involve an unwinding of the DNA over about 4 to 8 base pairs;[86](ii) elongation of the chain by the addition of ribonucleotides to the 3′-hydroxyl group of the incipient chain with the elimination of inorganic pyrophosphate;[87] (iii) progression of RNA biosynthesis forming an RNA strand which peels off the DNA template until the transcription enzyme reaches a termination point when the completed RNA is released and the enzyme is released from the DNA template.

Much information has been obtained about the promoter regions, for example, the sequence for the *E. coli* lac operon is known,[88] but relatively little is known about the process of termination and the structure of terminators. However, some information has been obtained for the possible terminator region of the *E. coli* tyrosine suppressor tRNA gene.[89] The products of transcription are polyribonucleotides complementary to regions of the transcribed DNA. These products may be rRNAs, tRNAs, or the mRNAs which then control protein biosynthesis. However, in many cases the transcription products are not themselves the active materials, but require post-transcriptional processing such as the loss of some nucleotides, the modification of some nucleotides, or the addition of specific nucleotide sequences.

For further details of the transcription process readers are directed to Refs. 8, 82, 85, to the references contained therein, and to standard texts on biochemistry. However, there is one further important point which should be mentioned. It has always been considered that the genes could be read in one way only and that they were non-overlapping, but in 1976 it was reported[90] that the DNA bacteriophage ϕX174 has one gene contained entirely within another so that transcription initiation sites cannot have that unique function. Later work by Sanger and his co-workers[91] has further demonstrated this point, and also contributed a great advance in DNA research by providing the complete sequence of this DNA bacteriophage.

The complete sequence and secondary structure of an RNA viroid has also now been published,[92] and Khorana has reviewed work leading to the total synthesis of a functional gene.[93]

Such advances indicate the intense activity in the field of nucleic acid research, an exciting area to which this chapter has tried to provide a very brief introduction.

REFERENCES

1. T. Caspersson, *Cell Growth and Cell Function*, Norton, New York (1950).
2. J. Brachet, *Chemical Embryology*, Interscience, New York (1950).
3. J. N. Davidson and C. Waymouth, *Biochem. J.*, **38**, 39. (1944).
4. J. N. Davidson and C. Waymouth, *Nutrition Abs. and Review*, **14**, 1 (1944–45).
5. J. D. Watson and F. H. C. Crick, *Nature*, **171**, 737, 964 (1953).
6. J. D. Robertus, J. E. Ladner, J. T. Finch, D. Rhodes, R. S. Brown, B. F. C. Clark, and A. Klug, *Nature*, **250**, 546 (1974).
7. (a) S. H. Kim, F. L. Suddath, G. J. Quigley, A. McPherson, J. L. Sussman, A. H. J. Wang, N. C. Seeman and A. Rich, *Science,* **185**, 435 (1974); (b) G. T. Robillard, C. E. Tarr, F. Vosman and H. J. C. Berendsen, *Nature*, **262**, 363 (1976); (c) M. C. Chen, R. Giegé, R. C. Lord and A. Rich, *Biochemistry*, **14**, 4385 (1975); (d) G. J. Quigley and A. Rich, *Science*, **194**, 796 (1976).
8. *Davidson's Biochemistry of the Nucleic Acids* (8th edn.), revised by R. L. P. Adams, R. H. Burdon, A. M. Campbell and R. M. S. Smellie, Chapman and Hall, London (1976).
9. E. Chargaff and J. N. Davidson (eds.) *The Nucleic Acids*, Vols 1–3, Academic Press (1955–60).
10. T. Caspersson and J. Schultz, *Nature,* **142**, 294 (1939); **143**, 602, (1939).
11. J. Brachet, *Enzymologia*, **10**, 87 (1941).
12. H. Fraenkel-Conrat, N. S. Snell and E. D. Ducay, *Arch. Biochem. Biophys.*, **39**, 80, (1952).
13. E. H. L. Chun, M. H. Vaughan and A. Rich, *J. Mol. Biol.*, **7**, 130 (1963).
14. C. J. Pollard, *Arch. Biochem. Biophys.*, **105**, 114 (1964).
15. A. Gibor and M. Izawa, *Proc. Natl. Acad. Sci.*, **50**, 1164 (1963).
16. M. Edelman, C. A. Cowan, H. T. Epstein and J. A. Schiff, *Proc. Natl. Acad. Sci.*, **52**, 1214 (1964).
17. G. Brawerman and J. W. Eisenstadt, *Biochem. Biophys. Acta.*, **91**, 47 (1964).
18. D. B. Roodyn and D. Wilkie, *The Biogenesis of Mitochondria*, Methuen, London (1968) and refs. therein.
19. C. G. Kurland, *J. Mol. Biol.,* **2**, 83 (1960).
20. A. S. Spirin in J. N. Davidson and W. E. Cohn (eds.) *Progress in Nucleic Acid Research*, Vol. 1, Academic Press, New York (1963) p. 301.
21. M. F. Singer and P. Leder, *Ann. Rev. Biochem.*, **35**, 195 (1966).
22. C. I. Davern, *Proc. Natl. Acad. Sci.*, **55**, 792 (1966).
23. P. A. Levene and T. Mori, *J. Biol. Chem.*, **83**, 803 (1929).
24. A. S. Jones and S. G. Laland, *Acta Chem. Scand.*, **8**, 603 (1954).
25. S. G. Laland and W. G. Overend, *Acta Chem. Scand.*, **8**, 192 (1954).
26. A. Bendich in S. P. Colowick and N. O. Kaplan (eds.) *Methods in Enzymology*, Vol. III (1957), p. 715.
27. E. Chargaff, *Experientia*, **6**, 201 (1950).
28. E. Chargaff, *Fed. Proc.*, **10**, 654 (1951).
29. E. Chargaff, *Essays on Nucleic Acids*, Elsevier, Amsterdam (1963).
30. P. Doty, *Harvey Lectures*, **55**, 103 (1961).
31. J. Marmur and P. Doty, *J. Mol. Biol.*, **4**, 430 (1962).
32. W. T. Astbury, *Symp. Soc. Exptl. Biol.*, **1**, 66 (1947).
33. R. Franklin and R. G. Gosling, *Nature*, **171**, 740 (1953); **172**, 156 (1953).

34. M. H. F. Wilkins and co-workers, *J. Mol. Biol.*, (1960); *ibid*, (1961); *ibid*, **12**, 60 (1965).
35. O. Sinonaglu, in B. Pullmann (ed.), *Molecular Associations in Biology*, Academic Press, New York (1968).
36. M. J. B. Tunis-Schneider and M. F. Maestre, *J. Mol. Biol.*, **52**, 521 (1970).
37. G. A. Rodley, R. S. Scobie, R. H. T. Bates, and R. M. Lewitt, *Proc. Natl. Acad. Sci.*, **73**, 2959 (1976).
38. J. F. Pardon, B. M. Richards, L. G. Skinner, and C. H. Ockey, *J. Mol. Biol.*, **76**, 267 (1973).
39. F. H. C. Crick and A. Klug, *Nature*, **255**, 530 (1975).
40. D. E. Pettijon and R. Hecht, *Cold Spring Harbor Symp. Quant. Biol.* **38**, 31 (1973).
41. A. S. Spirin, *Macromolecular Structure of Ribonucleic Acids*, Methuen, London, 1963.
42. R. W. Holley and co-workers, *Science*, **147**, 1462 (1965).
43. R. W. Holley, *Scientific American*, **214**, 30 (1966).
44. F. Sanger in L. Grossman and K. Moldave (eds.), *Methods in Enzymology*, Vol. 12, Academic Press, New York (1967).
45. M. Szekeley and F. Sanger, *J. Mol. Biol.*, **43**, 607 (1969)
46. G. G. Brownlee, *Determination of Sequences in RNA*, North-Holland, Elsevier, Amsterdam and New York (1972).
47. K. Burton, *Essays in Biochemistry*, Vol. 1, Academic Press, London (1965), p. 58.
48. H. Turler and E. Chargaff, *Biochim. Biophys. Acta*, **195**, 446 (1969).
49. A. S. Jones and R. T. Walker, *Nature*, **202**, 24, 1108 (1964).
50. D. Nathans and H. O. Smith, *Ann. Rev. Biochem.*, **44**, 273 (1975).
51. L. R. Lehman in *Progress in Nucleic Acid Research and Molecular Biology*, Vol. 2, Academic Press, New York (1961).
52. A. Hunter and I. Hlynka, *Biochem. J.*, **31**, 486 (1937).
53. W. Jones, *Amer. J. Physiol*, **52**, 203 (1920).
54. M. Kunitz, *J. Gen. Physiol.*, **24**, 15 (1940).
55. D. G. Smyth, W. H. Stein, and S. Moore, *J. Biol. Chem.*, **238**, 227 (1963).
56. S. Bernhard, *The Structure and Function of Enzymes*, Benjamin, New York, 1968.
57. G. Kartha, J. Bello and D. Harker, *Nature*, **213**, 862 (1967).
58. H. W. Wyckoff, K. D. Hardman, N. M. Allewell, T. Inagami, L. N. Johnson, and F. M. Richards, *J. Biol. Chem.*, **242**, 3984 (1967).
59. (a) B. Gutte and R. B. Merrifield, *J. Amer. Chem. Soc.*, **91**, 501 (1969); (b) G. Bernd and R. B. Merrifield, *J. Biol. Chem.*, **246**, 1922 (1971).
60. R. Hirschmann, R. F. Nutt, D. F. Veber, R. A. Vitali, S. L. Varga, T. A. Jacob, F. W. Holly, and R. G. Denkewalter, *J. Amer. Chem. Soc.*, **91**, 507 (1969).
61. T. Uchida and F. Egami, *J. Biochem.*, **61**, 44 (1967).
62. G. de Lamirande, C. Allard, H. C. DaCosta, and A. Cantero, *Science*, **119**, 351 (1954).
63. J. S. Roth, *J. Biol. Chem.*, **208**, 180 (1954).
64. E. Reid and J. T. Nodes, *Ann. N. Y. Acad. Sci.*, **81**, 618 (1959).
65. M. E. Mirault and K. Scherrer, *Eur. J. Biochem.*, **28**, 197 (1972).
66. A. L. M. Bothwell and S. Altman, *J. Biol. Chem.*, **250**, 1460 (1975).
67. S. K. Niyogi and A. K. Dalta, *J. Biol. Chem.*, **250**, 7307, 7313 (1975).
68. S. Linn and I. R. Lehman, *J. Biol. Chem.*, **240**, 1287, 1294 (1965).
69. I. R. Lehman, *J. Biol. Chem.*, **235**, 1474 (1960).
70. K. Shisido and T. Ando, *Biochim. Biophys. Acta,* **287**, 477 (1972).
71. A. J. Mikulski and M. Laskowski, *J. Biol. Chem.*, **245**, 5026 (1970).

72. M. Lasowski, in P. D. Boyer (ed.), *The Enzymes*, Vol. 4, (3rd edn.), Academic Press, New York (1971).
73. R. Bernadi and G. Bernadi, in P. D. Boyer (ed.) *The Enzymes*, Vol. 4 (3rd edn.), Academic Press, New York (1971).
74. M. Meselson and F. W. Stahl, *Proc. Natl. Acad. Sci.*, **44**, 671 (1958).
75. J. Cairns, *J. Mol. Biol.*, **6**, 208 (1963); *Sci. Amer.*, **214**, 36 (1966).
76. J. H. Taylor, *Molecular Genetics*, Academic Press, New York (1963).
77. R. Okazaki, T. Okazaki, K. Sakabe, K. Sugimoto, R. Kainurna, A. Sugino, N. Iwatsuki, *Cold Spring Harbour Symp. Quant. Biol.*, **33**, 129 (1968).
78. R. Okazaki, T. Okazaki, K. Sahabe, K. Sugimoto and A. Sugino, *Proc. Natl. Acad. Sci.*, **59**, 598 (1968).
79. K. Sugimoto, T. Okazaki, Y. Imac and R. Okazaki, *Proc. Natl. Acad. Sci.*, **63**, 1343 (1969).
80. I. R. Lehman, M. J. Bessman, E. S. Simms and A. Kornberg, *J. Biol. Chem.*, **233**, 163 (1958).
81. A. Kornberg, *Science*, **163**, 1410 (1969).
82. V. M. Ingram, *Biosynthesis of Macromolecules* (2nd edn.) Benjamin, Menlo Park, Cal. (1972).
83. A. Kornberg, *DNA Synthesis*, Freeman, San Francisco (1974).
84. A. Travers, *Cell*, **3**, 97 (1974).
85. R. H. Burdon, *RNA Biosynthesis*, Chapman and Hall, London (1976).
86. J. M. Saucier and J. C. Wang, *Nature New Biol.*, **239**, 167 (1972).
87. V. Maitra and S. Hurwitz, *Proc. Natl. Acad. Sci.*, **54**, 815 (1965).
88. R. C. Dickson, J. Abelson, M. W. Barnes, and W. S. Reznikoff, *Science*, **187**, 27 (1975).
89. P. C. Loewen, T. Sekiya and H. G. Khorana, *J. Biol. Chem.*, **249**, 217 (1974).
90. B. G. Barrell, G. M. Air, and C. A. Hutchison, *Nature*, **264**, 34 (1976).
91. F. Sanger, G. M. Air, B. G. Barrell, N. L. Brown, A. R. Coulson, J. C. Fiddes, C. A. Hutchison, P. M. Slocombe, and M. Smith, *Nature*, **265**, 687 (1977).
92. H. J. Gross, H. Domdey, C. Lossow, P. Jank, M. Raba, H. Alberty, and H. L. Sänger, *Nature*, **273**, 203 (1978).
93. H. G. Khorana, *Science*, **203**, 614 (1979).

Chapter 8

Events at the Ribosome—Translation

(A) INTRODUCTION

Having described the nucleic acids in the previous chapter and having indicated the role of DNA as the genetic material in providing a template for the biosynthesis of mRNA, in this chapter we are concerned with the role of mRNA and the ribosome in the biosynthesis of protein.

The genetic information in a cell is contained in the base sequences of the DNA. In the process of replication new DNA is biosynthesized, using the old DNA as template, for hereditary transmission to new cells. In the process of transcription the genetic information is transferred from the DNA to mRNA in a complementary manner. In the process of translation the four-letter language of the base sequences which comprise the genetic information is converted via mRNA and ribosomes into the twenty-letter language of the amino acids which make up proteins. Many reviews have been written on this topic, and here we will consider concisely the roles of the nucleic acids involved.

Significant advances in studies of the biosynthesis of proteins were not made until cell-free systems were developed and the use of synthetic messengers was initiated. The first cell-free system which was studied—the microsomal fraction from rat liver—was investigated by Siekevitz,[1] but later key experiments were carried out using rabbit reticulocytes[2] or *E. coli*.[3,4]

Such studies, which have been later confirmed by other workers, indicate that protein synthesis *in vitro* requires the following components:

(i) a pool of the twenty amino acids commonly found in proteins;
(ii) ATP, GTP, Mg^{2+};
(iii) appropriate activation enzymes for the amino acids;
(iv) a supply of functionally active ribosomal subunits;
(v) a pool of tRNA molecules for each amino acid;
(vi) a supply of the appropriate enzymes for peptide bond formation, chain initiation and chain termination;
(vii) mRNA or a synthetic substitute.

The following steps are envisaged to occur in protein synthesis:

(a) activation of the amino acids with the formation of aminoacyl-tRNA species;
(b) association of the ribosomal subunits on the mRNA;
(c) binding of an aminoacyl-tRNA to the ribosomal complex and chain initiation;
(d) sequential binding of subsequent aminoacyl-tRNAs with peptide bond formation;
(e) translocation of the ribosome relative to mRNA;
(f) chain termination and the release of completed chains from the complex.

(B) STEPS IN PROTEIN SYNTHESIS

(i) Activation of the Amino Acids

The amino acids are first phosphorylated by ATP under the catalytic action of an activating enzyme specific for the amino acid, and the aminoacyl group is transferred to the specific tRNA, the aminoacyl group being attached to the 3′-position of the terminal adenosine at the ACC end of the tRNA (See Section 7C). Thus the aminoacyl-tRNA synthetase enzymes must have binding sites particular to the amino acid and its appropriate tRNA.

In vitro, the isoleucyl-tRNA synthetase seems to be unusual in that in the absence of the natural $tRNA^{Ile}$ it will catalyse the formation of valinyladenylate.

The reactions which occur in the activation step are given in Fig. 8.1.

$$aa_1 + ATP \xrightleftharpoons{Mg^{2+}/E_1} aa_1 \sim AMP - E_1 + PP_i \rightleftharpoons aa_1 \sim AMP + E_1$$

$$aa_1 \sim AMP - E_1 + tRNA^1 \xrightleftharpoons{Mg^{2+}} aa_1 \sim tRNA^1 + E_1 + AMP$$

Fig. 8.1.

Although nineteen amino acids are activated directly by the process described above, the formation of the glutamine-$tRNA^{Gln}$ complex appears to be indirect and occurs via the following steps:[5]

$$ATP + \text{L-glutamate} + tRNA^{Gln} \xrightleftharpoons{Mg^{2+}} \text{Glu-}tRNA^{Gln} + AMP + PP_i \quad \text{(i)}$$

$$ATP + \text{L-glutamine} + \text{Glu-}tRNA^{Gln} \xrightleftharpoons{Mg^{2+}} \text{Gln-}tRNA^{Gln} + \text{L-glutamate} + ADP + P_i \quad \text{(ii)}$$

Reaction (i) requires Glu + RNA^{Gln} synthetase and reaction (ii) requires a specific aminotransferase.

The aminoacyl-tRNA synthetases show a very high selectivity which is of great importance in ensuring absolute reproducibility in protein synthesis.

Once the aminoacyl-tRNA has been formed, the information which guides the sequence of aminoacids in the protein to be assembled lies entirely in the principle of nucleic acid–base complementarity. This fact was demonstrated by a series of experiments in which cysteinyl-$tRNA^{cys}$, from *E. coli*, was reduced by Raney nickel to alanyl-$tRNA^{cys}$. It was found that alanine was incorporated specifically instead of cysteine in the appropriate positions of haemoglobin on introducing the modified loaded tRNA into a protein-synthesizing system. It was further found that ala-$tRNA^{cys}$ was unable to transfer the alanyl group to $tRNA^{ala}$ in the presence of either of the appropriate tRNA synthetases.[6]

By the above mechanism many tRNAs become loaded with their specific amino acids and carry them to the ribosomes where the process of polypeptide-chain synthesis occurs. The tRNA's are liberated and they can be reloaded to go through this process repeatedly.

In vitro amino acid activation is reversible, but *in vivo* the reverse reaction does not seem to be favoured and a high concentration of aminoacyl tRNA is maintained in the cytoplasm.

(ii) Initiation

The polypeptide chain is synthesized at the ribosome starting from the amino end and proceeding to the carboxyl end. The messenger RNA is 'read' in the direction $5' \rightarrow 3'$, and as this is the direction in which the mRNA chain is synthesized from the DNA template, it is possible that the first formed portion of mRNA may associate with a ribosome and start protein synthesis before transcription of the whole mRNA species is complete.

The mRNA attaches to the small ribosomal subunit and, at an 'initiation codon', the larger subunit associates with the smaller unit and a loaded tRNA, having the complementary anticodon to the initiation codon, takes up position. The larger ribosomal subunit then becomes associated with the mRNA—small ribosomal unit—aatRNA complex such that the initiation aatRNA occupies the donor (peptide) site of the larger subunit.

A number of initiation factors also seem to be necessary for protein synthesis to begin. Ochoa has purified three such factors—F_1, F_2 and F_3—from the 30S ribosomal subunits of *E. coli.*[7]

It is now well established (cf. Davidson[8]) that for *E. coli* in either whole cells or cell free systems prepared from this microorganism, the first amino acid to start the polypeptide chain is *N*-formylmethionine (**1**)

$$\begin{array}{l} \qquad\qquad\quad CH_2CH_2SCH_3 \\ \qquad\qquad\quad | \\ OHCNH-CHCO_2H \end{array}$$

(1)

The $tRNA^{fMet}$ from *E. coli* has been shown[9] to have a different primary structure from $tRNA^{Met}$ and to represent about 70% of the bacterial methionine tRNA. Formylation of methionine only takes place after $tRNA^{fMet}$ has become loaded whilst loaded $tRNA^{Met}$ cannot be formylated.

The anticodon for $tRNA^{fMet}$ is UAC, and the codon AUG, which also codes for $tRNA^{Met}$, is the initiation codon leading to the formation of a polypeptide with formylmethionine at the amino end. However, the codon GUG is also capable of acting as an initiation codon and can be read by $tRNA^{fMet}$, though not by $tRNA^{Met}$, leading to polypeptide starting with formylmethionine, When AUG is not sited at the initiation point of mRNA it codes for methionine whilst GUG in an internal position codes for valine.

To summarize the initiation sequence:

(i) The 30S ribosomal unit binds to mRNA in the presence of two initiation factors, one of which may act as a recognition site for the mRNA.
(ii) f-Met-$tRNA^{fMet}$ then binds to an AUG codon in the m-RNA-30S ribosomal subunit complex in the presence of the third initiation factor and in the presence of GTP which is hydrolysed to GDP and P_i.
(iii) Lastly the 50S ribosomal subunit binds and chain elongation proceeds.

The above processes are those which have been observed to occur in bacterial systems. However, the processes seem to be very similar in eukaryotic cells, although these cells do not have the transformylases of bacterial cells which can remove the *N*-formyl group from the polypeptide chain.[10] But eukaryotes have two species of methionine tRNA, one of which seems to have an initiator role and can be formylated by bacterial enzymes.[11,12] It is also known that some eukaryotic proteins, e.g. rabbit α- and β-globulin chains, are started with methionine.[12]

Four initiation factors have been isolated from rabbit reticulocyte and liver ribosomes[13].

(iii) Chain Elongation

The polypeptide chains are formed by the stepwise addition of individual amino acids, via aatRNAs, to the initiation amino acid or the incipient protein by peptide-bond formation and concomitant displacement of the tRNA from the peptide site. The peptide bond formation step is essentially a nucleophilic displacement of the tRNA at the peptidyl position by the amino group of the incoming aminoacyl tRNA. The reaction is catalysed by a peptidyl transferase which seems to be part of the larger ribosomal subunit. The polypeptide with its new amino acid is now bound to the ribosome by the tRNA of the new aminoacid. Each functioning ribosome thus carries a single growing polypeptide chain. However, each mRNA may carry a number of ribosomes at any one time, each of which will have a polypeptide chain at different stages of development.

These groups of ribosomes associated with a single strand of mRNA are known as polysomes or polyribosomes. The polysomes may contain up to

40 ribosomes, such as those isolated from *E. coli*,[14–17] or as few as five such as those from rabbit reticulocytes involved in the synthesis of haemoglobin.[18–22]

The ribosomes move relative to the mRNA—most probably the mRNA moves through the ribosomes—in a direction from the 5′- to the 3′- end of the mRNA chain. This movement of the ribosome relative to mRNA is termed *translocation*. This process requires GTP and two or three transfer factors. The peptide-bond formation is catalysed by a peptidyl transferase which is an intrinsic part of the larger ribosomal subunit.

The ribosome altogether spans about thirty nucleotide units[23] but two adjacent base triplet codons control the tRNAs held by the peptidyl and aminoacyl sites of the ribosome. The tRNAs which enter are those which have the anticodon complementary to the codon at the acceptor site. However, it is known that a single tRNA can recognize several codons as long as they differ in the last base in the codon. For example, serine with the anticodon AGI can recognize UCU, UCC and UCA.

This degeneracy of the genetic code has been suggested by Crick[24] to be due to a certain amount of latitude in the hydrogen-bonding requirements for the third base in the codon, an effect which he has called the 'wobble' hypothesis.

This observation has been supported by Khorana[25] and the wobble allowed seems to be (for the third base of a codon) U, C or A for I, and U or C for G.

A 'two out of three' system of recognition in ordinary translation has recently been suggested by Lagerkvist[26], whilst it has also been suggested that a quadruplet code provides the initiation signal[27]. But the principles of the classical genetic code are outlined below.

(C) RECOGNITION BETWEEN CODON AND ANTICODON—THE GENETIC CODE

It has just been mentioned that the binding of the loaded tRNAs at the peptidyl (P) and acceptor (A) sites of the ribosome are controlled by the complementarity of the bases in the anticodon loop of tRNA and the codon adjacent to the A and P sites.

It was early recognized by Crick and coworkers[28] that whereas the genetic information was carried by DNA and thence to the ribosomes by the mRNA, there must be a code for converting the sequences of four bases in the nucleic acids to the twenty odd possibilities for amino acids, and that this code was a triplet code—that is, a codon of three bases on the mRNA codes for a single amino acid in the polypeptide. Several people, notably Nirenberg, Ochoa and Khorana undertook the problem of assigning triplets of bases to each of the twenty amino acids and used several techniques. The first attempts were made using cell-free extracts from *E. coli* and synthetic nucleotide polymers—e.g. poly U was found to code for phenylalanine, poly A for lysine and poly C for proline. It was later found

Table 8.1 The genetic code

5′-OH terminal base	Middle base				3′-OH terminal base
	U	C	A	G	
U	Phe	Ser	Tyr	Cys	U
	Phe	Ser	Tyr	Cys	C
	Leu	Ser	term.	term.	A
	Leu	Ser	term.	Trp	G
C	Leu	Pro	His	Arg	U
	Leu	Pro	His	Arg	C
	Leu	Pro	Gln	Arg	A
	Leu	Pro	Gln	Arg	G
A	Ile	Thr	Asn	Ser	U
	Ile	Thr	Asn	Ser	C
	Ile	Thr	Lys	Arg	A
	Met (init.)	Thr	Lys	Arg	G
G	Val	Ala	Asp	Gly	U
	Val	Ala	Asp	Gly	C
	Val	Ala	Glu	Gly	A
	Val (init.)	Ala	Glu	Gly	G

that oligonucleotides containing three bases were the minimum that would be bound to ribosomes and would direct an incoming aminoacyl tRNA. All 64 possible triplets have been assigned by extension of this work.[29]

This ribosomal binding technique has been further extended using copolymers of defined sequence whilst alternative methods which have been investigated include evidence from bacteriophage RNAs of known sequence (or at least having regions of known sequence) and the use of mutations. Further details of these studies may be found in Davidson,[8] in the original literature, in *Progress in Nucleic Acid Research and Molecular Biology*, and in a number of other texts.

The three unassigned codons, UAA, UAG and UGA act as the chain-terminating signals. Thus the genetic code has been clearly demonstrated through a cosiderable amount of elegant work and it seems to be universal in that it seems to be obeyed by all types of living matter.[30,31]

The genetic code represented by the mRNA codons and the corresponding aminoacids is given in Table 8.1.

(D) CHAIN TERMINATION

The ribosome moves relative to the mRNA and the polypeptide chain contines to grow until the ribosome reaches one of the termination codons at the A site. Two releasing factors then bind to the ribosome and the com-

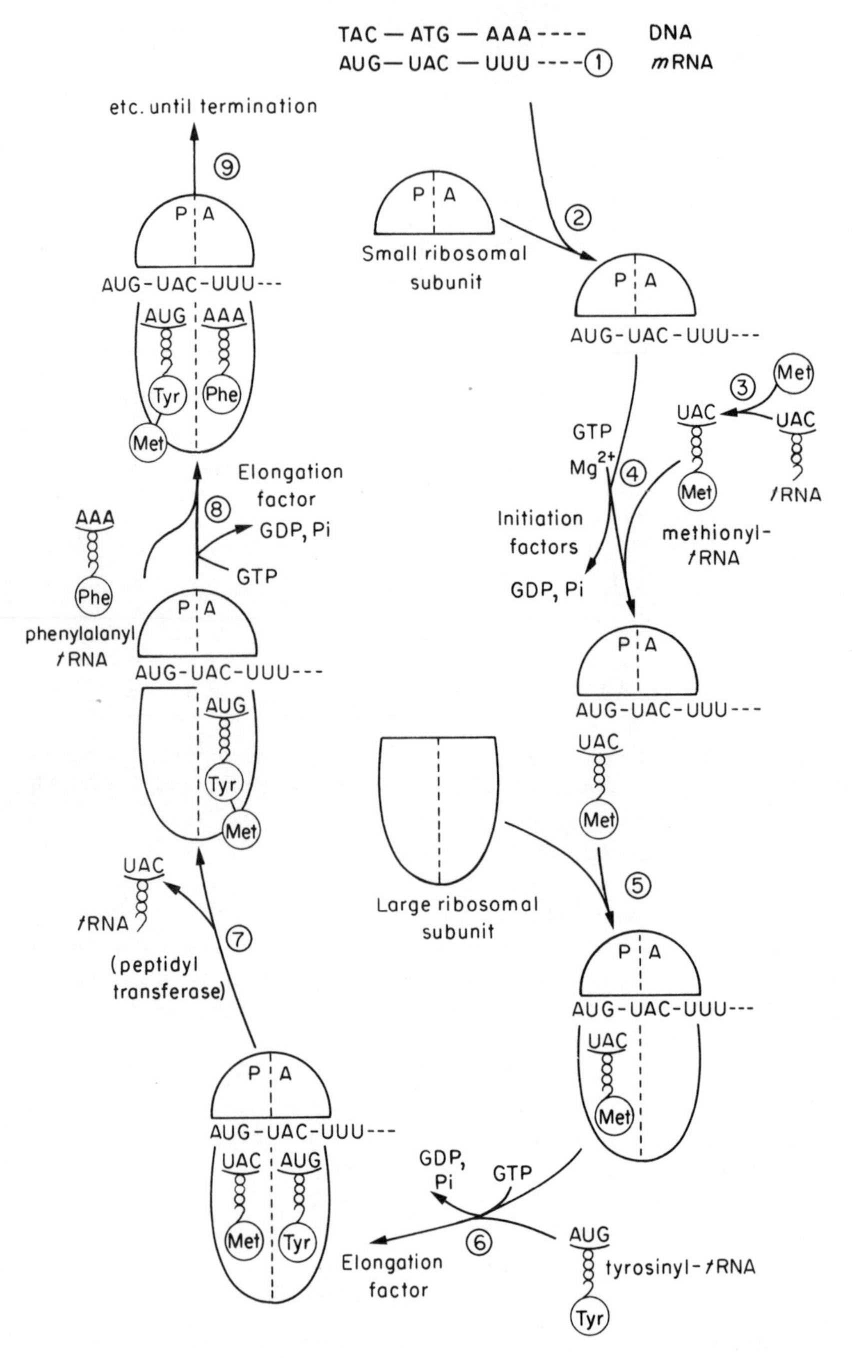

Fig. 8.2. Events at the ribosome.

pleted polypeptide chain is released from the ribosome. The ribosomal subunits also separate and are released from the mRNA. The precise details of all steps in chain termination are still not clear but it is evident that it is a multistep process requiring a number of proteins having specific catalytic activities, and it may be that of the order of five factors are required.[32]

The sequence of events which occur at the ribosome are indicated diagrammatically in Fig. 8.2.

(E) INHIBITION OF PROTEIN SYNTHESIS

For cell division to occur, thus for tumour development or bacterial growth to occur, it is necessary that there be DNA synthesis, RNA synthesis and the synthesis of active, viable, protein to serve in the appropriate catalytic and structural roles in the cell. Interference at any step in the process

$$\text{DNA} \xrightarrow{\text{replication}} \text{DNA} \xrightarrow{\text{transcription}} \text{mRNA} \xrightarrow{\text{translation}} \text{protein}$$

may thus result in useful anticancer or antibacterial action.

In previous sections (see Section 4F) reference has already been made to some aspects of the inhibitory actions of a number of compounds. In the last two chapters some further discussion of such topics will be made, and here the actions of some compounds as inhibitors of protein synthesis will be summarized. A selection of such compounds is given in Table 8.2.

Inhibitory action may occur at one of a number of sites and some compounds have several actions. For example, the antimetabolites may inhibit particular steps in the synthesis of nucleoside triphosphates but also exert a feed-back control in the general pathways of nucleic acid biosynthesis. The antibiotics frequently have particular sites of action, for example, Actinomycin D forms complexes with the guanine residues in DNA, probably intercalating the G–C base pair with the cyclic compound (Fig. 8.3) being situated in the groove and being hydrogen-bonded to guanine.[33] Actinomycin D inhibits both DNA polymerase and DNA-dependent RNA polymerase. However, its action on the transcription enzyme is very much more pronounced than its effect on the DNA polymerase.

Some other antibiotics seem to have specificity for the type of cell—for example cycloheximide seems to inhibit translocation specifically in eukaryotic cells, whilst a number of other compounds, e.g. tetracycline, erythromycin, chloramphenicol, have specificity for prokaryotic ribosomes.

Several reviews of antibiotic action and protein inhibition are available including refs. 34–36, and more detailed accounts of such action are given in the references cited both here and in these reviews.

Table 8.2 Some inhibitors of nucleic acid and protein synthesis

Compound	Action	Effect
Alkylating agents (bifunctional): e.g. Cyclophosphamide (Endoxan) Uracil mustard (Nor-dopa)	Cross link DNA	Inhibits replication and transcription
Antibiotics:		
e.g. Mitomycin C	Cross links G residues	Inhibits replication and transcription
Daunomycin Adriamycin	Binds to DNA	
Phleomycin Bleomycin	Attaches to T residues	Inhibits replication causes break in DNA
Actinomycin D	Binds to DNA, G residues	Much more effective inhibitor of transcription than of replication
Streptonigrin	Inhibits DNA and RNA polymerase	Inhibits transcription and replication
Rifampicin	Inhibits RNA polymerase	Inhibits transcription
Cordycepin Tubercidin	Inhibitors of purine biosynthesis	Inhibits nucleic acid biosynthesis
Streptomycin	Binds to smaller ribosomal subunit	Causes misreading of genetic code
Tetracycline	Binds to smaller ribosomal subunit	Blocks binding of aatRNA to A site
Erythromycin	Binds to larger ribosomal subunit	Blocks P site and inhibits translocation
Puromycin	Binds to larger ribosomal subunit	Causes termination of polypeptide chain

Intercalating agents		
e.g. Proflavine, ethidium bromide	Intercalate DNA causing some unwinding of the duplex helix	Causes errors in replication and transcription
Base analogues		
e.g. 8-Azaguanine	Inhibits GMP biosynthesis	Inhibits nucleic acid synthesis
6-Mercaptopurine	Inhibits conversion of IMP to AMP	
6-Azauracil	Inhibits OMP decarboxylase	
5-Fluorouracil	Inhibits thymidylate synthetase	Inhibits DNA biosynthesis Changes genetic code, hence changes protein synthesis
5-Bromouracil 5-Iodouracil	Incorporated into DNA in place of thymine	
2,6 Diaminopurine	Incorporated into nucleic acids in place of guanine	
Folate antagonists		
e.g. Aminopterin Methotrexate	Inhibit DHFR	Inhibit DNA synthesis
Other compounds		
e.g. Chloramphenicol	Binds to large ribosomal subunit in bacteria and mitochondria	Inhibits peptide bond formation
Cycloheximide	Acts at the ribosome	Inhibits translocation specifically in eukaryotic cells

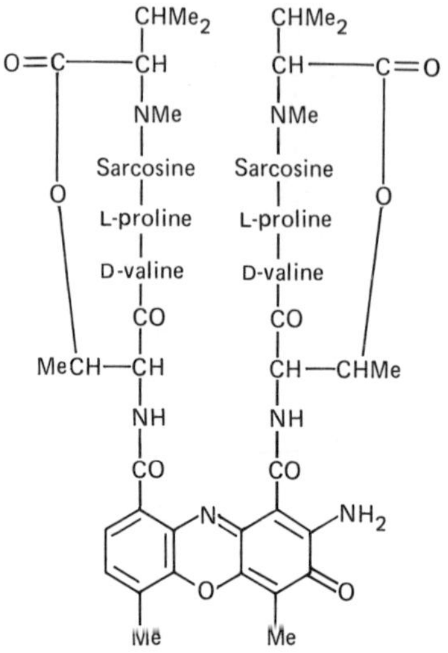

Fig. 8.3. Structure of actinomycin D

(F) MUTATION AND MUTAGENS

In Chapters 7 and 8 the roles of the nucleic acids in the transmission of genetic information from generation to generation and in the ordering of protein and enzyme synthesis have been indicated. It is the base sequences in a particular region of DNA which ultimately control the number and sequence of amino acids in a particular polypeptide. Such a region is a *gene* and, until recently (see Section 7G) it was believed that the information contained in the base sequence was non-overlapping and that one gene specified one polypeptide chain. Thus any change in the base sequence could result in a change in the protein coded for by that section of the genetic material. Such a change in the genetic information which results in a change in the proteins whose biosynthesis it governs is termed a mutation.

The genes are generally considered to constitute stable entities since hereditary characteristics are transmitted to successive generations and different species breed true. However, it is through mutation that evolution has occurred. Mutations can result from alterations in the shape, structure or constituent bases of DNA. Such alterations may be gross, such as chromosomal breaks, deletions, or translocation, or may be 'point mutations' in which only one or a small number of bases are altered.

Mutations may be induced by a variety of effects including:

(i) Substitution by base analogues
(ii) Chemical modifications of DNA or its constituent bases
(iii) Intercalation by compounds that bind to DNA

(iv) Depolymerization of DNA
(v) The action of u.v. or ionizing radiations.

Some compounds have useful antiviral, antimicrobial, or anticancer activity because of these properties (see Sections 9C, 10F) since mutations are frequently lethal to cells but the above items will be briefly mentioned here. Further details may be found in refs. 37 and 38.

(i) Substitution by Base Analogues

A number of synthetic pyrimidines and purines, when incorporated into DNA synthesizing systems, can become incorporated into the DNA. Such substitutions may lead to pairing mistakes when the newly synthesized mutant DNA undergoes replication or transcription. For example, 5-bromouracil can become incorporated in place of thymine in DNA, and this base has the property of pairing with either the natural base adenine or with guanine.

This occasional pairing error by the 5-bromodeoxyuridine site in DNA is thought to be due to the greater electronegativity of Br relative to Me tending to favour the enolic form to some extent rather than the keto form.[39] This enolic form is complementary to guanine instead of adenine (Fig. 8.4).

Other bases, e.g. 2-aminopurine, etc. also act in a similar way causing base-pairing errors.

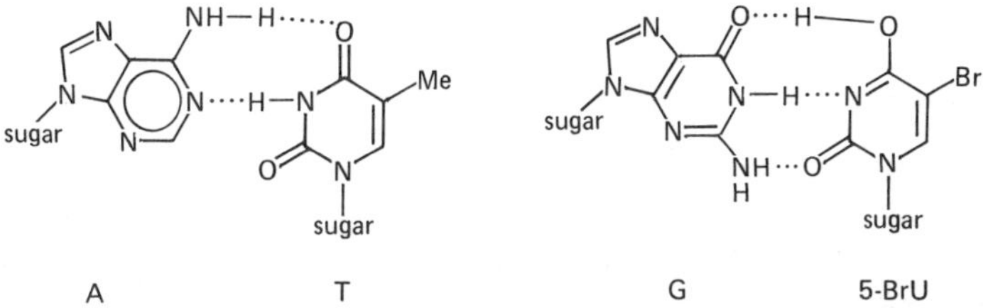

Fig. 8.4. Base-pairing of the enolic form of 5-BrU

(ii) Chemical Modifications of DNA and the Constituent Bases

A number of compounds fall into this class of potential mutagens, the most common type being alkylating agents which may be either mono- or poly-functional.

Monofunctional alkylating agents have a high specificity for alkylating the N-7 position of guanine which causes labilization of the N-9 sugar link, resulting in the loss of guanine residues from the DNA (see Section 3B(iii)d). The resulting gap in the base sequence results in either a deletion in the new strand or the introduction of a random base into the blank site.

It is also possible that alkylation of guanine leads to a change in the base-pairing effect, also resulting in a point mutation.

Polyfunctional alkylation agents (e.g. the nitrogen mustards, etc.) have the above effects but also can result in inter- or intra-chain bridging.[40]

It has been noted above that mitomycin C can also cross-link DNA (after enzymatic reduction).[41]

Such bridge formation causes inhibition of replication and transcription and causes base-sequence errors in these processes.

Several simple chemicals cause modifications in the nitrogenous bases of the nucleic acids. For example, nitrous acid can cause deamination of cytosine to uracil whilst it is present in a nucleic acid,[37] whilst sulphur dioxide, hydroxylamine and hydrazine also react with pyrimidine bases causing deamination of cytosine and/or ring opening of the pyrimidine ring.[37]

(iii) Intercalation of DNA

Some compounds are known to cause structural modifications to nucleic acids without undergoing reactions with the bases. Acridines, phenanthridines, a range of polycyclic aromatic hydrocarbons and some polycyclic antibiotics have been shown to intercalate adjacent base-pairs in helical regions of nucleic acids. This intercalation seems to produce frame-shift mutations (see ref. 37). It should be noted that many of the aromatic hydrocarbons which intercalate DNA (e.g. benzpyrene) are also carcinogenic (see Chapter 9).

Even caffeine has been reported to be mutagenic, although the mechanism by which it acts is uncertain. However, it has been suggested that caffeine interferes with DNA repair processes (see ref. 37).

(iv) Depolymerization of DNA

There is evidence that free radicals, for example, from hydrogen peroxide or from reactions which produce organic free radicals, reduce the u.v. absorption of DNA. It seems that the nitrogen bases are the site of attack and these are degraded to ring opened products. As a result the sugar-phosphate backbone of the DNA is labilised and strand breakage occurs.[42]

The β-glycosidic link is also susceptible to attack to release free bases and again the sugar-phosphate backbone becomes liable to cleavage.[43]

(v) The Action of u.v. and Ionizing Radiations

Ionizing radiations can also cause depolymerization of DNA strands and can induce modifications of the nitrogen bases. Ultraviolet radiation may also cause such effects, but the principal effect is the production of pyrimidine dimers. Of the possible types of pyrimidine dimer the thymine dimer is formed most readily. The production of such intrastrand dimers blocks the action of DNA polymerase and prevents replication.[44–47] However, there appear to be several enzymic mechanisms for the repair of such genetic damage. The type of thymine dimer which is formed is shown in Fig. 8.5:

Gross mutations or mutations that produce nonsense codes are most likely to be lethal and produce completely non-viable genetic material. However, some point mutations can produce genetic modifications. It is

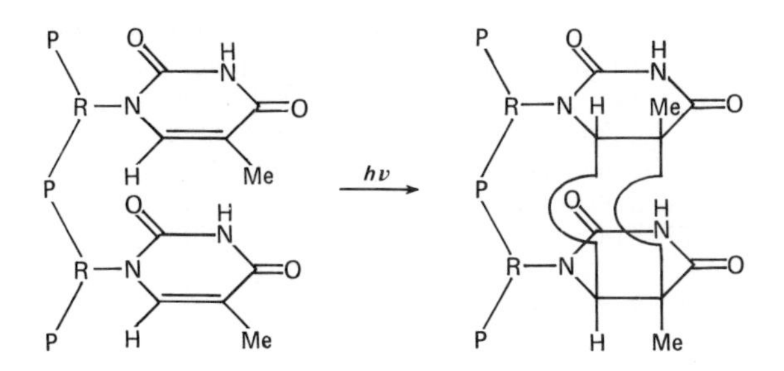

Fig. 8.5. Thymine dimer formation

through such changes that evolution has occurred. However, a discussion of genetics and heredity is not intended here and the reader is directed to specialist texts on this subject.

REFERENCES

1. (a) P. Siekevitz, *J. Biol. Chem.*, **195**, 549 (1952); (b) P. C. Zamecuik and E. B. Keller, *J. Biol. Chem.*, **209**, 338 (1954).
2. R. Schweet, H. Lamfrom, and E. Allen, *Proc. Natl. Acad. Sci. U.S.*, **44**, 1029 (1958).
3. M. R. Lamborg and P. C. Zamecuik, *Biochim. Biophys. Acta.*, **42**, 206 (1960).
4. (a) A. Tissières, D. Schlessinger, and F. Gros, *Proc. Natl. Acad. Sci. U.S.*, **46**, 1450 (1960); (b) A. Tissières and J. W. Hopkins, *ibid.*, **47**, 2015 (1961).
5. M. Wilcox and M. Nirenberg, *Proc. Natl. Acad. Sci. U.S.*, **61**, 229 (1968).
6. (a) F. Chaperille, F. Lipmann, G. Von Ehrenstein, B. Weisblum, W. J. Ray, and S. Benzer, *Proc. Natl. Acad. Sci. U.S.*, **48**, 1086 (1962); (b) G. Von Ehrenstein, B. Weisblum, and S. Benzer, *ibid.*, **49**, 669 (1963); (c) G. Herve and F. Chaperille, *Biochim. Biophys. Acta.*, **76**, 493 (1963).
7. (a) S. Sabol, M. A. G. Sillero, K. Iwasaki, and S. Ochoa, *Nature*, **228**, 1269 (1970); (b) S. Sabol and S. Ochoa, *Nature, New Biol.*, **234**, 233 (1971); (c) S. Lee-Huang and S. Ochoa, *Nature, New Biol.*, **234**, 236 (1971).
8. *Davidson's Biochemistry of the Nucleic Acids* (8th edn.), revised by R. L. P. Adams, R. H. Burdon, A. M. Campbell, and R. M. S. Smellie, Chapman and Hall, London (1976).
9. S. K. Dube, K. A. Marcker, B. F. C. Clark and S. Cory, *Nature*, **218**, 232 (1968).
10. J. M. Adams, *J. Mol. Biol.*, **34**, 131 (1968).
11. (a) J. C. Brown and A. E. Smith, *Nature*, **226**, 610 (1970); (b) A. A. Smith and K. A. Marcker, *Nature*, **226**, 607 (1970).
12. R. Jackson and T. Hunter, *Nature*, **227**, 672 (1970).
13. (a) D. A. Shafritz, P. M. Prichard, J. M. Gilbert, and W. F. Anderson, *Biochem. Biophys. Res. Commun.*, **38**, 721 (1970); (b) D. A. Shafritz, P. M. Prichard, J. M. Gilbert, W. C. Merrick, and W. F. Anderson, *Proc. Natl. Acad. Sci., U.S.*, **69**, 983 (1972); (c) D. J. Picciano, P. M. Prichard, W. C. Merrick, D. A. Shafritz, H. Graff, R. G. Crystal, and W. F. Anderson, *J. Biol. Chem.*, **248**, 204 (1973).
14. W. Gilbert, *J. Mol. Biol.*, **6**, 374 (1963); **6**, 389 (1963).
15. T. Staehelin, C. C. Brinton, F. O. Wettstein, and H. Noll, *Nature*, **199**, 865 (1963).
16. M. Schaecter, *J. Mol. Biol.*, **7**, 561 (1963).
17. Y. Kibo and A. Rich, *Proc. Natl. Acad. Sci.*, **51**, 111 (1964).

18. (a) J. R. Warner, A. Rich, and C. E. Hall, *Science,* **138**, 1399 (1962); (b) J. R. Warner, P. M. Knopt, and A. Rich, *Proc. Natl. Acad. Sci*., **49**, 122, (1963).
19. A. Gierer, *J. Mol. Biol*., **6**, 148 (1963).
20. P. A. Marks, E. R. Burka, R. Rifkind, and D. Danon, *Cold Spring Harbour Symp. Quant. Biol*., **28**, 269 (1963).
21. B. Hardesty, R. Miller, and R. Schweet, *Proc. Natl. Acad. Sci*., **50**, 924 (1963).
22. A. P. Mathias, R. Williamson, H. E. Huxley, and S. Page, *J. Mol. Biol*., **9**, 154 (1964).
23. M. Takanami and G. Zubay, *Proc. Natl. Acad. Sci*., **51**, 834, (1964).
24. (a) F. H. C. Crick, *Scientific American*, **215**, 55 (1966); (b) F. H. C. Crick, *J. Mol. Biol*., **19**, 548 (1966).
25. D. Söll, D. S. Jones, E. Ohtsuka, R. D. Faulkner, R. Lohrmann, H. Hayatsu, H. G. Khorana, J. D. Cherayil, A. Hampel, and R. M. Bock, *J. Mol. Biol*., **19**, 556 (1966).
26. U. Lagerkvist, *Proc. Natl. Acad. Sci., U.S.*, **75**, 1759 (1978).
27. U. Manderschied, S. Bertram, and H. G. Gassen, *FEBS Letters,* **90**, 162 (1978).
28. (a) F. H. C. Crick, *Cold Spring Harbour Symp. Quant. Biol*., **31**, 3 (1966); (b) F. H. C. Crick, *Proc. Roy. Soc.* (B), **167**, 331 (1967), (c) F. H. C. Crick, L. Barnett, S. Brenner, and R. J. Watts-Tobin, *Nature*, **192**, 1227 (1961); (d) F. H. C. Crick in J. N. Davidson & W. E. Cohn (eds.), *Progress in Nucleic Acid Research*, Vol. 1 Academic Press, New York (1963), p. 163.
29. H. Matthaei, G. Heller, H. P. Voigt, R. Neth, G. Schöch and H. Kübler, *Genetic Elements*, *FEBS Symposium* (D. Shugar), 233 (1966).
30. M. B. Matthews, *Essays in Biochemistry*, **9**, 59 (1973).
31. M. Ochoa and J. B. Weinstein, *Proc. Natl. Acad. Sci.*, **52**, 470 (1964).
32. (a) E. Scolnick, R. Tomkins, T. Caskey, and M. Nirenberg, *Proc. Natl. Acad. Sci.*, **61**, 768 (1968); (b) E. M. Scolnick and C. T. Caskey, *ibid*, **64**, 1235 (1969); (c) G. Milman, J. Goldstein, E. Scolnick and T. Caskey, *ibid*, **63**, 183 (1969); *ibid.,* **65**, 430 (1970); (d) C. T. Caskey, E. M. Scolnick, R. Tompkins, G. Milman and J. Goldstein, *Methods in Enzymology,* **20**, 367 (1971); (e) M. M. Rechler and R. G. Martin, *Nature,* **222**, 908 (1970); (f) Z. Vogel, A. Zamir and D. Elson, *Proc. Natl. Acad. Sci.,* **61**, 701 (1968).
33. H. M. Sobell, *Progr. Nucleic Acid Res. Mol. Biol*., **13**, 153 (1973).
34. T. J. Franklin and G. A. Snow, *Biochemistry of Antimicrobial Action* (2nd. edn.), Chapman and Hall, London (1975).
35. E. F. Gale, E. Cundliffe, P. E. Reynolds, M. A. Richmond, and M. J. Waring, *The Molecular Basis of Antibiotic Action*, Wiley, London, New York, Sydney, Toronto (1972).
36. I. H. Goldberg and P. A. Friedman, *Ann. Rev. Biochem*., **40**, 775 (1971).
37. L. Fishbein, W. G. Flamm, and H. G. Falk, *Chemical Mutagens*, Academic Press, New York and London (1970).
38. A. Goldstein, L. Aronow, and S. M. Kalman, *Principles of Drug Action* (2nd edn.), Wiley, New York, London, Sydney, Toronto, (1974), Chap. 10.
39. P. D. Lawley and P. Brookes, *J. Mol. Biol*., **4**, 216 (1962).
40. (a) K. Kohn, N. Steigfigel, and C. Spears, *Proc. Natl. Acad. Sci*., **53**, 1154 (1965); (b) K. Kohn, D. Green and P. Doty, *Fed. Proc*., **22**, 582 (1963).
41. V. N. Iyer and W. Szybalski, *Proc. Natl. Acad. Sci*., **50**, 355 (1963).
42. H. Pries and W. Zillig, *Z. Physiol. Chem*., **342**, 73 (1965).
43. J. A. V. Butler and B. E. Conway, *Proc. Roy. Soc.,* **B141**, 562 (1953).
44. R. E. Rasmussen and R. B. Painter, *Nature*, **203**, 1360 (1964).
45. D. Pettijohn and P. Hanawalt, *J. Mol. Biol*., **9**, 395 (1964).
46. R. B. Setlow, *Progr. in Nucleic Acid Res. and Mol. Biol*., **9**, 257 (1968).
47. P. Hanawalt, *Endeavour*, **31**, 83 (1972).

Chapter 9

Carcinogens and Cancer Chemotherapy

(A) INTRODUCTION

Cancer is not a single disease but is a general name used to describe a number of related conditions each of which is characterized by the uncontrolled growth of certain cells. The tumour which forms in these neo- or hyperplastic conditions, which is malignant, is differentiated from benign tumours by four principal criteria:

(i) they are not encapsulated and can invade and destroy the tissue in which they arise;
(ii) they are characterized by unlimited power of disorderly reproduction quite different from the orderly way in which the normal cells of an organ are reproduced;
(iii) malignant tumour cells show loss of differentiation;
(iv) they are capable of producing secondary growths remote from the primary tumour.

Leukaemia is a form of cancer in which there is uncontrolled over-production of white corpuscles, and the number of these in the blood is permanently increased. The condition is also characterized by enlargement of the spleen and lymph glands, and changes in bone marrow.

The precise nature of cancerous growth is still unclear, and a considerable research effort is being prosecuted into the causes of cancer as well as methods for the control of cancer. There is a considerable amount of literature now available, including the journals *Cancer Research*, *Advances in Cancer Research*, *British Journal of Cancer*, *European Journal of Cancer*, as well as many papers in other journals, reviews in the popular science journals such as *New Scientist*, *Scientific American*, as well as in *Nature* and *Science*, and there are many books on different aspects of cancer and anti-cancer agents.

Cancer is an important medical phenomenon that seems to be contributing to the deaths of increasing numbers of people every year in virtually all countries of the world.[1] This chapter aims to review briefly some of the current knowledge of carcinogens and cancer chemotherapy in relation to the contribution of the nitrogen heterocycles and nucleic acids to such topics.

It seems that cancer results from a sequence of biological alterations which may occur over a long period of time, but there is no type of tumour in which the sequence of steps of its origin and development from a normal cell is fully understood. The evidence for this multifactoriality is mainly statistical in man but is experimental in animals,[2] and in the human disease retinoblastoma at least four factors seem to be in operation—DNA alteration, cell differentiation, cell proliferation, and chance, which is perhaps error during DNA synthesis.

It was recognized quite some time ago that certain occupations carried a high cancer risk for certain types of disease, and it is now well known that a wide variety of chemicals can induce cancer—for example, vinyl chloride, β-naphthylamine, benz[*a*]pyrene, asbestos, etc. As time passes, yet more relatively common compounds are being identified as having carcinogenic activity in animals and must therefore also be suspect in humans.

In addition to chemical carcinogenesis, it is also established that other agents—for example u.v. light, X-rays, and nuclear radiation—are also carcinogenic, and more recently, oncogenic viruses have attracted much attention. A discussion of the role of viruses in tumour initiation is beyond the scope of this chapter, but further information may be found in the current literature and in the references contained in ref. 2.

The factors in the development of cancer which may be influenced by chemical carcinogens are: mutations, DNA repair, cell toxicity, cell proliferation, cell differentiation, and the induction of activating enzymes. Other factors which are important in the development of cancer are immunity, tumour progression, and (perhaps) viruses.[2]

(B) CHEMICAL CARCINOGENESIS

Although a wide variety of chemicals have been found to have carcinogenic action,[3,4] generally speaking, the important examples fall into five groups:

(i) polycyclic aromatics
(ii) biological alkylating agents
(iii) aromatic amines and azo-compounds
(iv) *N*-nitrosamines and -amides
(v) metallic compounds and miscellaneous carcinogens.

Some examples of the above types of compound are given in Fig. 9.1, and some mechanisms by which they may exert their action are given below. A recent comprehensive review of this topic is available from the American Chemical Society.[4]

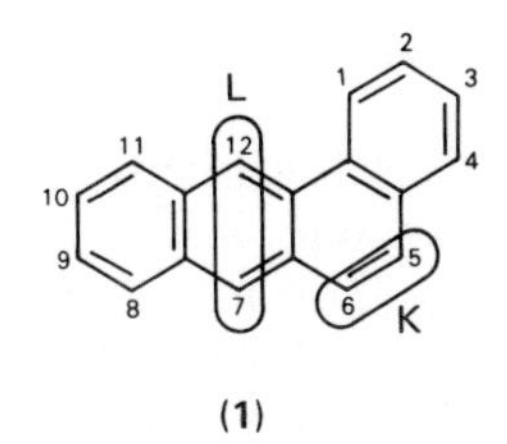

(1)

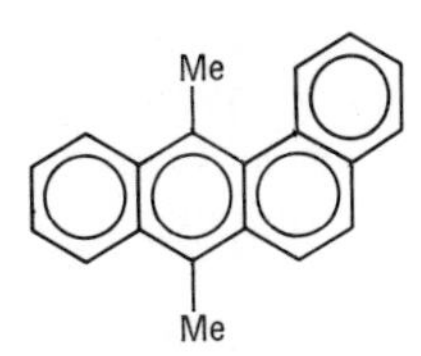

7,12-dimethylbenz[a] anthracene

benz[a] pyrene

(i) Polycyclic aromatics

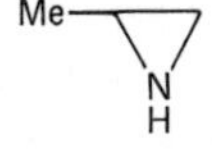

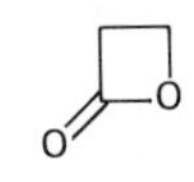

$S(CH_2CH_2Cl)_2$

propyleneimine

β-propiolactone

bis-β-chloroethyl sulphide

(ii) Biological alkylating agents

H_2N–C_6H_4–C_6H_4–NH_2

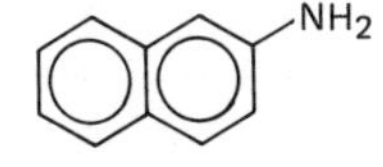

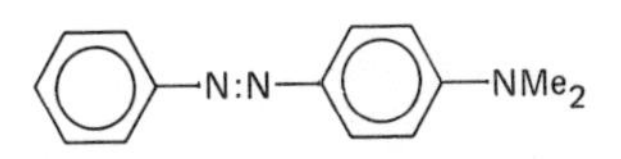

benzidine

β-naphthylamine

4-dimethylaminoazobenzene

(iii) Aromatic amines and azo-compounds

Me_2NNO

$NH_2CON(Me)NO$

N–NO

N-nitrosodimethylamine

N-methyl-*N*-nitrosourea

N-nitrosopiperidine

(iv) *N*-nitrosamines and amides

$Ni(CO)_4$

Nickel carbonyl

O O O O O OMe

aflatoxin B_1

$CH_2{=}CHCl$

Vinyl chloride

(v) Miscellaneous.

Fig. 9.1. Some examples of known carcinogens.

(i) Polycyclic aromatics

A variety of polycyclic aromatic hydrocarbons have been shown to have carcinogenic activity, but almost certainly it is not the parent hydrocarbon which is carcinogenic, but they are carcinogen precursors which are activated *in vivo* by the microsomal mixed-function oxidases. Probably most chemical carcinogens are similarly activated by oxidizing enzymes to the active carcinogen.[3,4]

The structures of the critically active carcinogens of the polycyclic aromatic type are now known. However, the Pullmans[5] have proposed that the carcinogenic activity depends on the presence of a K region having high alkenic character and an L region of low reactivity, indicated in benz[*a*]anthracene (**1**) in Fig. 9.1.

It has been proposed that the carcinogenic process involves enzymatic epoxidation of the K bond and subsequent reaction with a protein or nucleic acid,[6] the K-region oxides usually being more carcinogenic than the parent hydrocarbons, although there are exceptions. There seems to be poor correlation between the binding of polycyclic aromatics to DNA and their carcinogenic activity[7,8] but there are good correlations of carcinogenicity with hydrophobicity, K-region reactivity, and charge-transfer complex-forming ability.[9]

It has been suggested that the polycyclic aromatics may be carcinogenic by virtue of their mutagenicity. The mutational theory of carcinogenesis was originally proposed by Boveri[10] and general support for the operation of mutational factors in carcinogenesis arises from the fact that frequently mutagens are also carcinogens.[11] However, there are difficulties in attempting to correlate mutagenicity with carcinogenicity since mutation studies are usually carried out on microbial systems whilst carcinogenic studies are usually carried out on mammals.

In the case of the polycyclic aromatics there does not seem to be a correlation between carcinogenic activity and the ability to bind to DNA *in vitro*,[12] but a positive correlation has been shown between the *in vivo* binding of hydrocarbons to DNA and their carcinogenic potency.[13] This fact does support the involvement of metabolized hydrocarbons, rather than the hydrocarbons themselves, as the carcinogenic species, and the covalent attachment of the polycyclic aromatic hydrocarbons to the nitrogen bases of DNA has been demonstrated.[14,15] Thus the carcinogenic activity seems to arise from a chemical reaction with DNA rather than physical intercalation of DNA and charge-transfer complexation.

DNA repair has been demonstrated to have a role in carcinogenesis in three separate studies—experimental studies on repair of carcinogen-induced damage, the effect of repair on viral tumour induction, and tumour-incidence studies of people having hereditary DNA-repair defects.

Lawley[16,17] has suggested that this could be the major role of the polycyclic aromatics. If these compounds inhibit DNA repair mechanisms, this would mean that either a greater number of DNA lesions went unrepaired, hence to be expressed as spontaneous mutations, or to result in defective recombination and so to give rise to mutated genomes.

(ii) Biological Alkylating Agents

The largest class of potential mutagens which man is likely to encounter is probably the biological alkylating agents,[17] and as many mutagens are also carcinogens, this probably also represents a large carcinogen hazard. Many alkylating agents have been shown to be carcinogenic (e.g. see refs. 2 and 3), these being simple alkyl halides, nitrogen or sulphur 'mustards', or other alkylating species. Such alkylating agents may be carcinogenic because of their mutational properties or effects on DNA repair mechanisms, but because of these same effects the alkylating agents are also used in cancer chemotherapy—to control the growth of the tumour—and as such are important anti-cancer drugs. A discussion of the effects of alkylating agents on DNA is therefore deferred until the section on cancer chemotherapy (see Section 9C). However, a comprehensive review of carcinogenesis by alkylating agents has been written by Lawley (see ref. 4, p. 83).

(iii) Aromatic Amino and Azo Compounds

A number of aromatic amines and azo compounds have been shown to be carcinogenic, the involvement of β-naphthylamine (**2**) in causing cancer of the bladder in workers in the rubber industry being a classic case, but as early as 1895 Rehn[18] described cancer of the bladder in workers engaged in the manufacture of aniline dyes.

The carcinogenic form of the aromatic amines is the *N*-hydroxy derivative.[19,20] Similarly, 3-hydroxyanthine (**3**) and 3-hydroxyguanine (**4**) are also oncogenic after further activation by sulphotransferase, these N-hydroxylated species being enzymatically converted *in vivo* to cationic or free radical species.[21]

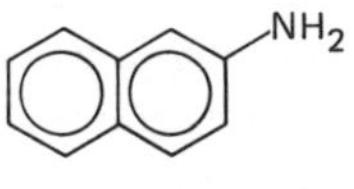

NH_2 HN N O N N H OH HN N HN N N H OH O O

(2) 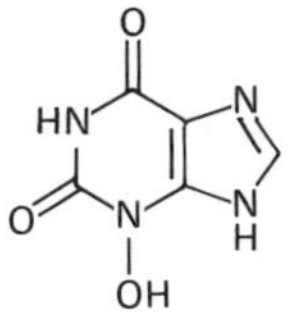 **(3)** 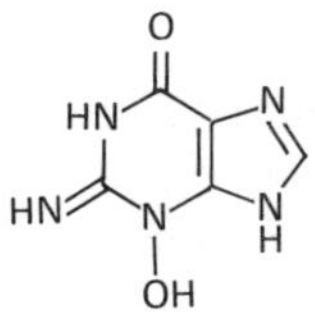 **(4)**

These products then attack methionine residues in proteins and guanine residues in DNA and so are both mutational and interfere with DNA repair.

(iv) N-Nitrosamines and Amides

The *N*-nitrosamines and amides are also both highly carcinogenic and mutagenic after *in vivo* enzymatic reduction to diazoalkanes which give rise to highly reactive carbonium ions or free radicals.[3,17] These products then interfere with DNA repair mechanisms and possibly other oncogenic mechanisms in a similar way to the metabolic products of the aromatic amines and azo compounds.

(v) Miscellaneous Carcinogens

In addition to the above compounds of the type (i)–(iv), a wide arrange of other examples of carcinogen have been observed, including some naturally occurring compounds such as pyrrolizidine alkaloids, aflatoxins and some other compounds. Materials such as asbestos, nickel carbonyl, and chromium compounds have also been implicated as carcinogens.[3] It is only comparatively recently that the carcinogenic activity of metal ions has received much attention,[22,23] but it is likely to achieve greater importance in the future, since humans are exposed to a variety of such compounds.

The mechanisms by which these compounds are carcinogenic are currently only speculative, but most studies to date have considered the cellular DNA as the most likely site of attack. A suggested scheme for the mechanism of action of chemical carcinogens is given in Fig. 9.2. However, with the considerable effort being carried out on cancer research, we shall no doubt see further developments as time passes. Some further accounts of possible carcinogenic mechanisms are given in the series *Progress in Nucleic Acid Research and Molecular Biology* (Academic Press).

To summarize, we can say that a number of factors contribute to the development of neoplasia, although carcinogenesis is presently only beginning to be understood, and in no case is it known as yet how a tumour develops in the detailed molecular biology. Sometimes a single agent can cause neoplasia if it has different effects whilst in other cases a tumour arises due to a succession of apparently innocent events.

Alterations in the genomes have attracted considerable interest since most known carcinogens seem to cause such alterations. However, the other events such as DNA repair, cell proliferation, differentiation, the immune mechanism, and the role of carcinogen-activating enzymes are also attracting interest, whilst further factors to be considered include host susceptibility, the generation of carcinogens in the body, and possibly further factors related to the structure and function of cell surface membranes.

The fully rational treatment of cancer will not be possible until the mechanisms of carcinogenesis and tumour development are understood, but the approaches to cancer chemotherapy currently used are discussed below.

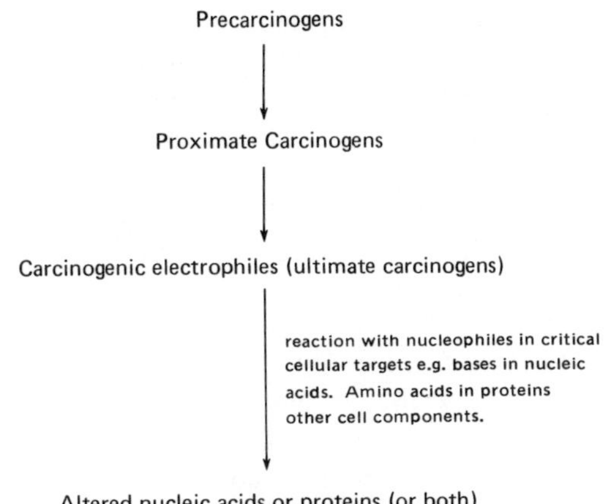

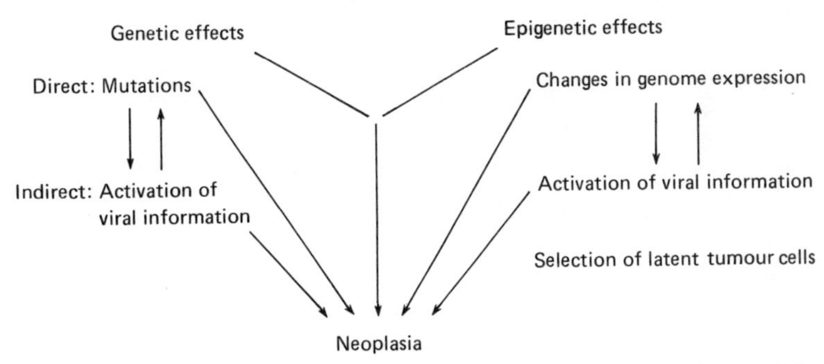

Fig. 9.2. Metabolic activation of chemical carcinogens and possible mechanisms of action of the agents (after E.C. Miller and J.A. Miller, ref. 4, p. 754)

(C) THE CHEMOTHERAPY OF CANCER

Study of the chemotherapy of cancer is one of the important aspects of medicinal chemistry today, and there is an extensive literature including *Cancer Chemotherapy Reports*, *Advances in Cancer Research*, and other journals, and including reviews by Ferguson,[3] Knock,[24] and in Cole *et al.*[25], Sartovelli,[26] and Holland and Frei.[27]

The first recorded example of useful cancer chemotherapy is that of Huggins and Hodges (1941)[28] who reported that the sex hormone oestrogen was useful in the treatment of cancer of the prostrate gland in men. Later work showed that the chemical warfare agents, the nitrogen mustards, had potential antileukaemic activity, and also that folate antagonists also showed such activity.

A recent review[3] quotes that there are about 45 anticancer drugs in current use on the official list of the National Cancer Institute with about 40 more at, or close to, clinical trials.

Knock[24] has categorized the anticancer agents into the following groups: (a) alkylating agents, (b) sulphydryl inhibitors (c) antimetabolites, (d) antibiotics and alkaloids, (e) steroids and hormones, and (f) miscellaneous drugs. Other cancer therapy such as radiation and surgery will not be covered in this account. A point to be noted is that many of the compounds used as anticancer agents are the same types of compound that are carcinogenic and that both types of compound are generally found to interact with the genetic material of the cell. The factors which initiate uncontrolled growth are similar to those which control cell growth.

In attempting to devise agents for the chemical control of diseases it is necessary to find some specificity particular to the disease-causing agent so that it may be counteracted without adverse affects on the host or healthy tissue. In the case of an invading organism such as a protozoon or bacterium, this is relatively easy, since the host and the parasite are different species and have different metabolic capabilities. In the case of cancer, it is the tissue of the host which is undergoing uncontrolled growth giving rise to the cancerous state, and differences between cancerous and normal cells have to be found and exploited.

A number of differences have been found between cancerous and normal cells, cancer cells generally having:[3]

(a) lower pH;
(b) greater free-radical character;
(c) tumour-produced hormone peptides;
(d) tumour-associated antigens;
(e) lower calcium-ion and higher potassium-ion concentrations;
(f) different potassium-isotope ratios;
(g) elevated amounts of methylated nucleosides;
(h) higher concentrations of plasma mucoproteins and mucopolysaccharides;
(i) greater need of exogenous zinc;
(j) higher biowater content;
(k) a need, in the case of leukaemic cells, for exogenous *L*-asparagine.

However, as yet is seems that these differences are more useful for detecting cancerous cells, rather than being therapeutically exploitable.

Most of the anticancer agents in use exert their action by interfering with the biosynthesis or the role of nucleic acids, and those heterocyclic derivatives of the purines, pyrimidines, and pteridines which find use in cancer chemotherapy are considered below:

Alkylating agents: Even simple alkylating agents such as methyl iodide, dimethylsulphate, etc. are mutagenic, and other alkylating agents can exert a variety of biological effects including mutagenesis, carcinogenesis, and tumour inhibition.

Although bifunctionality of the alkylating agent does not seem to be required for mutagenic and carcinogenic action,[29] it does seem to be required for good anticancer activity. Simple alkylating agents are capable of reacting with the amino group nitrogen, ring nitrogen, and oxygen functions of the pyrimidine and purine bases in nucleic acids. For DNA the order of reactivity with alkylating agents is guanine N-7 > adenine N-3 > adenine N-1 > cytosine N-1 > adenine N-7, with the other possibilities coming after that.[24]

These single alkylations of DNA can cause alterations in the hydrogen-bonding characteristics of the bases, e.g. a methylation at O-6 of guanine would reduce the hydrogen-bonding capability from three to two, and this can result in errors or replication or transcription (Fig. 9.3(a)).

In the case of guanine N-7 alkylation, the 7,9-disubstituted guanine is unstable at neutral pH and results in loss of the guanine residue from the DNA chain.[30] Such an effect also leads to a point mutation with the possibility of an error in replication or transcription.

With a bifunctional alkylating agent, in addition to it acting as a single functional compound, it may cause dialkylation. If this occurs in the same DNA strand, then two point mutations may be introduced. However, if interstrand dialkylation occurs, this may lead to inhibition of strand separation which would inhibit replication and may inhibit transcription.

A large number of derivatives of the original nitrogen mustard (**5**) and of other alkylating agents of the ethylenimine and epoxide types have been synthesized as potential anticancer agents.[31] Of these compounds, uracil

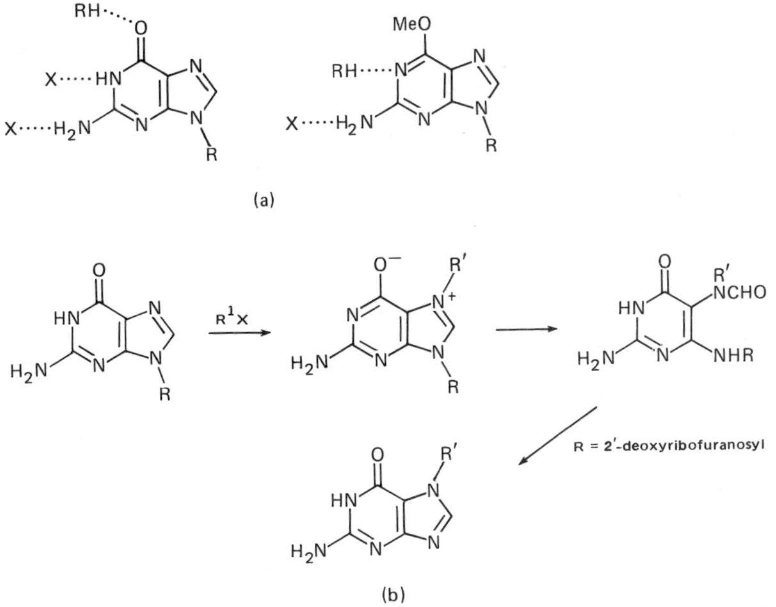

(a)

R = 2′-deoxyribofuranosyl

(b)

Fig. 9.3. (a) Effect of O-6 Methylation of Guanosine and H bonding. (b) Excision of guanine residue by alkylation

mustard (5-(bis-(β-chloroethyl)amino)uracil) (Uramustine) (**6**a) has undergone extensive clinical trials and has been found to be effective in a number of cancers including chronic lymphocytic leukaemia, Hodgkinson's disease, and lymphosarcoma. It is less toxic than the drug cyclophosphamide (cytoxan) (**7**).

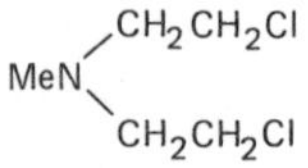

(**5**)

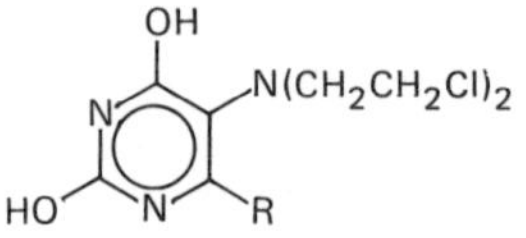

(**6**) **a**, R = H
b, R = Me

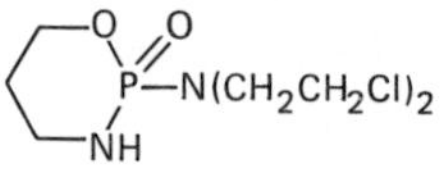

(**7**)

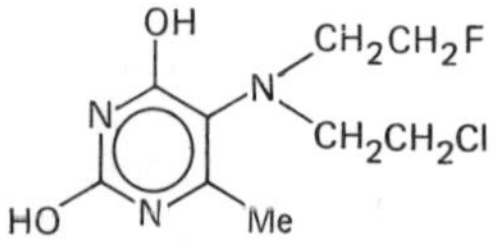

(**8**)

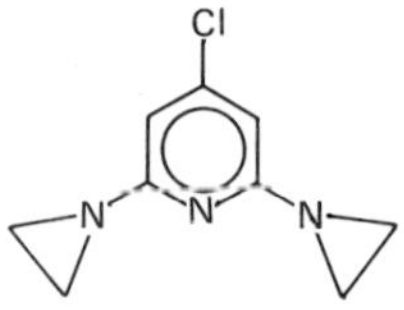

(**10**)

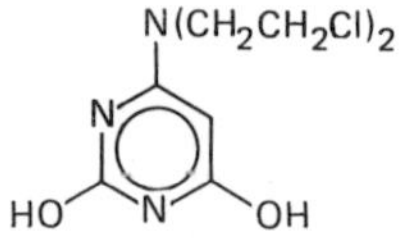

(**9**)

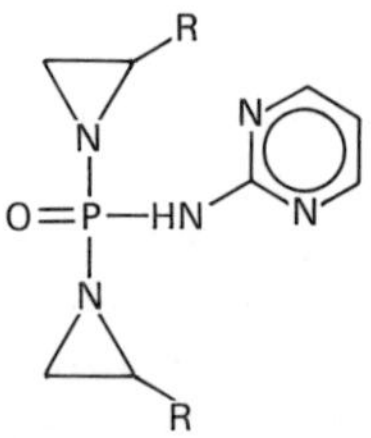

(**11**) **a**, R = H
b, R = Me

The 6-methyl derivative (**6**b) (Dopan) and a fluoro analogue (fluoropan) (**8**) have also been found to be effective in a number of cases. However, the (bis-β-chloroethylamino)uracil (**9**) has been found to have no anticancer activity.[32]

Some ethylenimino derivatives of pyrimidine have also been found to be active against some cancers. 2,6-Bis(aziridinyl)-4-chloropyrimidine (ethymidine, etimidin) (**10**) has been found to have beneficial effect on lung cancer, ovarian cancer, and lymph nodes on the neck. Whilst some phosphamide derivatives (phosphazine (**11**a) and methylphosphazine have also been found to be active in animal systems.

Sulphydryl inhibitors: Although a sulphydryl inhibitor was the first type of compound to be used for the chemotherapy of cancer,[33] when Lissauer reported the use of potassium arsenite for the treatment of leukaemia in 1865, this method of approach has not found favour due to the ubiquity of sulphydryl groups in cellular proteins and the lack of specificity of sulphydryl reagents. However, a number of compounds capable of reacting with sulphydryl groups have been investigated clinically for their ability to control tumour growth,[24] but none of these has been of extensive use nor have any pyrimidine or purine derivatives been used in such a role.

Antimetabolites: Since for cell division to occur there must be successful replication and transcription, attack at the nucleic acids frequently leads to inhibition of tumour growth. The antimetabolites interfere with normal nucleic acid biosynthesis by either competing with the usual metabolites for incorporation into the nucleic acids or by acting as inhibitors of the enzymes involved in the biosynthesis of nucleic acids or their precursors. A wide variety of analogues of the pyrimidines and purines, and their nucleosides, which occur in nucleic acids, have been investigated for antitumour activity[34] and some of these heterocyclic antimetabolites play an important part in modern anticancer chemotherapy.

Some of the important antimetabolite anticancer agents in clinical use are shown in Fig. 9.4. Unfortunately, of course, the antimetabolites interfere with cell division of all cells and they do not usually show selectivity between normal and cancer cells. As a result tumour regression by the antimetabolites is rarely achieved without manifestations of relatively severe toxicity to normal tissues. However, where localized treatment is possible, then the antimetabolites can be very useful and in a number of cases regional chemotherapy can be used.

The antimetabolites represent an important group of anticancer agents and derivatives of pyrimidines, purines, and pteridines are used clinically, the mode of action of some of the important examples being given below.

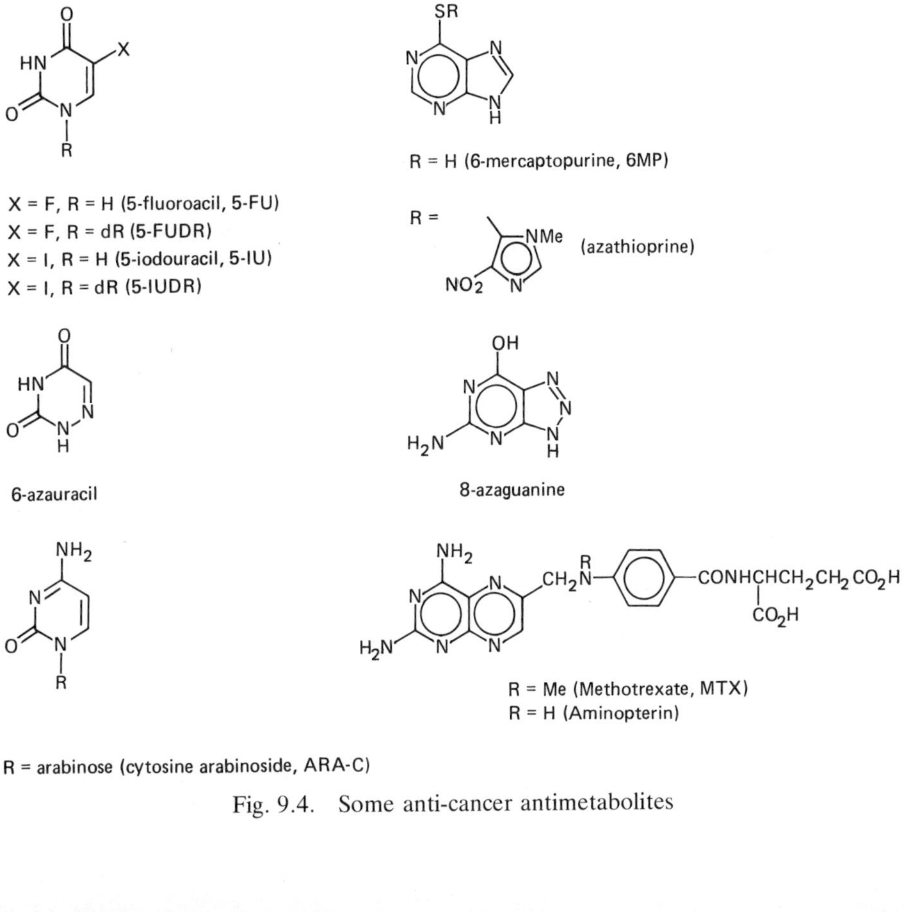

X = F, R = H (5-fluoroacil, 5-FU)
X = F, R = dR (5-FUDR)
X = I, R = H (5-iodouracil, 5-IU)
X = I, R = dR (5-IUDR)

R = H (6-mercaptopurine, 6MP)

R = (azathioprine)

6-azauracil

8-azaguanine

R = Me (Methotrexate, MTX)
R = H (Aminopterin)

R = arabinose (cytosine arabinoside, ARA-C)

Fig. 9.4. Some anti-cancer antimetabolites

(i) Folate Antagonists (see Section 6E)

Derivatives of tetrahydrofolic acid play a key role in the biosynthesis of purines, when they are involved in the introduction of C2 and C8 into the purine ring, and in the biosynthesis of thymidylate from dUMP. In 1950 it was observed that the compounds aminopterin and amethopterin (methotrexate) inhibited the incorporation of ^{14}C formate into the purines of both DNA and RNA, and in 1952 it was observed that these same compounds inhibited the incorporation of ^{14}C formate into the thymine of DNA in leukaemic cells.[35,36]

Methotrexate (MTX) has now been used in cancer chemotherapy for more than 20 years and has been found to be useful against a wide variety of malignancies.[37] Although many analogies have been synthesized and tested, MTX is still the single drug of choice for the treatment of meningeal leukaemia.[38]

In combination cancer chemotherapy *p* (2,4,6-triamino-5-pyrimidinylazo) benzenesulphonic acid (**12**) has been shown to have high synergistic effects in the treatment of patients with acute leukaemia.[39] However, in general, although 5-arylazopyrimidines having 2,4-diamino substitution have been found to have antifolate activity in animals and test systems, they have not found much use in clinical situations.

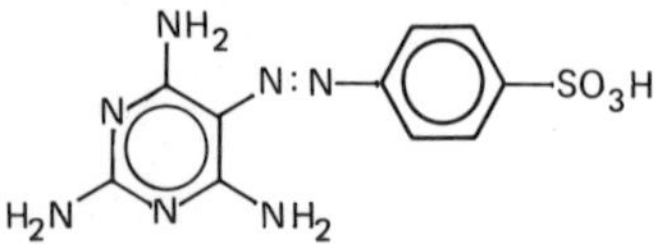

(**12**)

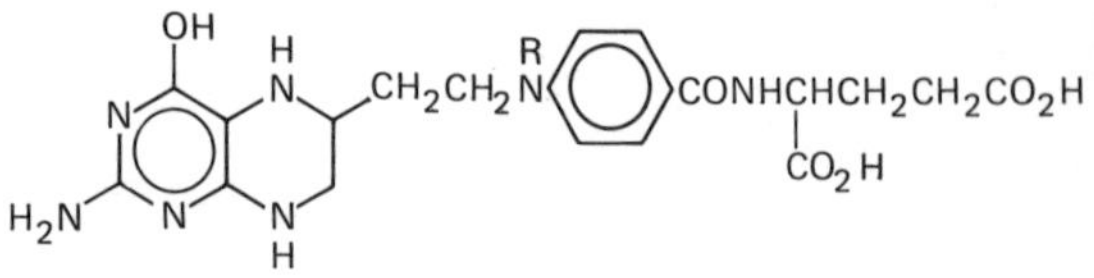

(**13**) **a**, R = H
b, R = Me

Methotrexate and related folate antagonists exert their action in inhibiting nucleic acid biosynthesis by powerfully inhibiting the enzyme dihydrofolate reductase. The kinetics of the inhibition are *pseudoirreversible*, i.e. the methotrexate cannot be significantly displaced from the enzyme by physiological levels of folic acid but the effects of methotrexate overdose can be alleviated by the administration of folic acid.

The following generalities have been proposed[3] for the anticancer effectivity of antifolate folic acid analogues:

(a) for significant activity there should be a free 4-NH_2 group, the exceptions being compounds (**13**a) and (**13**b);
(b) the introduction of halogens into the benzene ring of MTX increases activity;
(c) modifications of the glutamic acid moiety of MTX do not alter activity;
(d) the 1- and 3-deaza-analogues of MTX and their dihydro derivatives retain some activity;

(e) most pyrimidine derivatives and simple pteridine derivatives are inactive.

Professor E. C. Taylor has now synthesized the folate antagonist 5-deazafolic acid, which seems to be the most potent dihydrofolate reductase inhibitor yet made, but details of this study are not yet (March 1979) available.

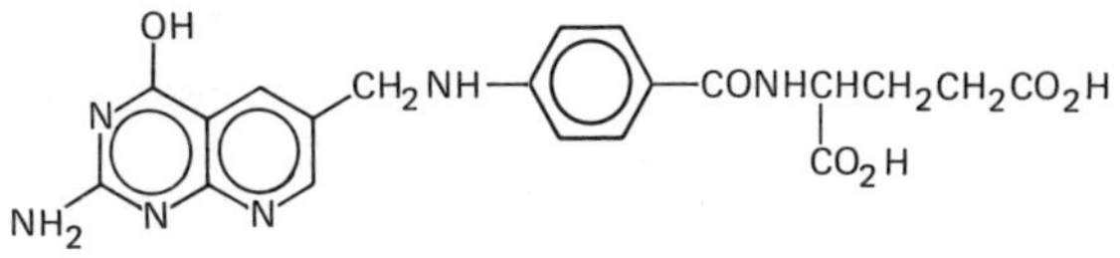

5-deazafolic acid

(ii) Pyrimidine Antimetabolites

The idea of using analogues of uracil in cancer studies stemmed from the observation that uracil is utilized in nucleic acid biosynthesis to a greater extent in a chemically induced rat liver tumour than in the normal rat liver.[40,41] 5-Fluorouracil (5-FU), its riboside (5-FUR), and deoxyriboside (5-FUDR) were synthesized and were found to have considerable activity against a range of tumours.[42,43] The other 5-halogenated uracils also showed antitumour activity, but 6-FU and other 6-halogenated uracils do not show such activity. However, only 5-FU and its derivatives are widely used in cancer chemotherapy, and although 5-iodouracil has anticancer activity it is short-acting, but 5-IUDR is of great importance as an antiviral agent for the treatment of conjunctivitis due to herpes simplex or vaccinia viruses.[44,24]

5-FU and 5-FUDR are converted, in vivo, to the nucleotide 5-FUDRP, which is a strong inhibitor of thymidylate synthetase. The size of the fluorine atom is similar to that of hydrogen and 5-FUDRP competes with dUMP at the enzyme site, but the 5-position cannot be methylated. Thus DNA synthesis is inhibited, this effect explaining the anticancer action of 5-FU. Although 5-FUDR does not become incorporated into DNA the corresponding riboside does incorporate into RNA,[45] this effect causing the high toxicity of 5-FU.[24]

5-FUDR also inhibits orotic acid metabolism and the incorporation of phosphate into DNA.[45] Thus 5-FU has multiple biochemical effects, but probably the most significant in cancer chemotherapy is inhibition of thymidylate synthetase.

5-IU is converted in vivo to 5-IUDR, which can be incorporated into DNA. 5-IUDR thus differs in its mechanism of action from 5-FUDR in that it acts in place of thymidine, whereas 5-FUDR inhibits the formation of thymidine.

6-Azauridine, the nucleoside of 6-azauracil, is a pyrimidine analogue antimetabolite which has also found use in the clinical chemotherapy of cancer.[46,47] 6-AzaUMP acta in the de novo synthesis of pyrimidine nucleotides by inhibiting the conversion of OMP to UMP by the enzyme orotidylate decarboxylase (see Section 6C).

Another pyrimidine antimetabolite which has found use as an anticancer agent is cytosine arabinoside (ARA-C) (see Section 4F). This compound inhibits the reduction of CMP to dCMP and thus inhibits DNA synthesis. Like 5-IU it has found use as an antiviral agent for the treatment of DNA viruses[48,49] such as herpes simplex, but it has also found use for the treatment of leukaemias and other neoplastic diseases.[24,50]

(iii) Purine Analogues

The most extensively used purine antimetabolite is 6-mercaptopurine (6MP) (**14**a) which is converted in vivo to the ribonucleotide. 6MP inhibits the conversion of IMP to AMP[51] (see Section 6B) and also inhibits the de novo synthesis of purines by a feedback-inhibition mechanism. However, the physiologically significant mechanism of action of the drug is still not known.

Although a number of analogues of 6MP have been tested[52] they do not seem to have improved effectivity and 6MP still seems to be the purine antimetabolite of choice in the treatment of a number of cancers. The compound Azathioprine (Imvran) (**14**b) was developed and this had a more favourable therapeutic index in animals. However, this has not been reflected in a significant clinical advantage over 6MP for the treatment of leukaemia, but it has become quite widely used in cases where it is necessary to suppress the immune response.[24,52]

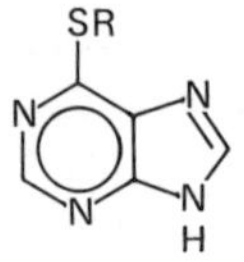

(14) 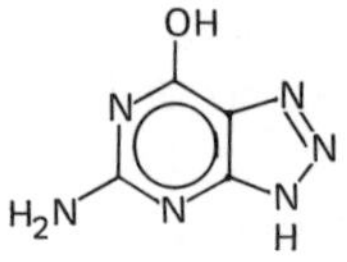 **(15)**

a, R = H

b, R =

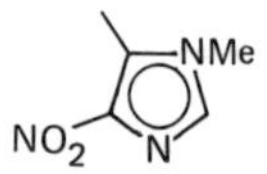

8-Azaguanine (5-amino-7-hydroxy-*v*-triazolopyrimidine) (**15**) is an antifungal antibiotic (Pathocidin) produced by a strain of *Streptomyces albus*. It inhibits the biosynthesis of GMP and also has been found to incorporate into nucleic acids in many biological systems.[45] It has found some use clinically for the treatment of cancer.[45]

(iv) Hormones and Miscellaneous Anticancer Agents:

In addition to the above derivatives of the nitrogen heterocycles, a variety of other anticancer agents have found use ranging from simple derivatives such as nitrosoureas and guanidines (e.g. CCNU (**16**)) which are alkylating agents, to hormones, some metal complexes (e.g. *cis*-$Pt(NH_3)_2Cl_2$), and to a variety of antibiotics. Of the antibiotics, the compound Adriamycin (**17**), has been shown to have a wide spectrum of clinical activity and is currently of great interest.[53]

In Russia there has been great interest in the antibiotic Variomycin A (**18**), whilst in Japan Mitomycin C (**19**) has attracted interest.[3]

These compounds may act at various points in the sequence of DNA → RNA → protein, but they seem to bind to DNA and inhibit the transcription process.

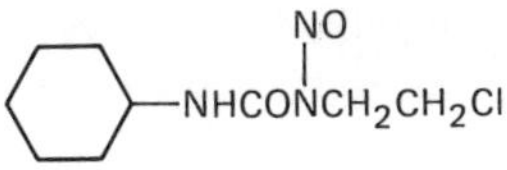

(**16**)

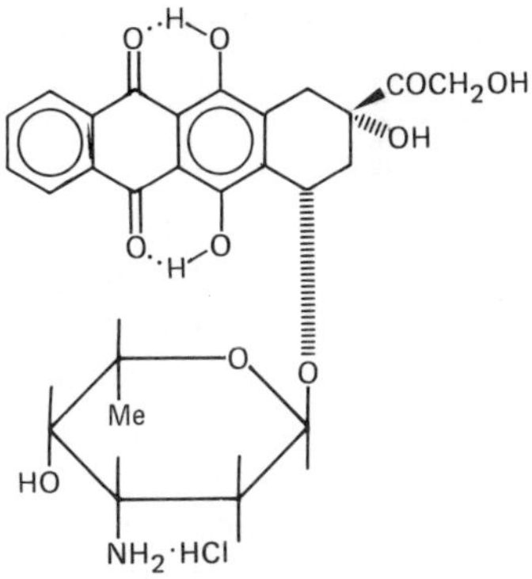

(**17**)

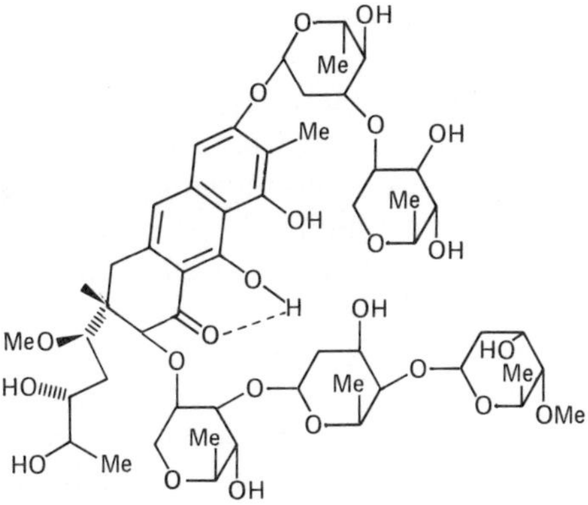

(**18**)

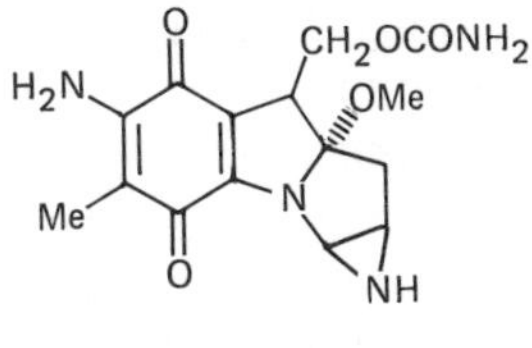

(**19**)

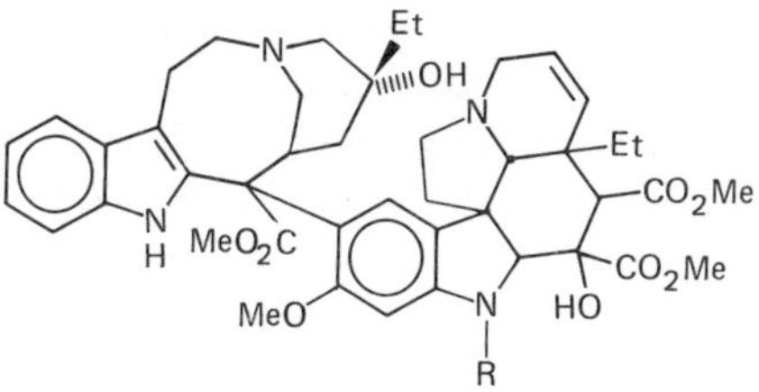

(**20**) **a,** R = CHO
b, R = Me

Other natural products which have been found to have anticancer activity in some cases include the periwinkle alkaloids vincristine (**20**a) and vinblastine (**20**b). However, for a further survey of such compounds the reader is directed to refs. 3 and 23 and to the current literature on cancer chemotherapy.

In addition to the use of the compound *cis*-dichlorodiammine platinum (**II**) as an antitumour agent, a recent report[54] has shown that some 'platinum–pyrimidine blue' complexes have significant antitumour effects in experimental systems. These complexes are deep blue soluble complexes of varying structure and composition but having some simple hydroxypyrimidines as ligands.

(D) CONCLUSION

The field of cancer chemotherapy is very active and almost daily new compounds are being shown to have some activity, although compounds coming to clinical trials are still comparatively small in number. However by the time that this book is completed some new compounds will doubtless have been put forward.

There are about ten human cancers which are currently responsive to therapy such that 50 per cent of patients should achieve normal life expectancy,[55] although some major fatal cancers such as lung cancer are not generally responsive to chemotherapy. The incidence of lung cancer still seems to be increasing.

The purpose of cancer chemotherapy is to enable the patient to achieve normal life expectancy or at least to provide extended remission—the current survival rate of all lung-cancer patients from diagnosis to death remains less than six months.[56] In order to improve the methods of chemotherapy obviously an understanding of the biochemical mechanisms of carcinogenesis and neoplasia is needed, and also a better understanding of the mechanisms of action of known anticancer agents is required.

Improved remissions are sometimes obtained by using a combination of drugs or a number of drugs in succession and also by a combination of chemotherapy with radiation therapy or surgery. However, in addition to the four widely used methods for the treatment of cancer, namely surgery, radiation, chemotherapy, and immunotherapy, several less developed methods are being investigated and may prove useful, for example, electrosurgery, thermotherapy and cryosurgery.[3]

It has been suggested that rather than the prevention of cancer, it may be that the best that can be hoped for in cancer treatment is its control, and that the greatest long-range chances of overcoming cancer lie in immunotherapy and the avoidance or elimination of environmental carcinogens.[3]

However, cancer chemotherapy is, and will remain, a useful method of control for the forseeable future, and the search for new and better anticancer agents will remain a valuable and lively field of study.

REFERENCES

1. (a) *National Cancer Institute 1973 Fact Book*, US Dept. of Health, Education and Welfare, Washington, D.C.; (b) *75 Cancer Facts and Figures*, American Cancer Society, New York (1975).
2. J. V. Frei, *Chem. Biol. Interactions*, **13**, 1 (1975).
3. L. N. Ferguson, *Chem. Soc. Rev.*, **4**, 289 (1975).
4. C. E. Searle (ed.), *Chemical Carcinogens*, A.C.S. Monograph 173, A.C.S., Washington (1976).
5. A. Pullman and B. Pullman, *Adv. Cancer Res.*, **3**, 129 (1971).
6. E. Boyland in *Biochem. Soc. Symp.*, **5**, 40 (1950); P. O. P. Ts'O and J. A. DiPaolo, (eds.), *Chemical Carcinogenesis* Part A, Marcel Dekker, New York (1974).
7. D. M. Jerina and J. W. Daly, *Science*, **185**, 573 (1974).
8. T. Kuroki, E. Huberman, H. Marquardt, J. K. Selkirk, C. Heidelberger, P. L. Grover, and P. Sims, *Chem. Biol. Interactions*, **4**, 389 (1972).
9. R. Franke, *Chem. Biol. Interactions*, **6**, 1 (1973).
10. T. Boveri, *The Origin of Malignant Tumours*, Fischer, Jena (1914) (translated by M. Boveri; Ballière, Tindall and Cox, London (1929)).
11. A. G. Knudson, *Adv. Cancer Res.*, **17**, 317 (1973).
12. E. Boyland and B. Green, *Biochem. J.*, **96**, 15P (1965).
13. P. Brookes and P. D. Lawley, *Nature*, **202**, 781 (1964).
14. P. Brookes and P. D. Lawley, *J. Cellular Comp. Physiol.*, **64**, Suppl. 1, 111 (1964).
15. (a) H. W. S. King, M. H. Thompson, and P. Brookes, *Cancer Res.*, **35**, 1263 (1975); (b) M. H. Thompson, M. R. Osborne, H. W. S. King, and P. Brookes, *Chem. Biol Interactions,* **14**, 13 (1976).
16. P. D. Lawley, 17th Mosbacher Symp., Dental Ges. Physiol. Chem. (1966), quoted in ref. 17.
17. L. Fishbein, W. G. Flamm, and H. L. Falk (eds.), *Chemical Mutagens*, Academic Press, New York and London (1970).
18. L. Rehn, *Arch. Klin. Chir.*, **50**, 588 (1895).
19. J. L. Radomski, G. M. Conzelman, A. A. Rey, and E. Brill, *J. Nat. Cancer Inst.*, **50**, 989 (1973).
20. P. D. Lotlikar, L. Luka, and K. Zaleski, *Biochem. Biophys. Res. Comm.*, **59**, 1349 (1974).
21. C. B. Brown, M. N. Teller, I. Smullyan, N. J. M. Birdsall, T.-C. Lee, J. C. Parham, and G. Stohrer, *Cancer Res.*, **33**, 1113 (1973).
22. D. R. Williams, *Chem. Rev.*, **72**, 203 (1972).
23. A. Furst and R. T. Haro, *Progr. Exp. Tumor Res.*, **12**, 102 (1969).
24. F. E. Knock, *Anticancer Agents*, Thomas, Springfield, Ill. (1967).
25. W. H. Cole (ed.) *Chemotherapy of Cancer*, Lea and Febiger, Philadelphia (1970).
26. A. C. Sartorelli (ed.), *Cancer Chemotherapy*, A.C.S. Symposium No. 30, A.C.S., Washington (1977).
27. J. F. Holland and E. Frei (eds.), *Cancer Medicine*, Lea and Febiger, Philadelphia (1973).
28. C. Huggins and C. V. Hodges, *Cancer Res.*, 293 (1941).
29. G. P. Warwick, *Cancer Res.*, **23**, 1315 (1963).
30. P. D. Lawley, *Proc. Chem. Soc. (London)*, **1957**, 290.
31. *Cancer Chemotherapy Rept.*, **26**, (1963).
32. C. C. Cheng and B. Roth in G. P. Ellis and G. B. West (eds.), *Progress in Medicinal Chemistry*, Vol. 8, Butterworths, London (1971).
33. D. A. Karnofsky in P. A. Plattner (ed.), *Chemotherapy of Cancer*, Elsevier, Amsterdam (1964).

34. P. Langen *Antimetabolites of Nucleic Acid Metabolism*, Gordon and Breach, New York, London, Paris (1975).
35. H. E. Skipper, J. H. Mitchell, and L. L. Bennett, *Cancer Res.*, **10**, 510 (1950).
36. H. E. Skipper, L. L. Bennett, and L. W. Law, *Cancer Res.*, **12**, 667 (1952).
37. R. B. Livingstone and S. K. Carter, *Single Agents in Cancer Chemotherapy*, Plenum Press, New York (1970).
38. L. E. Broder and S. K. Carter, *Meningeal Leukaemia*, Plenum Press, New York (1972).
39. S. Hibino, *Cancer Chemotherapy Rept.*, **13**, 141 (1961).
40. R. J. Rutman, A. Cantarow, and K. E. Paschkis, *Cancer Res.*, **14**, 119 (1954).
41. K. C. Liebman and C. Heidelberger, *Fed. Proc.*, **14**, 243 (1955).
42. R. Duschinsky, E. Pleven, and C. Heidelberger, *J. Amer. Chem. Soc.*, **79**, 4559 (1957).
43. C. Heidelberger and F. J. Ansfield, *Cancer Res.*, **23**, 1226 (1963).
44. P. Calabresi and A. D. Welch in L. S. Goodman and A. Gilman (eds.), *The Pharmacological Basis of Therapeutics*, Macmillan, New York (1965).
45. C. C. Cheng in G. P. Ellis and G. B. West (eds.), *Progress in Medicinal Chemistry*, Vol. 6., Butterworths, London (1969).
46. H. J. Fallon, E. Frei, and E. J. Freidreich, *Amer. J. Med.*, **33**, 526 (1962).
47. R. C. DeConti, R. W. Turner, and P. Calabresi, *Proc. Amer. Ass. Cancer Res.*, **6**, 14 (1965).
48. G. E. Underwood, *Proc. Soc. Exp. Biol. Med.*, **111**, 660 (1962).
49. H. E. Kaufman, *Ann. N.Y. Acad. Sci.*, **130**, 168 (1965).
50. *Martindale, The Extra Pharmacopoeia*, (26th edn.) The Pharmaceutical Press, London (1973).
51. J. Lasnitski, R. E. F. Matthews, and J. D. Smith, *Nature*, **173**, 346 (1954).
52. J. A. Stock in R. J. Schnitzer and F. Haeking (eds.), *Experimental Chemotherapy*, Vol. 4, Academic Press, New York and London (1966).
53. F. Arcamore, G. Cassinelli, G. Franceschi, R. Mondelli, P. Grezzi, and S. Penco, *Gazzetta,* **100**, 949 (1970).
54. J. P. Davidson, P. J. Faber, R. G. Fischer, S. Mansey, H. J. Peresie, B. Rosenberg, and L. VanCamp, *Cancer Chemotherapy Rpts.,* **59**, 287 (1975).
55. C. G. Zubrod, *Cancer*, **30**, 1474 (1972).
56. O. S. Selawry, *Cancer Chemotherapy Rpts., Part 3*, **4**(2), 5, (1973).

Chapter 10

Nitrogen Heteroaromatic Pharmaceuticals

(A) INTRODUCTION

Perusal of the national pharmacopoeiae (for example the British or the US pharmacopoeia) or the *Merck Index*[1] will show the large contribution that heterocyclic compounds make towards currently prescribed drug substances, and of these chemotherapeutically useful heterocyclic compounds the pyrimidines, purines, and pteridines, and related compounds, comprise a substantial proportion.

To cover in one chapter the whole range of biochemical and medicinal properties of these groups of compounds and to include each compound that is potentially useful in this context would be impossible. This chapter intends to briefly review the more important or more interesting examples of nitrogen heteroaromatic pharmaceuticals and agricultural chemicals rather than to attempt to provide an extensive list of compounds and an extensive list of uses.

The naturally occurring antibiotics (and ARA-C) have been referred to in Chapter 4, the sulphonamides have been referred to in Chapter 6, and cancer chemotherapy has been dealt with in the previous chapter, but a whole variety of other activities including diuretic, antithyroid, fungicidal, antihelmintic, and surface anaesthetic properties have also been associated with nitrogen heterocyclic compounds. Reviews of pyrimidines[2] of biological and medicinal interest, of purines and pyrimidines,[3] and purines[4] are available, whilst some information concerning pteridines is also available.[5] In addition to these reviews further information on the chemotherapeutic action of these heterocycles will be found in the series *Progress in Medicinal Chemistry*, (Butterworths), *Advances in Pharmacology and Chemotherapy* (Academic Press), *Advances in Drug Research* (Academic Press), and *Progress in Drug Research* (Birkhäuser Verlag). A bibliography of reviews in medicinal chemistry is also available.[6]

Because the purines, pyrimidines, and pteridines are widely distributed in natural materials and living organisms it is not surprising that so many derivatives of these compounds have biological activity, but the range of activity reaches beyond analogue and antimetabolite action and it seems that in some cases the heterocyclic ring acts as a useful carrier for biologically active substituents. A selection of chemotherapeutically useful pyrimidines, purines and pteridines is given below.

It is not intended that this chapter should give a complete review of pharmaceuticals, but it is designed to point out the contribution of the pyrimidine-related heterocyclic compounds to medicine. In many cases there are many structurally unrelated compounds which have similar or superior properties to those mentioned here.

(B) THE BARBITURATES

Barbituric acid **(1a)** has been known for well over 100 years and the first hypnotic barbiturate 5,5-diethylbarbituric acid (barbitone, barbital, veronal) **(1b)** was introduced into medicine by Fischer and von Mering in 1903. Since that time more than 2500 barbiturates have been synthesized but only about a dozen are widely prescribed today. They are usually prescribed in combination with other drugs for use as sedatives, tranquillizers, hypnotics, anticonvulsants and antiepileptics, and sometimes as analgesics.[2c,7] A review of the barbiturates is available[8] whilst a short account is given in the *Kirk—Othmer Encyclopedia of Chemical Technology*.[9] The mechanism and sites of action of the barbiturates have also been reviewed.[10]

The principal requirement for barbiturates to show activity is that there is 5,5-disubstitution by two lipophylic groups, both of which have at least two carbon atoms. Neither barbituric acid nor 5,5-dimethylbarbituric acid have significant hypnotic activity. Variation in the structure and chain length of the substituents causes changes in the activity of the barbiturates. For example, the introduction of hydrophylic groups such as hydroxyl or amino into the 5- substituents causes loss of hypnotic action, the introduction of a single halogen (preferably bromine) sometimes increases the hypnotic potency. The introduction of unsaturated substituents also usually increases hypnotic potency but sometimes these changes may produce convulsant activity.

The hypnotic properties generally increase with the length of the 5-substituents and the derivatives of major commercial importance usually have up to 7 carbon atoms. Above about 7 or 8 carbon atoms the activity of the barbiturates decreases, convulsant properties do not appear and the toxicity becomes a critical factor. 5,5-Dibenzylbarbituric acid **(1c)** is, for example, devoid of hypnotic activity. In general barbiturates having two dissimilar substituents at the 5-position are more active than the symmetrically substituted ones.

The barbiturates can be classified according to the time for which they

Table 10.1 Duration of action of some barbiturates (data from ref. 9)

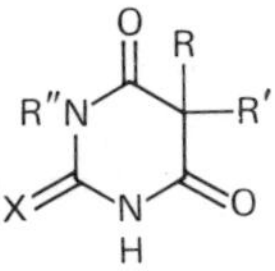

Name	R	R′	R″	X	Length of action
Barbital	C_2H_5	C_2H_5	H	O	long
Mephobarbital	C_2H_5	C_6H_5	CH_3	O	long
Methabarbital	C_2H_5	C_2H_5	CH_3	O	long
Phenobarbital	C_2H_5	C_6H_5	H	O	long
Amobarbital	C_2H_5	$CH_2CH_2CHMe_2$	H	O	intermediate
Aprobarbital	$CH_2CH{=}CH_2$	$CHMe_2$	H	O	intermediate
Butabarbital	C_2H_5	$CH(CH_3)CH_2CH_3$	H	O	intermediate
Allobarbital	$CH_2CH{=}CH_2$	$CH_2CH{=}CH_2$	H	O	intermediate
Talbarbital	$CH_2CH{=}CH_2$	$CH(CH_3)CH_2CH_3$	H	O	intermediate
Vinbarbital	C_2H_5	$C(CH_3){=}CHCH_2CH_3$	H	O	intermediate
Cyclobarbital	C_2H_5	1-(cyclohex-1-enyl)	H	O	short
Heptabarbital	C_2H_5	1-(cyclohept-1-enyl)	H	O	short
Pentobarbital	C_2H_5	$CH(CH_3)CH_2CH_2CH_3$	H	O	short
Secobarbital	$CH_2CH{=}CH_2$	$CH(CH_3)CH_2CH_2CH_3$	H	O	short
Hexobarbital	CH_3	1-(cyclohex-1-enyl)	CH_3	O	ultrashort
Methohexital	$CH_2CH{=}CH_2$	$CH(CH_3)C{\equiv}CCH_2CH_3$	CH_3	O	ultrashort
Thiamylal	$CH_2CH{=}CH_2$	$CH(CH_3)CH_2CH_2CH_3$	H	S	ultrashort
Thiopental	C_2H_5	$CH(CH_3)CH_2CH_2CH_3$	H	S	ultrashort

have effect into long, intermediate, short, and ultra-short-acting drugs. The duration of action of some barbiturates is indicated in Table 10.1. In general it seems that the presence of phenyl groups or unsaturated substituents gives longer-acting drugs, simple aliphatic substituents give intermediate drugs, whilst chain-branched substituents lead to barbiturates which have a short time of action but heightened activity. Ultra-shortacting barbiturates are obtained usually by alkylation of the ring nitrogen, whilst the replacement of the 2-oxygen by sulphur to give, for example, thiopental (**1d**) also usually provides highly potent, short- or ultra-shortacting compounds.

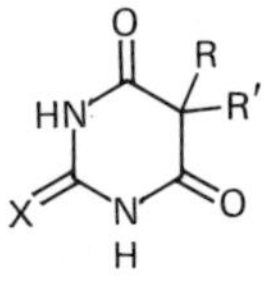

(**1**)

a, R = R′ = H, X = O

b, R = R′ = C_2H_5, X = O

c, R = R′ = $CH_2C_6H_5$, X = O

d, R = C_2H_5, R′ = $CH(CH_3)C_3H_7$, X = S

In general, there seems to be a correlation between the lipid solubility of the barbiturates and their hypnotic action and duration of action. The barbiturates are acidic, and the preferred form of the drugs for therapeutic use is as the mono sodium salts which are water soluble, whilst it is the un-ionized form which is considered to be the active form. Thus the dissociation constants of the barbiturates are also important in controlling their hypnotic activity.

In addition to their roles mentioned above, barbiturate derivatives have also been claimed to have activity as anticholinergics for the treatment of Parkinson's disease, as anti-inflammatory agents, and there have been claims that some compounds have hypotensive, diuretic, and antiviral activity.[2c]

Although they have been known for many years and have been extensively studied, the biochemical mechanisms by which the barbiturates act are still unclear. They are known to have profound effect upon the central nervous system but for the most part studies have only lead to descriptions of their actions. The barbiturates are known to elevate the threshold of neurons by stabilization of the cell membrane and prolongation of the time for recovery from excitation. But how this effect is mediated is unknown.[10] The barbiturates are also known to have an 'enzyme-inducing' action and to cause increased activity in the hepatic microsomal polyfunctional oxidases.[11]

The barbiturates are metabolized principally by four methods:

(i) oxidation of a 5-substituent;
(ii) dealkylation of ring nitrogens;
(iii) desulphurization of thiobarbiturates to barbiturates;
(iv) cleavage of the barbiturate ring to give ureas.

Dealkylation at the 5-position and glucuronide formation are also routes by which some barbiturates are metabolized. The metabolites, or unchanged barbiturates, are excreted in the urine, the extent of metabolism and the nature of the metabolites depending on the structure of the barbiturate. Barbital, for example, is mainly excreted unchanged (95%) in rats together with the glucuronide (3%) and 5-ethylbarbituric acid (2.5%).[12]

Thiopental is almost completely metabolized in the rat, the routes being indicated in Fig. 10.1.

Fig. 10.1. Metabolites of thiopental[12]

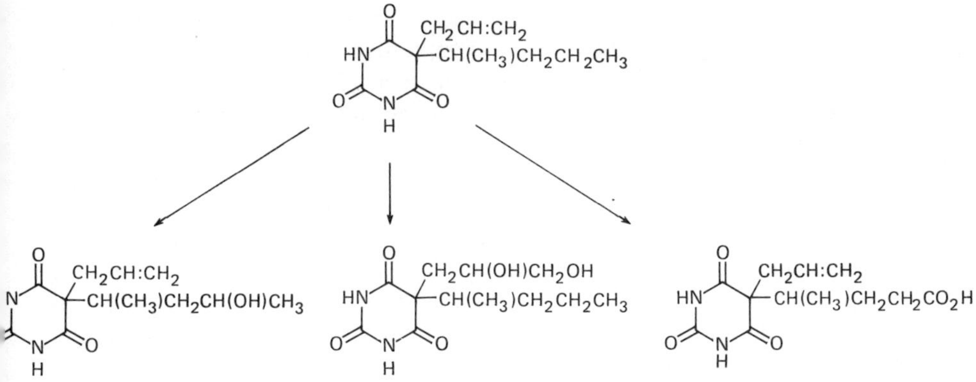

Fig. 10.2. Metabolites of seconal[12]

In addition to side-chain oxidation to the carboxylic acid, hydroxylation reactions may also occur. For example, the metabolism of the barbiturate seconal is shown in Fig. 10.2.

A considerable quantity of barbiturates is manufactured and prescribed each year and they represent the major 'drugs of abuse'. For example in the U.S.A. in 1961 407,000 lbs. of barbiturates were sold (Ref. 9, p. 72) this is equivalent to about 500 million hypnotic doses.[10] A similar situation exists in Britain where in 1963, 1,200 million barbiturate tablets were prescribed to N.H.S. patients and, in 1965, a survey of 305 cases of drug overdoses in the South of England showed that barbiturates were responsible for 48% of the cases. However the barbiturates play an important role in the treatment of many conditions and their value should not be underestimated.

(C) ANTIMALARIALS

In the early 1940s, when there was a considerable research effort into combating bacterial and other infections which affected troops, a number of pyrimidine derivatives were investigated for such activity following the success of the sulphonamides (see Section 6F). Compounds containing the 2,4-diaminopyrimidine structure were found to inhibit the utilization of folic acid in some microorganisms, this action being due to the fact that many such compounds bound to the enzyme dihydrofolate reductase. However, the first compound which was an important antimalarial was paludrine (proguanil, chloroguanide) (**2**).[13]

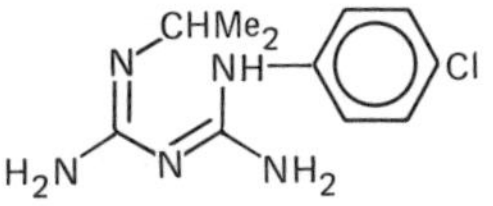

(2) 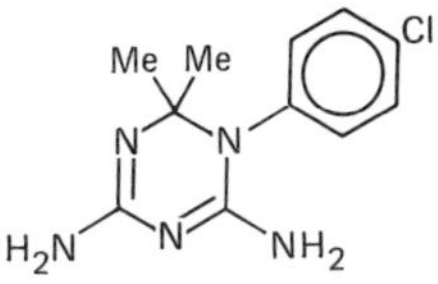 (3) 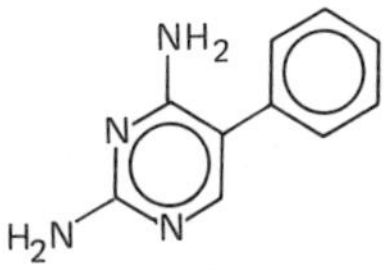 (4)

This compound was reported to be active due to its *in vivo* conversion to the dihydrotriazine (**3**), which was a competitive inhibitor of dihydrofolate reductase. The similarity in structure between (**3**) and (**4**) suggested that derivatives of 2,4-diamino-5-phenylpyrimidine might also have antimalarial activity.

High antimalarial activity in experimental animals was found with 2,4-diamino-5-phenylpyrimidines, although 5-benzyl, 5-phenoxy, and 5-arythiopyrimidines were very much less active. It was found that substitution of electronegative groups such as halogen or nitro into the *para* position of the benzene ring gave increased activity, that alkylation at position 6 also increased activity, but alkylation of the amino groups greatly reduced activity.

The most potent antimalarial agent of this type was found to be 2,4-diamino-5-(*p*-chlorophenyl)-6-ethylpyrimidine (**5**) (Daraprim, pyrimethamine).

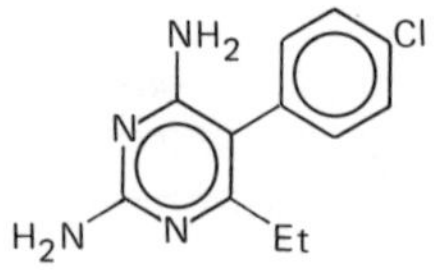

(**5**)

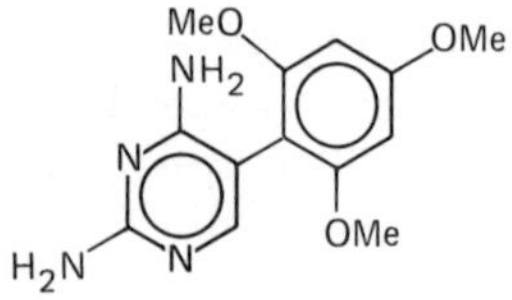

(**6**)

Daraprim was found to be 60 times as active as paludrine against *Plasmodium gallinaceum* and 200 times as active against *P. berghei*.[14] Later, another 2,4-diaminopyrimidine-trimethoprim (Syraprim) (**6**) was found to give rapid clinical remission in humans infected with *P. falciparum*.[15]

Daraprim is used clinically as a prophylactic drug against *P. falciparum*. It is rapidly absorbed and slowly excreted and is probably the most persistent antimalarial drug. When given in weekly doses of 25 mg it is sufficient to prevent the appearance of parasites in the blood.[16] The antimalarial action of Daraprim can be potentiated by the simultaneous administration of sulphonamides and sulphones,[2a] and also the drug is frequently administered with schizontocidal drugs such as chloroquine.[16] However resistant parasite strains can develop when Daraprim alone is used, but this problem can usually be overcome by treatment with a combination of Daraprim and sulphormethoxine.

(D) DIURETICS

The methylated purines caffeine (**7**), theophylline (**8**) and theobromine (**9**) occur naturally in the leaves and fruit of a number of plants, in particular those of the tea, coffee, and cola species. Commercially caffeine is obtained by extraction of tea dust which can contain up to 3.5% by weight of the compound.

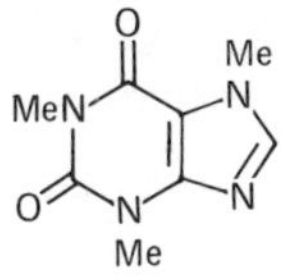

(7) 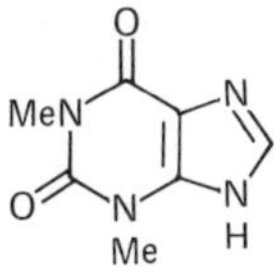 (8) 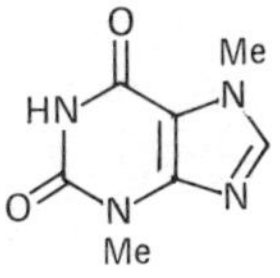 (9)

Theophylline and theobromine, as the sodium or calcium salts, form double salts with the corresponding salts of organic acids, e.g. salicylate, acetate, glycinate, these combinations being therapeutically useful diuretics and also having cardiac stimulant and vasodilatory action.

Also of therapeutic importance as diuretics are derivatives of theophylline containing one or two molecules of an alkylamine, the most widely used being aminophylline, which is the compound having two molecules of theophylline to one of ethylenediamine.

Caffeine also possesses some diuretic activity but its main therapeutic uses are as a cerebral stimulant or cardiac stimulant. It is often used as a constituent of many tablets for the relief of headaches, usually combined with aspirin and phenacetin. Caffeine citrate is a compound which effervesces on dissolution in water and is a useful, non-intoxicating stimulant in headaches due to tiredness.

In attempts to obtain analogues of the methylated xanthines it was found that some of the 6-aminouracils showed considerable activity as oral diuretics in experimental animals.[17]

From such studies the drug amisometradine (Rolicton) **(10)** was developed which had better activity and considerably less toxicity than aminophylline. Aminometridine **(11)** has also found use clinically but is much more prone to cause side effects than amisometradine.

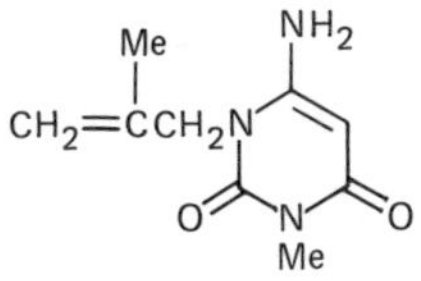

(10) 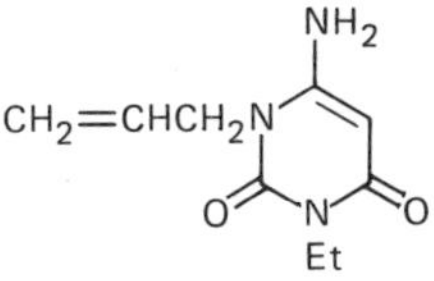 (11) 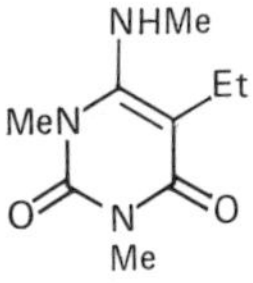 (12)

The mode of action of these compounds in man seems to be specific inhibition of tubular reabsorption of $Na^{\oplus}$ and $Cl^{\ominus}$, but does not seem to be accompanied by an increase in glomular filtration rate nor by changes in factors regulating the acid–base balance. Amisometradine is a clinically useful, but only moderately potent, oral diuretic which is indicated for the treatment of patients with mild heart conditions but not for initial diuresis in severe heart cases.

Some other pyrimidines of this type including the compound **(12)** have also been found to have diuretic activity, but so far these have not found use in the clinic. However the pteridine triamterene (2,4,6-triamino-5-phenylpteridine) is marketed in some countries, in combinations with other substances, as a diuretic.

(E) ANTITHYROID PYRIMIDINES

2-Thiouracil (2-TU, **13a**) and its 6-alkylated derivatives have been used extensively for the treatment of hyperthyroidism to depress the clinically overactive thyroid in thyrotoxicosis.[18] However, these compounds have not been completely successful in giving permanent remission of hyperthyroidism, and they are also liable to produce adverse reactions in susceptible patients, e.g. depression of erythrocyte formation, agranulocytosis, etc.

It has been suggested that the mechanism of action of these antithyroid drugs is to react with iodine liberated within the thyroid cells to form a disulphide and thus to prevent iodination of tyrosine.[19–21] Such a reaction is known to occur with the 2- and 4-mercaptopyrimidines. This explains why the S-substituted thiouracils show no antithyroid activity until they have undergone enzymic cleavage to the free thiouracil.

Substitution at the 1- or 5-position of 2-thiouracil results in decreased antithyroid activity, but alkyl substitution at position 6 generally increases the activity whereas the introduction of electronegative substituents, e.g. CF_3, decreases activity. 4-Thiouracil has no antithyroid activity and probably the most widely used compound is 6-propyl-2-thiouracil (**13b**) (Propacil) although 5-iodo-2-thiouracil (**13c**) is also used.

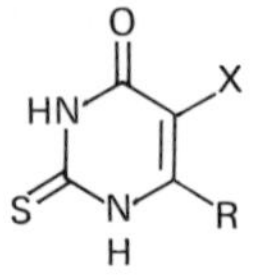

(13)

a, R = H, X = H
b, R = *n*Pr, X = H
c, R = H, X = I

(F) ANTIVIRAL AGENTS

In economic terms viral infections are important since days lost in industry due to absenteeism because of them are probably greater than days lost for any other single cause. However, the problems of producing a highly effective, non-toxic, antiviral agent are considerable, because a virus is so much simpler than a bacterium that the points of attack are very much less and there is the difficulty of producing a virus-specific agent which will not attack the host cell. In many cases, for example the common cold, influenza, the illness is quite mild and the patient will recover without undergoing treatment, so any antiviral agent which is marketed must be one with virtually no side effects or toxicity. However, some conditions, e.g. ocular herpes keratitis, which requires topical administration of the antiviral agent, and grave conditions such as Herpes virus encephalitis have proved to be amenable to antiviral chemotherapy. Recently the compound virazole[22] (**14**) has been marketed in some South American countries for the treatment of viral hepatitis, but it has not found universal acceptance.

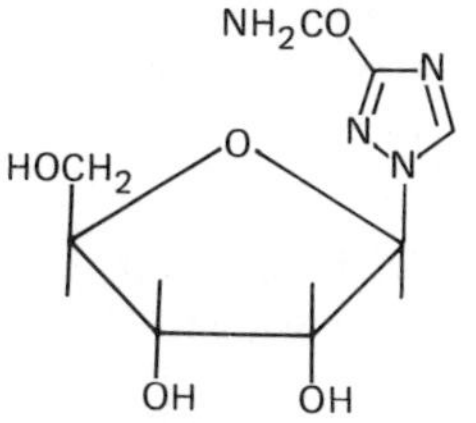

NH_2CO N N N HOCH$_2$ O OH OH

(14)

Of the antiviral compounds in current use, several of the more important clinically useful compounds are pyrimidine or purine derivatives, the best known being 5-iodo-2′-deoxyuridine (IDUR **15**), cytosine arabinoside (Ara-C **16**) and adenine arabinoside (Ara-A **17**).

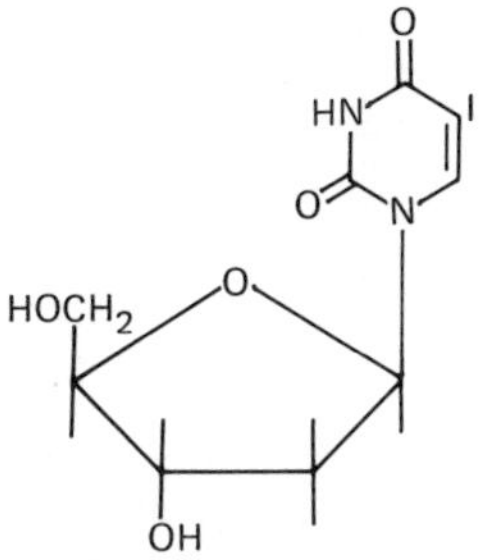

O HN I O N HOCH$_2$ O OH

(15)

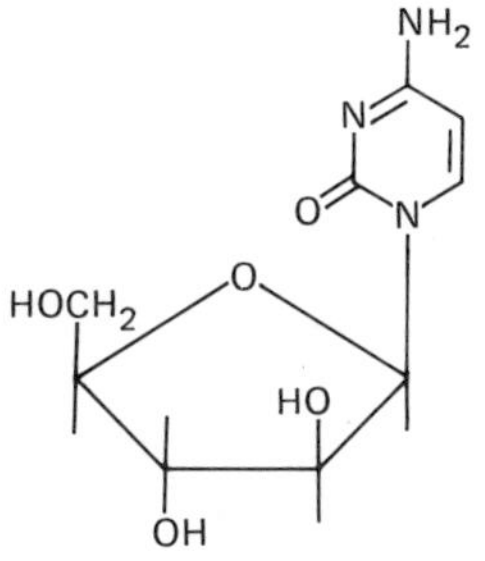

NH_2 N O N HOCH$_2$ O HO OH

(16)

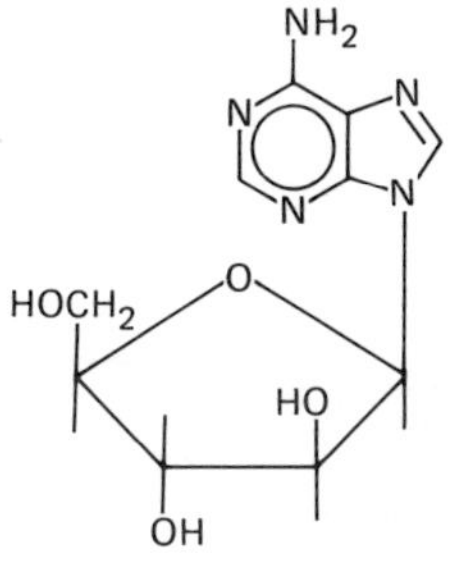

NH_2 N N N N HOCH$_2$ O HO OH

(17)

Ara-C and Ara-A have been discussed in Section 4F. In addition to showing antiviral activity these compounds also show anticancer activity as they interfere with nucleic acid biosynthesis.

Ara-A is active against a variety of DNA viruses but not RNA viruses *in vitro*. It has been shown to have significant therapeutic action against ocular herpes keratitis, and also to be active against herpes simplex and vaccinia infections of mice. The properties and activity of Ara-A have been reviewed.[23]

Ara-C has a similar spectrum of activity to Ara-A but has, in addition to activity against herpes keratitis in man, been found to be active against severe generalized varicella (chickenpox).[24]

IDUR has been found to be active against herpes zoster, particularly for older people, when applied topically as a 40% solution in dimethylsulphoxide, and when the treatment was continued for more than 4 days no further lesions appeared. The duration of the pain associated with the infection was reduced from 30 days to about 2.5 days.[25]

A number of other pyrimidines and purines, and their nucleosides or nucleotides, have been investigated for antiviral activity, and although a number have been found to be active *in vitro* or in some model systems, none have been found to be very useful in clinical situations. 5-Bromouracil, 5-trifluoromethyldeoxyuridine (5-F_3TDR), and a number of analogues have been synthesized and investigated: of these 5-F_3TDR was found to be more effective than IUDR against dendritic herpes infections in man.[27]

(G) SOME OTHER PYRIMIDINES OF PHARMACEUTICAL INTEREST

In addition to the above-mentioned uses of pyrimidine derivatives as pharmaceuticals these heterocyclic compounds show a wide range of other activities. For example, some pyrimidine sulphonamides show orally active hypoglycaemic or diabetic action, e.g. compounds of type (**18**).[28] Pyrantel (**19**), as the pamoate salt, has been used successfully in the clinic as an anti-helmintic,[9] whilst a number of pyrimidines of type (**20**) have been patented as antitrichomonal agents.

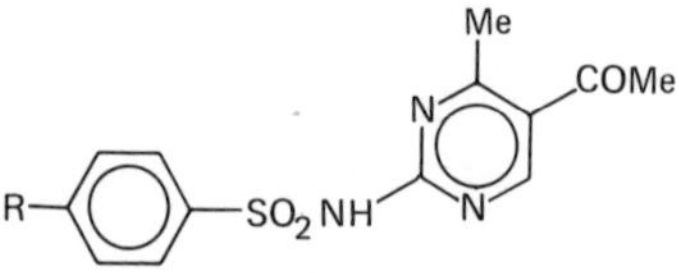

(18)

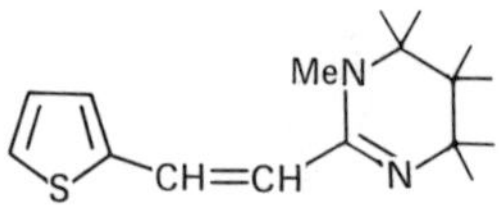

(19)

The compound thonzylamine (**21**) has found use as an antihistamine agent, whilst the compound 2-dimethylamino-4-chloro-6-methylpyrimidine (**22**) possesses tranquillizing properties, the activity being comparable with meprobamate.[33]

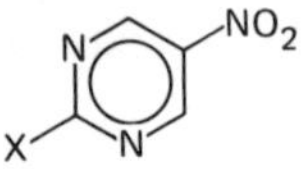

(20) 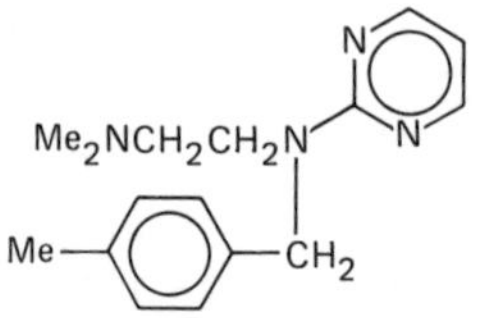 (21) 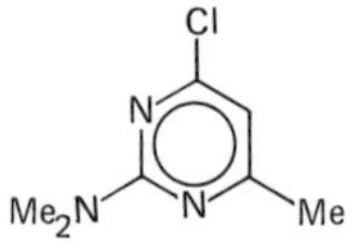 (22) 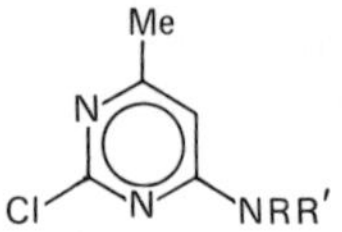 (23) **a**, R = R′ = Me; **b**, R = H, R′ = Me

Castrix (**23a**) is a positional isomer of (**22**) and is a well-known rodenticide[31] whilst the related compound 2-chloro-4-methyl-6-methylaminopyrimidine (**23b**) has also been used in such a role.[32]

Further examples of the biological actions of pyrimidines may be found in refs. 2 (a–c).

(H) SOME PYRIMIDINE AND RELATED NATURAL PRODUCTS

In addition to naturally occurring pyrimidine and purine antibiotics mentioned in Section 4F, and in addition to the nucleic acid bases, the presence of the methylated xanthines in tea, coffee, etc., and the B vitamins, pyrimidine and related heterocycles have an even wider distribution in living organisms.

From the species *Vicia* the pyrimidine-5-yloxyl-β-glucosides vicine (**24a**) and convicine (**24b**) are obtained.

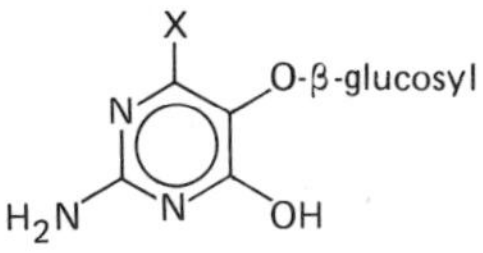

(**24**) **a**, X = NH_2
b, X = OH

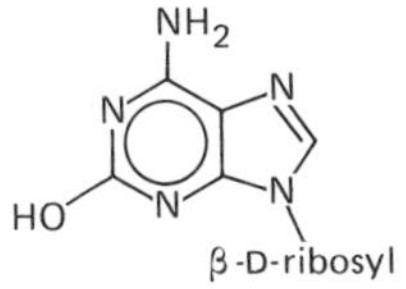

(**25**)

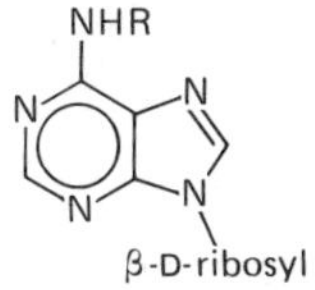

(**26**) **a**, R = 2-furylmethyl
b, R = $CH_2CH{=}C(CH_3)CH_2OH$
c, R = $CH_2CH{=}C(CH_3)CH_3$

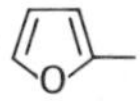

The presence of these glycosides in fava beans is causative of the condition *favism*, anaemia resulting from a combination of inherited low glucose-6-phosphate dehydrogenase and 6-phosphogluconate dehydrogenase activity in erythrocytes and oxidation of glutathione because of utilisation of NADPH by the fava glycosides in the erythrocytes of affected persons who eat such beans.

The β-ribofuranoside of isoguanine (6-amino-2-hydroxypurine) (**25**) has been isolated from croton beans whilst a number of 6-substituted aminopurines are known to be present in the plant and animal kingdoms. These compounds seem to be associated with growth control—for example kinetin (**26a**), zeatin (**26b**) and triacanthine (**26c**).

Pteridines are found in the wings of butterflies and other insects as well as being a part of the vitamin folic acid. The most commonly occurring pteridines are the aminohydroxy compounds listed in Table 10.2, although several other compounds of this type are also found.

In addition to these compounds, some triazinopyrimidines have been found in some micro-organisms. Two which have been found to show some antibiotic action are also listed in Table 10.2.

(I) PYRIMIDINE DERIVATIVES AS AGRICULTURAL CHEMICALS

The use of Castrix (2-chloro-4-dimethylamino-6-methylpyrimidine) and the 4-methylamino analogue as rodenticides has been mentioned above. However, in more recent years a range of pyrimidines has been developed which have wide usage in agricultural chemistry as fungicides, broad-range insecticides, and herbicides.

The substituted uracils Isocil (**27a**), Bromacil (**27b**) and Terbacil (**28**)

Table 10.2 Some naturally occurring pteridines and pyrimidotriazines

Systematic name	Trivial name
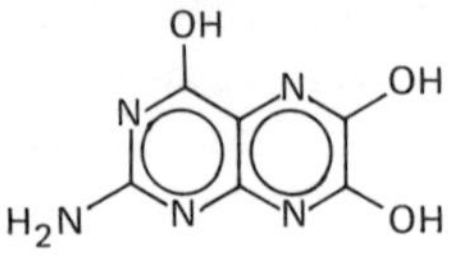 2-Amino-4,6,7-trihydroxypteridine	Leucopterin
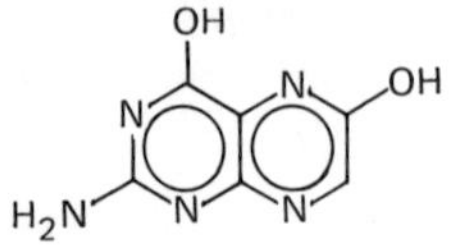 2-Amino-4,6-dihydroxypteridine	Xanthopterin
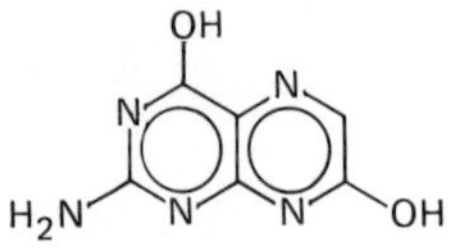 2-Amino-4,7-dihydroxypteridine	Isoxanthopterin
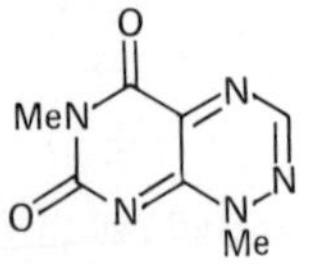 1,6-Dimethyl-5,7-dioxo-1,5,6,7-tetrahydropyrimidine (5,4-e-)-as-triazine	Toxoflavin (Xanthotricin)
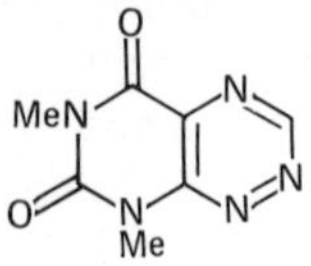 6,8-Dimethyl-5,7-dioxo-5,6,7,8-tetrahydropyrimido (5,4-e)-as-triazine	Fervenulin (Planomycin)

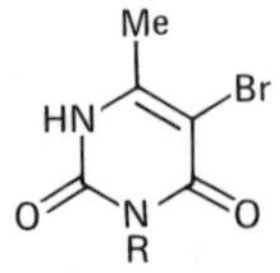

(27) a, R = $CHMe_2$
b, R = $CHMeCH_2CH_3$

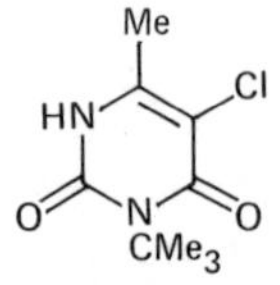

(28)

were developed by Du Pont and introduced in the early 1960s as broad-spectrum herbicides having high phytotoxicity and low mammalian toxicity.[33,34] Isocil has since been superseded by the other two compounds. They are useful against many annual and some perennial weeds in a number of crops, for example, sugar cane, apples, citrus crops, etc.[35] It is generally accepted that the substituted uracils are direct inhibitors of photosynthesis at the chloroplast level and they are thought to act by inhibiting photosynthetic electron transport.[35–37]

Some simple pyrimidine derivatives have also been found to have very useful systemic activity against powdery mildew in cucumbers and in barley,[36,38] two compounds of particular note being Dimethirimol (**29a**) and Ethirimol (**29b**) (ICI, Plant Protection Ltd.) The *N*, *N*-dimethylsulphamate ester of ethirimol is Bupirimate **(30)** which shows extremely useful activity against powdery mildews on a number of crops. Another recent pyrimidine fungicide, triarimol **(31)**, has been shown to be a broad-spectrum fungicide but to be of limited use due to undesirable toxological effects.[36]

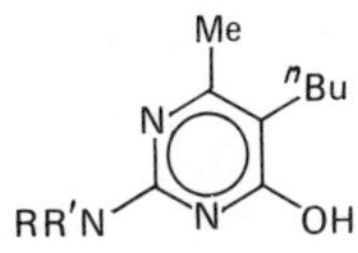

(29) **a**, R = R′ = Me; **b**, R = H, R′ = Et **(30)** 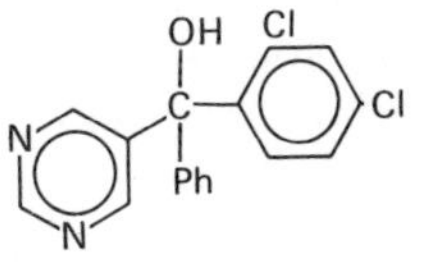 (31)

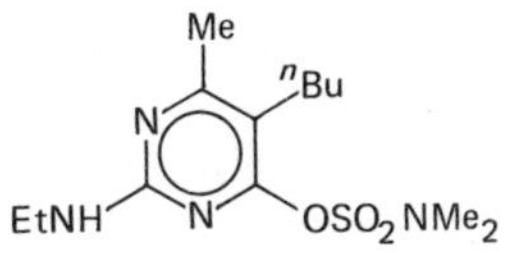

The precise mode of action of these pyrimidines is uncertain, although it has been suggested that they interfere with purine biosynthesis, perhaps by acting as pyridoxal antagonists.[36,39,40]

A number of pyrimidines having phosphorous ester substituents have been found to be useful against a variety of insect pests. The *Pesticide Manual*[37] lists 502 pesticides in current use, of which 92 are organophosphorous pesticides including the pyrimidines given in Fig. 10.3. Also listed is the pyrimidine carbamate shown. Each of the pesticides shown in Fig. 10.3 is thought to inhibit the enzyme acetylcholinesterase.[41]

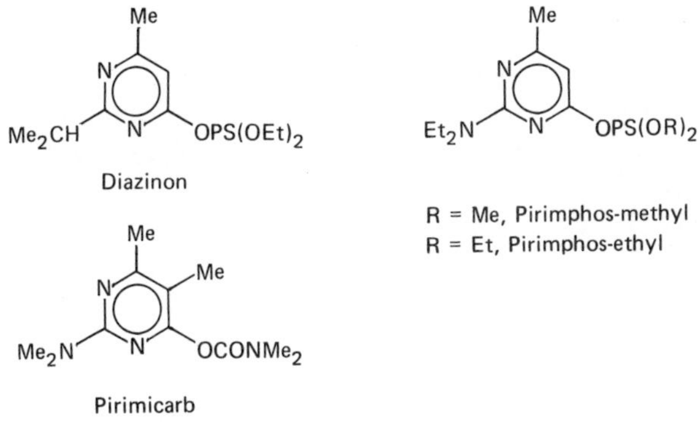

Diazinon

R = Me, Pirimphos-methyl
R = Et, Pirimphos-ethyl

Pirimicarb

Fig. 10.3. Pyrimidine pesticides

(J) CONCLUSION

In this chapter, and in some earlier chapters of this book, reference has been made to a number of synthetic pyrimidines, purines, and pteridines which have useful physiological action, and many of the compounds mentioned are marketed because of that action. Those compounds reaching the general public are a small fraction of the many heterocyclic derivatives that are synthesized and tested every year. The importance which workers in the pharmaceutical and crop protection fields place upon the investigation of the heterocyclic compounds in the title of this book, and on analogues of them, is evident from the volume of work published on such studies in the current literature.

A cursory glance at some of the journals on the shelves of the library at the time of writing shows papers having such titles as '. . . pyrazolo (1,5-a) pyrimidines, a nonbenzodiazepinoid class of antianxiety agents devoid of potentiation of central nervous system depressant effects of ethanol or barbiturates',[42] and 'Inhibitors of *B. subtilis* DNA polymerase III, 6-(arylalkylamino) uracils and 6-anilinouracils'.[43] A number of other such titles have appeared in recent years and the range of activities reported for nitrogen heterocycles is a very wide one.

At a recent meeting, Dr. A. Holý[44] listed a number of possible areas in which nucleotides and related compounds were of potential interest. The main activities of potential interest are listed in Table 10.3.

Such a list is not intended to be exhaustive, nor indeed is it meant to be an authoritative list of the types of action expected of nitrogen heterocycles, but it does indicate the range of activities for which these compounds show promise. Already the compound (**32**) has been shown to have high antiviral activity. It also shows synergism with Ara-A in the treatment of some viruses and is active in the early stages of viral replication, although it is not antibacterial since it does not penetrate the bacterial cell wall.[44] Another compound, acycloguanosine (**33**), has also recently been shown to have antiviral activity[45] and a report of clinical trials on this compound has now appeared (*The Lancet*, February 1979).

Antiviral nucleoside analogues have recently been reviewed.[46]

The C-nucleoside (**34**) has been shown to be antileukaemic, and with almost each issue of the major journals new heterocyclic compounds having biological action, or new actions for previously described heterocycles, are being reported.

In future years we should expect to see derivatives of the pyrimidines, purines and pteridines playing an even greater role in man's fight for survival than they do now. The message is clear. Let us continue to prosecute research on all aspects of the chemistry and biochemistry of the nitrogen heterocycles and their nucleosides, nucleotides and polynucleotides with vigour, for such compounds hold the key of life.

Table 10.3 Main activities of potential interest of nucleotides and related compounds

Antibacterial	Immunosuppressive
Antiviral	Blood platelet aggregating
Antifungal	Hypocholesterolemic
Antileukaemic	Neurostimulating
Antitumour	
Antiprotozoal	

R = S-2,3-dihydroxypropyl

(**32**)

R = D-ribosyl

(**34**)

(**33**)

REFERENCES

1. *The Merck Index* (9th edn.), Merck, Rahway, N. J. (1976).
2. (a) C. C. Cheng in G. P. Ellis and G. B. West (eds.) *Progress in Medicinal Chemistry*, Vol. 6, Butterworths, London, (1969), p. 67, (b) C. C. Cheng and B. Roth, *ibid*., Vol. 7 (1970), p. 285, (c) C. C. Cheng and B. Roth, *ibid*., Vol. 8, (1971), p. 61.
3. G. B. Elion and G. H. Hitchings in A. Goldin, F. Hawking, and R. J. Schnitzer (eds.) *Advances in Chemotherapy*, Vol. 2, Academic Press, New York and London (1965), p. 91.
4. J. A. Montgomery in G. P. Ellis and G. B. West (eds.) *Progress in Medicinal Chemistry*, Vol. 7, Butterworths, London (1970), p. 69.
5. (a) W. Pfleiderer, (ed.) *Chemistry and Biology of Pteridines*, de Gruyter, Berlin and New York (1975); (b) *Ciba Symposium on the Chemistry and Biology of Pteridines*, Churchill, London (1954).
6. G. P. Ellis, *Medicinal Chemistry Reviews*, Butterworths, London, (1972).
7. A. Goth, *Medical Pharmacology*, C. V. Mosby, St. Louis (1966).
8. W. J. Doran in F. F. Blicke and H. Cox (eds.), *Medicinal Chemistry*, Vol. 4., Wiley, New York and London (1959).
9. *Kirk–Othmer Encyclopedia of Chemical Technology* Vol. 3 (2nd edn.), Wiley–Interscience, New York, London, Sydney, (1964), p. 60.
10. L. S. Goodman and A. Gilman (eds.), *The Pharmacological Basis of Therapeutics*, (3rd edn.), Macmillan, London and New York, (1965), p. 105.
11. (a) W. J. Doran, *Barbituric Acid Hypnotics*, Vol. 4 of F. F. Blicke and H. Cox (eds.) *Medicinal Chemistry*, Wiley, New York (1959); (b) A. H. Conney, *Pharmacol. Rev*., **10**, 171 (1967).
12. D. V. Parke, *The Biochemistry of Foreign Compounds*, Pergamon Press, Oxford, London (1968).
13. F. H. S. Curd, D. G. Davey and F. L. Rose, *Ann. Trop. Med. Parasitol*., **39**, 157 (1945).
14. E. A. Falco, L. G. Goodwin, G. H. Hitchings, I. M. Rollo, and P. B. Russell, *Brit. J. Pharmacol*., **6**, 185 (1951).
15. D. C. Martin and J. D. Arnold, *J. Clin. Pharmacol*., **7**, 336 (1967).
16. WHO Technical Meeting Report, *Chemotherapy of Malaria*, Geneva (1961).
17. V. Papesch and E. F. Schroeder, *J. Org. Chem*., **16**, 1879 (1951).

18. E. B. Astwood, *J. Pharmacol. Exptal. Therap.*, **78**, 79 (1943), *J. Am. Med. Soc.*, **122**, 78 (1943).
19. W. H. Miller, R. O. Roblin, and E. B. Astwood, *J. Amer. Chem. Soc.*, **67**, 2201 (1945).
20. J. Calvo and J. Goemine, *Arch. Biochem.*, **10**, 531 (1946).
21. D. Campbell, F. W. Landgrebe, and T. H. Morgan, *Lancet*, **246**, 630 (1944).
22. (a) R. W. Sidwell, J. H. Huffman, G. P. Kharl, L. B. Allen, J. T. Witkowski, and R. K. Robins, *Science,* **177**, 705 (1972); (b) Ger. Öffen Pat. 2,220,246, J. T. Witkowski and R. K. Robins to I. C. N. (Chem. Abs., **78**, 84766 (1973)).
23. F. M. Schabel, *Chemotherapy*, **13**, 321 (1968).
24. T. C. Hall, J. Griffith, G. Watters, R. Baringer, and S. Katz, *Pharmacologist*, **10**, 171 (1968).
25. B. E. Juel-Jensen, *Ann. N. Y. Acad. Sci.*, **173**, 74 (1970).
26. Brit. P. 1170565 (1969) to Robugen G.m.b.h.
27. C. Heidelberger, D. L. Dexter, and W. H. Wolberg, *Proc. Amer. Ass. Cancer Res.*, **11**, 35 (1970), Abstr. 135.
28. Belg. P. 622,086, to Farbenfabriken Bayer (*Chem. Abs.*, **59**, 10078 (1963)).
29. H. L. Howes, J. E. Lynch, and G. F. Smith, *Amer. Soc. Trop. Med. Hyg.*, 17th Meeting, Atlanta, Georgia, (November 1968) (cf. ref. 2(b)).
30. Belg. P. 657,135 to Smith, Kline and French (*Chem. Abs.*, **64**, 19637 (1966).
31. U.S.P. 2, 219, 858 to Winthrop Chemical Co. (Chem. Abs., **35** 1805 (1941)).
32. J. F. Bates, *Proc. Intern. Soc. Sugar-Cane Technologists*, **11**, 695 (1962) (cf. ref. 2 (a)).
33. R. W. Varner, *Abstr. Weed Soc. Am.*, **1961**, 19.
34. H. C. Bucha, W. E. Cupery, J. E. Harrod, H. M. Loux, and L. M. Ellis, *Science*, **137**, 537 (1962).
35. P. C. Kearney and D. D. Kaufman (eds.), *Herbicides, Chemistry, Degradation and Mode of Action* (2nd edn.), Marcel Dekker, New York and Basel, (1975).
36. N. R. McFarlane, (ed.) *Herbicides and Fungicides—Factors Affecting Their Activity*, Spec. Publ. No. 29, Chemical Society, London (1977).
37. H. Martin (ed.), *Pesticide Manual*, British Crop Protection Council (1972).
38. *Approved Products for Farmers and Growers*, MAFF, London (1976).
39. A. Calderbank, *Acta Phytopathol.*, **6**, 355 (1971).
40. P. Slade, B. D. Cavell, R. J. Hemingway, and M. J. Sampson, in A. S. Tahori, (ed.), *Herbicides, Fungicides, Formulation Chemistry*, Proc. 2nd Int. IUPAC Congress, Vol. V, Gordon and Breach, New York, (1972).
41. J. R. Corbett, *The Biochemical Mode of Action of Pesticides*, Academic Press, London and New York (1974).
42. W. E. Kirkpatrick, T. Okabe, I. W. Hillyard, R. K. Robins, A. T. Dren, and T. Novinson, *J. Medicin. Chem.*, **20**, 386 (1977).
43. N. C. Brown, J. Gambino, and G. E. Wright, *J. Medicin Chem.*, **20**, 1186 (1977).
44. A. Holý, *Nucleotide Group, 10th Annual Symposium*, Chemical Society Birmingham (20–21 December 1977).
45. H. J. Schaeffer, L. Beauchamp, P. de Miranda, G. B. Elion, D. J. Bauer, and P. Collins, *Nature*, **272**, 583 (1978).
46. E. De Clercq and P. F. Torrence, *J. Carbohydrates, Nucleosides and Nucleotides*, **5**, 187 (1978).

Addendum: Some Biological and Biochemical Terms used in this book

CELL, AND SUB-CELLULAR COMPONENTS

The *cell* is the unit of living matter. Some microorganisms are unicellular but others are multicellular, whilst in higher organisms different types of cell are present corresponding to the different types of tissue. However, basically a cell is the simplest organism capable of separate existence. Different cells have different shapes and sizes but they have the same general features:

An outer *cell membrane* (plant cells may also have an extra *cell-wall* structure) which is a phospholipid bilayer enclosing the cell.

The medium within the cell which is a complex aqueous solution and suspension of material is the *cytoplasm* (or cell sap).

Within the cell is a body enclosed by a double membrane which is the *nucleus*. The nucleus contains the DNA (associated with protein) in discrete entities which are the *chromosomes* (or the chromatin). In the resting state the chromosomes are associated together in a dense region (or regions)—the *nucleolus*. An extensive membrane system links the nucleus to the cell membrane and passes through the cytoplasm: this is the *endoplasmic reticulum.* On disruption of the cell the endoplasmic reticulum tends to form itself into spherical membrane-bound particles—the *microsomes*.

Free in the cytoplasm, or attached to the endoplasmic reticulum, are found dense particles composed of protein and RNA, which are the sites of protein biosynthesis. These are the *ribosomes*. Groups of ribosomes attached to mRNA are called *polysomes*.

Further double-membrane bounded sub-cellular species which are the sites of 'energy metabolism' are the *mitochondria*. Plant cells may also have components containing chlorophyll which are the *chloroplasts.*

Both animal and plant cells also contain other sub-cellular components (organelles) which contain digestive enzymes. These are the *lysosomes*.

Prokaryotic cells are those which have no well-defined nucleus nor other membrane-bound organelles characteristic of normal cells. They are the primitive form of cells and it seems that the organelles of eukaryotic cells have evolved from them. Prokaryotes also have their genetic material (DNA) in the form of simple filaments.

Eukaryotic cells have a discrete nucleus bounded by a double membrane from the rest of the cell. The *genes* are contained in a number of *chromosomes* consisting of DNA with complexed protein.

The *gene* is the unit of the hereditary material (DNA) of a cell. It is a segment of DNA which will code for the biosynthesis (via the intermediacy of mRNA) of one particular protein chain. Until recently it was thought that the genes were non-overlapping.

The *genome* is the set of all the different *chromosomes* found in each nucleus of a given species.

The *chromosomes* are the thread-like systems of DNA and protein which occur in the nucleus of all animal and plant cells.

ORGANISMS AND ORGANS

Viruses are species at the interface between organic and inorganic matter. They are species which consist of nucleic acid (RNA or DNA) with associated protein. Many viruses can be crystallized. They have no metabolic function of their own but invade cells and use the host cell machinery to produce viral specific components using the viral nucleic acid as template.

Bacteriophages are types of virus which are parasitic to bacteria.

Bacteria are microscopic organisms which may be unicellular or multicellular and which have prokaryotic cells. They usually multiply by simple cell division or by some other form of asexual reproduction. They are usually classed according to their general shape—e.g. spherical, rod-shaped, filamentous, etc.

Bacteria mentioned in this book are:

Escherichia coli

Bacillus subtilis

Steptococcus lactis

Streptomyces albus

Yeasts are unicellular *fungi* (mycophyta) which are simple eukaryotic organisms without chlorophyll. The fungi include mushrooms, moulds, but also species such as *Penicillium* which has been important in the development of *antibiotics* (i.e. antimicrobial agents produced by microorganisms).

Aspergillus oryzae is a fungus.

Protozoa are unicellular species classed as animals differing from bacteria in that they have a well-defined nucleus. The malaria parasites are protozoa. Those mentioned in this book are:

Plasmodium gallinaceum

P. berghei

P. falciparum

Trypanosomes are a type of protozoon which are transmitted to man, cattle and other higher animals by insects. They cause serious diseases such as sleeping sickness.

Trichomonas is another protozoon which causes genital infections.

Helminths is a name for worms which are higher parasitic organisms which can affect man and other higher animals.

In man and other higher animals there are a variety of organs, tissues and body fluids. Many of these will be known to the reader. However, some terms which have been used in this book and which may not be familiar are given below:

Plasma is the fluid part of the blood (serum and fibrinogen) from which the blood cells have been separated.

Erythrocytes are the red blood cells. *Reticulocytes* is the name given to newly formed red blood cells.

Lymphocytes are a type of white blood cell produced in the lymphatic glands of the body. The *lymph* is a fluid, somewhat similar to blood plasma, which circulates through lymphatic vessels in the body.

The *thymus gland* lies in the lower part of the neck and the upper part of the chest. Its function has still not been fully defined but there is increasing evidence that it plays an important part in the immune responses of the body.

The *kidneys* are a pair of glands which filter fluid from the blood so that compounds may be excreted via the urine. The *glomeruli* filter the non-protein portion of the plasma and this filtrate is then passed through *tubules* where some of it is re-absorbed. Compounds such as urea, uric acid, etc. are thus excreted by the body. Urinary excretion is also important for the excretion of many drug metabolites.

Hormones are substances secreted by certain glands within the body which influence the action of other tissues and organs within the body.

Intestinal flora refers to the normal bacterial population of the intestine.

MISCELLANEOUS BIOCHEMICAL AND OTHER TERMS USED IN THIS BOOK

cytotoxic—lethal to cells.
intraperitoneal—injection into the peritoneal cavity of the body.
intravenous—injection into a vein.
LD_{50}—the median lethal dose, i.e. the dose which causes 50% fatalities in experimental animals.
phytotoxic—lethal to plants.
prophylactic—treatment required to prevent the onset of disease or symptoms.
The Svedberg unit of sedimentation, $S = 1 \times 10^{-13}$ s. Used in centrifugation to define the weight of particles since the rate of sedimentation is proportional to the molecular weight of the particles (for particles of similar shape).
Therapeutic index—a measure of the safety of a drug. The ratio of the median lethal dose to the median effective dose.
X-linked recessive trait—a characteristic associated with the X sex chromosome which manifests itself only if both parents have the associated genes.

Index